Reagents for Organic Synthesis

Reagents for Organic Synthesis

VOLUME 7

Mary Fieser

Louis F. Fieser

A WILEY-INTERSCIENCE PUBLICATION
JOHN WILEY & SONS
NEW YORK · CHICHESTER · BRISBANE · TORONTO

Library of Congress Catalog Card Number: 66-27894

ISBN 0-471-02918-1

Printed in the United States of America

10 9 8 7 6 5 4 3 2 1

PREFACE

This volume of *Reagents* is the last to bear both our names. Because of unforeseen difficulties, it only covers literature of 1976 and the publication has been delayed.

We are again particularly grateful for the continuing help of the following chemists: Professors John A. Secrist, Robert H. Wollenberg, William Roush, and Rick Danheiser, Dr. William Moberg, Dr. Mark A. Wuonola, Dr. James V. Heck, Dr. Ving Lee, Dr. May Lee, Dr. Janice Smith, David Wenkert, Homer Pearce, Howard Simmons, III, Donald W. Landry, John Lechleiter, Stephen Kamin, and Dale L. Boger. We acknowledge with gratitude the help for the first time of Anthony Feliu, Jeffrey C. Hayes, Dr. Leslie A. Browne, Marcus A. Tius, Dr. Bruce H. Lipshutz, Dr. William C. Vladuchick, John Maher, Dr. Pierre Lavallée, Dr. Daniel Sternbach, John Munroe, Paul Hopkins, Donald Wolanin, Jay W. Ponder, and Stephen Brenner.

The picture was taken by Dr. Melvin Bellott at a reunion luncheon for *Reagents* proofreaders at the Harvard Club on the occasion of the celebration of the birthday of Professor R. B. Woodward.

MARY FIESER
LOUIS F. FIESER

Cambridge, Massachusetts
January 1979

CONTENTS

A

Acetic acid, 2, 5–7; **5,** 3.

Cyclization. Treatment of the germacrane sesquiterpene agerol (1) with 80% aqueous acetic acid affords a 2:1 mixture of (2) and (3) in high yield rather than

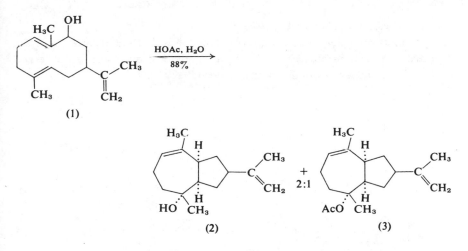

a selinane derivative.[1] These products are the only known isovetivane derivatives. This framework was originally assigned erroneously to natural products shown subsequently to be spiranes.

[1] F. Bellesia, U. M. Pagnoni, and R. Trave, *J.C.S. Chem. Comm.*, 34 (1976).

Acetic anhydride, 1, 3; **2,** 7–10; **5,** 3–4; **6,** 1–2.

Review.[1] Synthetic uses of acetic anhydride have been reviewed, particularly as applied to the synthesis of heterocycles (127 references).

[1] D. H. Kim, *J. Heterocyclic Chem.*, **13,** 179 (1976).

Acetic anhydride–Acetyl chloride.

Reaction with ketoximes.[1] Cyclohexanone oxime (1) is converted into (2)

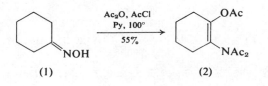

when heated in pyridine with these reagents. The reaction was also used to convert testosterone acetate oxime (3) into the diosphenol (5).

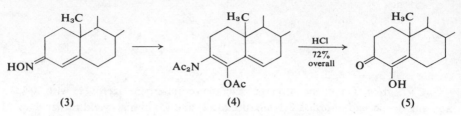

(3)   (4)   (5)

[1] M. V. Bhatt, C. G. Rao, and S. Rengaraju, *J.C.S. Chem. Comm.*, 103 (1976).

Acetic anhydride–Sodium acetate, 6, 5–6.

Claisen rearrangements.[1] Karanewsky and Kishi[2] have reported that the normal Claisen rearrangement of allyl aryl ethers can be effected in acetic anhydride and sodium or potassium acetate (150–200°).

Examples:

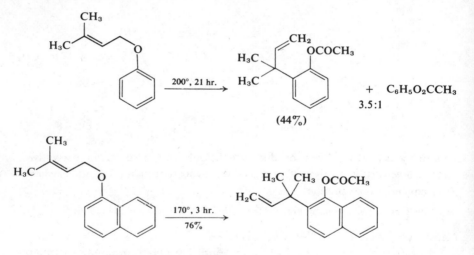

[1] S. J. Rhoads and N. R. Raulins, *Org. Reactions*, **22**, 1 (1975).
[2] D. S. Karanewsky and Y. Kishi, *J. Org.*, **41**, 3026 (1976).

Acetic anhydride–Trifluoroacetic anhydride.

Pummerer rearrangement (6, 2, 5–6). α-Acetoxy sulfides can be obtained by rearrangement of sulfoxides in TFAA–Ac$_2$O in the presence of 2,6-lutidine at 20°. The rearrangement is more rapid than in Ac$_2$O–NaOAc.[1]

(1) $RCH_2CH_2\overset{\overset{O}{\uparrow}}{S}C_6H_5 \xrightarrow[\text{60–85\%}]{\text{TFAA–Ac}_2\text{O}} RCH_2\overset{OAc}{\overset{|}{C}H}SC_6H_5$

[1] R. Tanikaga, Y. Yabuki, N. Ono, and A. Kaji, *Tetrahedron Letters*, 2257 (1976).

Acetyl chloride, 1, 11; **4,** 5–6; **5,** 4–5.

Reaction with imines. The reaction of some acid chlorides with an imine in the presence of a tertiary amine often leads to a β-lactam.[1] The reaction of acetyl chloride, however, follows a different course, and an oxazinone (2) is

(I) $2CH_3COCl + N(C_2H_5)_3 \longrightarrow$

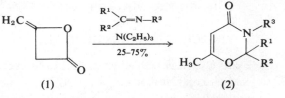

(1)   (2)

obtained. Apparently acetyl chloride is converted into diketene (1), which is cleaved by the amine to $CH_3COCH=C=O$. This oxoketene then undergoes cyclo-addition to the imine to form (2).[2]

1,3-Oxazetidines. The reaction of acetyl chloride with N-isopropylidene-anilines is also different in that it can lead to three products, depending on substituents in the phenyl ring (equation II).[3]

(II) $Ar—N=C(CH_3)_2 + CH_3COCl \xrightarrow{N(C_2H_5)_3}$

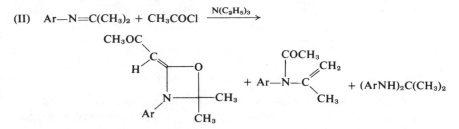

[1] H. Staudinger, *Ann.,* **356,** 51 (1907).
[2] A. Maujean and J. Chuche, *Tetrahedron Letters,* 2905 (1976).
[3] H. Iwamura, M. Tsuchimoto, and S. Nishimura, *ibid.,* 1405 (1975).

Acetyl nitrate, $CH_3CO_2NO_2$. Mol. wt. 105.05. The nitrate is prepared *in situ* from HNO_3 (98%) and Ac_2O (1:2–3) at $<0°$.[1]

Oxidation of sulfides to sulfoxides.[2] This well-known nitrating agent has been found to be a suitable reagent for oxidation of sulfides to sulfoxides without further oxidation to sulfones. The reaction is carried out at $0°$; yields are generally in the range of 80–100%. Surprisingly the reaction is faster than that with peroxy compounds. In the case of sulfoxides that tend to give Pummerer-type reactions with Ac_2O, benzoyl nitrate is the preferred oxidant. This acyl nitrate is prepared from benzoyl chloride and silver nitrate; it is somewhat more reactive than acetyl nitrate, and oxidations can be conducted even at $-76°$.

[1] S. Sifniades, *J. Org.,* **40,** 3562 (1975).
[2] R. Louw, H. P. W. Vermeeren, J. J. A. van Asten, and W. J. Ultée, *J.C.S. Chem. Comm.,* 496 (1976).

Acrylonitrile, 1, 14.

   Nitriles. Japanese chemists have reported a new synthesis of nitriles by the reaction of acrylonitrile with cuprous methyltrialkylborates (equations I and II).

(I) $R_3B + CH_3Li \longrightarrow [R_3BCH_3]Li \xrightarrow[-\text{LiBr}]{\text{CuBr}} [R_3BCH_3]Cu$

(II) $[R_3BCH_3]Cu + CH_2{=}CH{-}CN \longrightarrow \xrightarrow[88-93\%]{H_2O} RCH_2CH_2CN$

(III) $[R_3BCH_3]Cu + CH_2{=}CHCOOC_2H_5 \longrightarrow RCH_2CH_2COOC_2H_5$

These salts also react with ethyl acrylate to form carboxylic esters (equation III), but yields are only mediocre.[1]

[1] N. Miyaura, M. Itoh, and A. Suzuki, *Tetrahedron Letters*, 255 (1976).

Adogen 464, 6, 10.

   1,3-Benzoxathioles. These heterocycles can be prepared conveniently by reaction of 2-hydroxybenzenethiols with methylene bromide (phase-transfer conditions).[1]

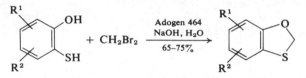

[1] S. Cabiddu, A. Maccioni, and M. Secci, *Synthesis*, 797 (1976).

"π-Allylpalladium acetate dimer," $(\eta^3\text{-}C_3H_5PdOAc)_2$. Mol. wt. 413, yellow decomposes at 100–130°. Prepared by reaction of π-allylpalladium chloride dimer with silver acetate (65% yield).[1]

   Head-to-tail dimerization of isoprene. Heck *et al.*[2] have found that dimerization of isoprene to head-to-tail dimers can be controlled by use of palladium catalysts in combination with organophosphine ligands. Thirty-six experiments are reported in which the various important variables were explored. The most satisfactory results were obtained with $(\eta^3\text{-}C_3H_5PdOAc)_2$, 2 $P(o\text{-}C_6H_4CH_3)_3$, $N(C_2H_5)_3$, and formic acid, with THF as solvent. In this way the two head-to-tail dimers can be obtained in 79% yield (equation I). These dimers were separated

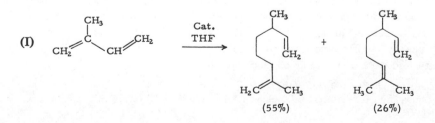

as the same hydrogen chloride adduct, 7-chloro-3,7-dimethyl-1-octene, from the other dimers formed. This adduct was converted into α- and β-citronellol and linalool.

[1] S. D. Robinson and B. L. Shaw, *J. Organometal. Chem.*, **3**, 367 (1965).
[2] J. P. Neilan, R. M. Laine, N. Cortese, and R. F. Heck, *J. Org.*, **41**, 3455 (1976).

π-Allylpalladium chloride dimer, $(C_3H_5PdCl)_2$. Mol. wt. 365.85, m.p. 120° dec. Suppliers: Alfa, ROC/RIC.

Thiepin system. Dibenzo[*bd*]thiepin (4) has been prepared recently by ring enlargement accomplished with this Pd(II) species (equation 1). Mild conditions are required because of the ease of sulfur extrusion.[1]

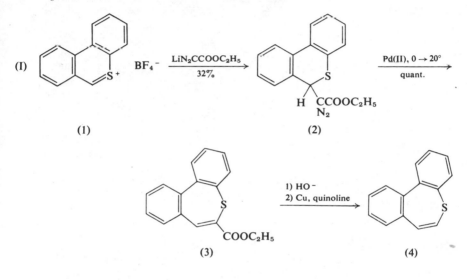

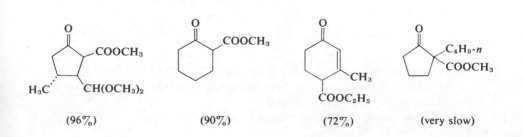

[1] K. Nakasuju, K. Kawamura, T. Ishihara, and I. Murata, *Angew. Chem. Int. Ed.*, **15**, 611 (1976.)

Alumina, Al_2O_3, **1**, 19–20; **2**, 17; **3**, 6; **4**, 8; **6**, 16–17.

Decarboalkoxylation of β-keto esters. β-Keto esters are decarboalkoxylated by basic alumina (Merck type T) in refluxing dioxane containing 1.5% water. The reaction is slow in dry solvent. β-Keto acids are also decarboxylated under these conditions. Some typical β-keto esters are shown together with the yields of the corresponding ketones.[1]

Elimination reactions. Woelm W-200 alumina dehydrated at 400° and 0.06 torr effects conversion of acyclic and cyclic secondary tosylates and sulfamates to olefins and alcohols (approximately 10:1 mixtures for acyclic substrates). An *anti* elimination predominates.[2]

Examples:

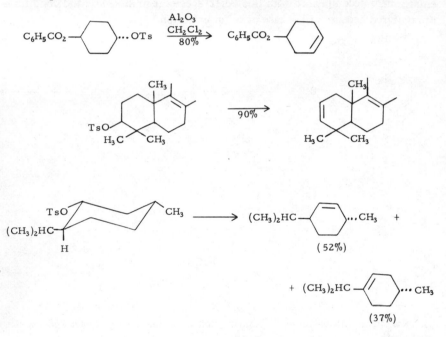

Oxidation of **sec-alcohols.** Posner *et al.*[3] have reported another reaction possible on alumina surfaces (**6,** 16–17): Selective oxidation of some diols with trichloroacetaldehyde (or benzaldehyde) as hydride acceptor (equation I).

(I) $RCHOH(CH_2)_nCH_2OH \xrightarrow[\sim 65\%]{\underset{Al_2O_3}{Cl_3CCHO}} RC(CH_2)_nCH_2OH$
$\overset{O}{\overset{\|}{}}$

Likewise 5α-androstane-3β,17β-diol is oxidized to 5α-androstane-3β-ol-17-one (68% yield). The selectivity in these oxidations is greater than that obtained with the more expensive Fetizon reagent.

An even more useful oxidation is that of cyclobutanol to cyclobutanone by trichloroacetaldehyde on activated alumina at 25° in over 99% yield[4]; this reaction generally proceeds in low yield by usual methods.

[1] A. E. Greene, A. Cruz, and P. Crabbé, *Tetrahedron Letters*, 2707 (1976).
[2] G. H. Posner and G. M. Gurria, *J. Org.*, **41**, 578 (1976).
[3] G. H. Posner, R. B. Perfetti, and A. W. Runquist, *Tetrahedron Letters*, 3499 (1976).
[4] G. H. Posner and M. J. Chapdelaine, *Synthesis*, 555 (1977).

Aluminum chloride, 1, 24–34; **2**, 21–23; **3**, 7–9; **4**, 10–15; **5**, 10–13; **6**, 17–19.

Ene reactions (**5,** 11). The reaction of alkenes with methyl acrylate catalyzed by AlCl₃ results in ene adducts.[1] Cyclobutanes were not observed, as reported earlier by Sands.[2]

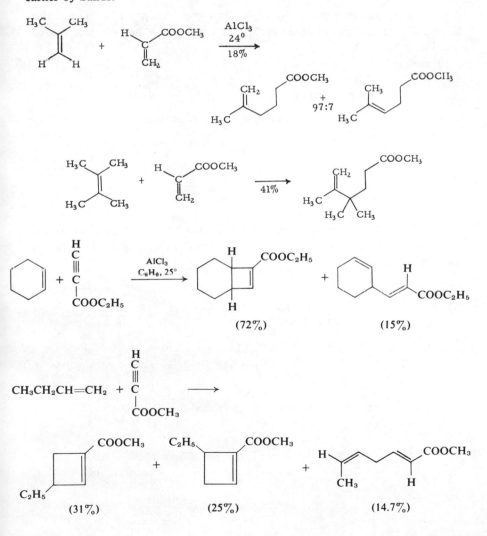

Surprisingly, the catalyzed reaction of alkenes with propiolate esters results in [2 + 2] cycloadducts, as well as ene adducts, in a manner that is related to the substitution on the alkene. No cycloadducts are formed from alkenes with two substituents on one end of the double bond.[3]

Reduction of alkyl halides.[4] Alkyl halides are reduced readily to hydrocarbons by organosilanes under catalysis with aluminum chloride. This hydrocarbon synthesis is limited to alkyl halides that are not rearranged by the catalyst. Primary, secondary, and tertiary halides are reduced. Reduction with triethyl-

$$RX + R'_3SiH \xrightarrow{AlCl_3} RH + R'_3SiX$$

deuteriosilane has shown that in some cases rearrangements to more stable intermediate secondary or tertiary carbenium ions precede the hydrogen transfer reaction.

Examples:

$$\text{1-Bromoadamantane} + (C_2H_5)_3SiH \xrightarrow[79\%]{AlCl_3} \text{adamantane}$$

$$\text{2-Bromoadamantane} \xrightarrow[84\%]{} \text{adamantane}$$

$$\text{Benzhydryl chloride} \xrightarrow[100\%]{} \text{diphenylmethane}$$

$$\text{Bromocyclohexane} \xrightarrow[90\%]{} \text{cyclohexane}$$

$$\text{1-Bromohexane} \xrightarrow[64\%]{} \text{hexane}$$

γ,δ-*Unsaturated alcohols; propargylic alcohols.*[5] The former alcohols can be obtained by addition of allylsilanes to aldehydes (equation I) catalyzed by aluminum chloride.

(I) $RCHO \quad | \quad R^1CH{=}CHCH_2Si(CH_3)_3 \xrightarrow[CH_2Cl_2, -25^0]{AlCl_3}$
$$[R^1 = H, C_2H_5, (CH_3)_3SiCH_2]$$

$$\underset{(CH_3)_3SiO \quad R^1}{RCH{-}CH{-}CH{=}CH_2} \xrightarrow[40-45\%]{\underset{CH_3OH}{H_2O}} \underset{R^1}{RCHOHCHCH{=}CH_2}$$

Propargylic alcohols are obtainable by a related reaction (equation II).

(II) $RCHO \quad + \quad (CH_3)_3SiC{\equiv}CSi(CH_3)_3 \xrightarrow[CH_2Cl_2, -40^0]{AlCl_3}$

$$\underset{OSi(CH_3)_3}{RCHC{\equiv}CSi(CH_3)_3} \xrightarrow[40-65\%]{\underset{CH_3OH}{H_2O}} RCHOHC{\equiv}CSi(CH_3)_3$$

α,β-*Unsaturated ketones*. Organomercurials ordinarily do not react with acid halides unless aluminium halides are used as catalysts (equation I). The

(I) $$R_2Hg + R^1COCl \xrightarrow[CH_2Cl_2]{AlBr_3} RCOR^1 + RHgCl$$

reaction is not particularly useful since only one R group reacts. However, vinylmercuric chlorides undergo this reaction rapidly at 20° to give α,β-unsaturated ketones with retention of geometry (equation II). The R group can

(II)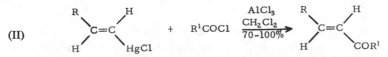

be alkyl or aryl, but the R^1 group must be an alkyl group for satisfactory results.[6]

β,γ-*Unsaturated ketones*. The reaction of acid chlorides with allylic silanes in the presence of aluminum chloride results in β,γ-unsaturated ketones, formed with allylic rearrangement (equations I and II).[7]

(I) $$(CH_3)_3SiCH_2CH{=}CH_2 + RCOCl \xrightarrow[\sim 75-80\%]{AlCl_3} RCOCH_2CH{=}CH_2 + (CH_3)_3SiCl$$

(II)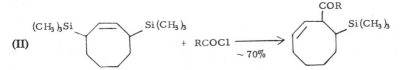

This method was used for an attractive synthesis of artemisia ketone (1) as shown in equation (III).[8]

(III) $$(CH_3)_2C{=}CHCH_2Cl \xrightarrow[58\%]{\substack{1)\ \ Mg,\ ether,\ -5^0 \\ 2)\ \ (CH_3)_3SiCl}} (CH_3)_2C{=}CHCH_2Si(CH_3)_3$$

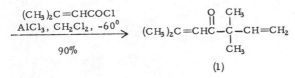

(1)

[1] H. Greuter and D. Belluš, *Syn. Comm.*, **6**, 409 (1976).

[2] R. D. Sands, *ibid.*, **3**, 81 (1973).

[3] B. B. Snider, *J. Org.*, **41**, 3061 (1976).

[4] M. P. Doyle, C. C. McOsker, and C. T. West, *ibid.*, **41**, 1393 (1976).

[5] G. Deleris, J. Dunogues, and R. Calas, *Tetrahedron Letters*, 2449 (1976).

[6] R. C. Larock and J. C. Bernhardt, *ibid.*, 3097 (1976).

[7] R. Calas, J. Dunogues, J.-P. Pillot, C. Biran, F. Pisciotti, and B. Arreguy, *J. Organometal. Chem.*, **85**, 149 (1975).

[8] J.-P. Pillot, J. Dunogues, and R. Calas, *Tetrahedron Letters*, 1871 (1976).

N-Amino-4,6-diphenylpyridone, (1). Mol. wt. 262.28, m.p. 166°.
Preparation[1]:

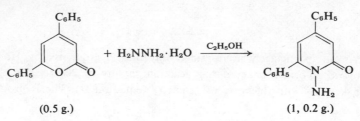

(0.5 g.)   (1, 0.2 g.)

Nitriles from aldehydes.[2] Aldehydes can be converted into nitriles by the sequence depicted in (I). The reaction is useful for nitriles that are stable to heat.

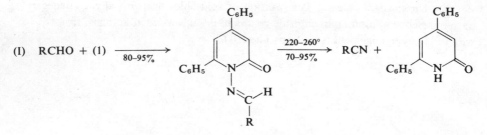

(I)   RCHO + (1) $\xrightarrow[80-95\%]{}$ ... $\xrightarrow[70-95\%]{220-260°}$ RCN + ...

---

[1] I. El-S. El-Kholy, F. K. Rafla, and M. M. Mishrikey, *J. Chem. Soc.* (*C*), 1578 (1970).
[2] J. B. Bapat, R. J. Blade, A. J. Boulton, J. Epsztajn, A. R. Katritzky, J. Lewis, P. Molina-Buendia, P.-L. Nie, and C. A. Ramsden, *Tetrahedron Letters*, 2691 (1976).

(S)-1-Amino-2-methoxymethylpyrrolidine, (1). Mol. wt. 130.19, b.p. 56–57°/3 torr, $\alpha_D - 50°$. Stable to storage at 10°.
Preparation from (S)-proline:

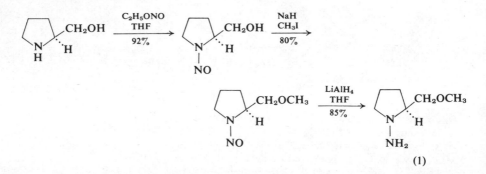

(1)

Enantioselective α-alkylation of ketones.[1] This reaction can be carried out by conversion of ketones into chiral hydrazones on reaction with (1), metallation with LDA, alkylation, and ozonolysis (equation I). The enantiomeric purity of the products varies over a wide range (30–87%).

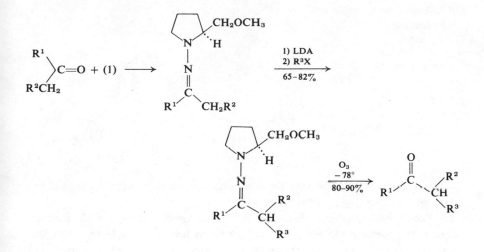

[1] D. Enders and H. Eichenauer, *Angew. Chem. Int. Ed.*, **15**, 549 (1976).

Ammonium acetate, 1, 38.

   2-Piperidones.[1] The Knoevenagel reaction of aryl or heterocyclic aldehydes with ethyl 4-nitrobutanoate[2] in the presence of ammonium acetate has been used for the synthesis of 2-piperidones (equation I).

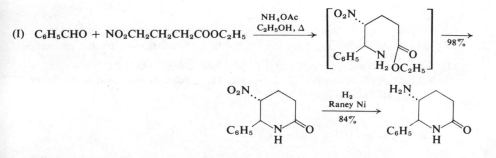

(I) $C_6H_5CHO + NO_2CH_2CH_2CH_2COOC_2H_5$

[1] H. Bhagwatheeswaran, S. P. Gaur, and P. C. Jain, *Synthesis*, 615 (1976).
[2] N. J. Leonard and D. L. Telly, *Am. Soc.*, **72**, 2537 (1956).

Antimony(V) chloride, 1, 42; **4,** 21; **5,** 18–19; **6,** 22–23.

Chlorination of alkylphenylacetylenes. These alkynes react with $SbCl_5$ to give dichloroalkenes in fair yields; the (Z)-isomers are formed predominantly and in some cases exclusively.

Examples:

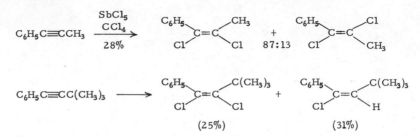

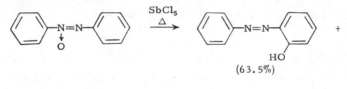

The reaction of chlorine with these alkynes also leads to dichloroalkenes, but this reaction shows slight stereoselectivity.[1]

Rearrangement of azoxybenzenes. $SbCl_5$ and azoxybenzene in CCl_4 form an orange 1:1 complex (95.5% yield), which when heated for 5 hours at 86–88° rearranges mainly to *o*-hydroxyazobenzene. Rearrangement of azoxybenzenes

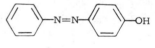

with sulfuric acid or chlorosulfuric acid (Wallach rearrangement) leads mainly to *p*-hydroxyazobenzenes. The new conditions were shown to be general.[2]

[1] S. Uemura, A. Onoe, and M. Okano, *J.C.S. Chem. Comm.,* 145 (1976).
[2] J. Yamamoto, Y. Nishigaki, M. Imagawa, M. Umezu, and T. Matsuura, *Chem. Letters,* 261 (1976).

B

Benzenesulfonylhydrazine, $C_6H_5SO_2NHNH_2$. Mol. wt. 172.21, m.p. 101–103°. Supplier: Aldrich.

Vinylsilanes. Benzenesulfonylhydrazones of carbonyl compounds can be converted into vinylsilanes by deprotonation followed by quenching with chlorotrimethylsilane. Yields tend to be only fair for aliphatic ketones, but are generally improved by use of TMEDA as solvent.[1]

Examples:

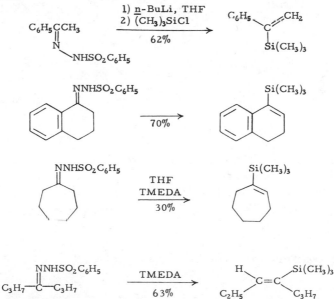

[1] T. H. Chan, A. Baldassarre, and D. Massuda, *Synthesis*, 801 (1976).

Benzenesulfonyltetrazole (1). Mol. wt. 210.22, m.p. 88–90°, decomposes within 10 days in a dessicator.

Preparation:

$$C_6H_5SO_2Cl + HN{\overset{N}{\underset{N=N}{\big|}}} \xrightarrow[\substack{60-70\%}]{\substack{N(C_2H_5)_3 \\ Dioxane}} C_6H_5SO_2-N{\overset{N}{\underset{N=N}{\big|}}}$$

(1)

13

Polynucleotides. Arylsulfonyltetrazoles,[1] especially benzenesulfonyltetrazole, are highly effective reagents for synthesis of the phosphotriester bond between protected oligonucleotides. They are considerably more active than aryl-sulfonyltriazoles.[2]

[1] J. Stawinski, T. Hozumi, and S. A. Narang, *Canad. J. Chem.*, **54**, 670 (1976).
[2] N. Katagiri, K. Itakura, and S. A. Narang, *Am. Soc.*, **97**, 7332 (1975).

Benzotriazolyloxytris(dimethylamino)phosphonium hexafluorophosphate (BOP), (1), 6, 34–35.

Improved preparation:

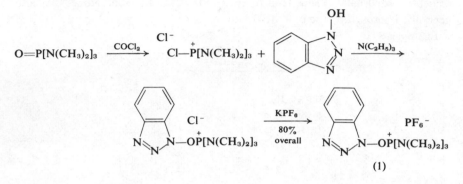

Peptide synthesis.[1] The paper includes preparations of two peptides using this reagent: BOC-L-Thr-L-Phe-OCH₃ (98% yield) and BOC-L-Cys(S-Benzyl)-L-Lys(NεCb)-L-Asn-L-PheOCH₃ (96% yield).

[1] B. Castro, J.-R. Dormoy, B. Dourtoglou, G. Evin, C. Selve, and J.-C. Ziegler, *Synthesis*, 751 (1976).

2-Benzoyloxymethylbenzoyl chloride,

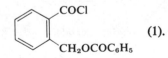

(1).

Mol. wt. 274.70, m.p. 51–52°.
Preparation:

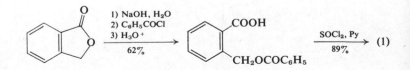

Amine protection. The reagent reacts with amines (Py, 0°) to form amides (2). These derivatives are cleaved rapidly by sodium methoxide in methanol at 20° to form phthalide and the amine.[1]

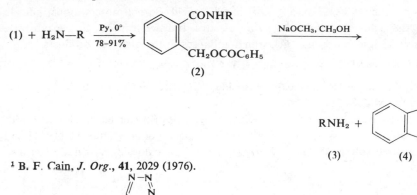

(2)

RNH₂ +

(3)   (4)

[1] B. F. Cain, *J. Org.*, **41**, 2029 (1976).

Benzoyltetrazole,   Mol. wt. 174.16.

The reagent is prepared *in situ* from benzoyl chloride, tetrazole, and triethylamine in dioxane.

Benzoylation of nucleosides.[1] Nucleosides are benzoylated in >90% yield by this reagent. Some selectivity between primary and secondary hydroxyl groups is noted. The primary amino group of deoxycytidine is benzoylated, but no N-benzoylation is detected with deoxyguanosine or deoxyadenosine.

[1] J. Stawinski, T. Hozumi, and S. A. Narang, *J.C.S. Chem. Comm.*, 243 (1976).

N-Benzyl-1,4-dihydronicotinamide, (1), 6, 36–37.

Reduction of **gem-***bromonitro compounds.* 2-Bromo-2-nitropropane (2) is

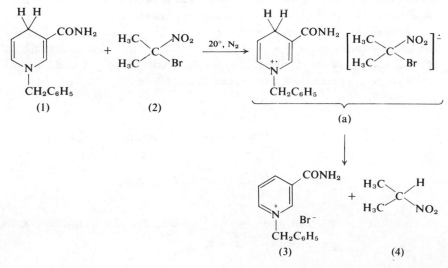

reduced by (1) at room temperature to 2-nitropropane (4) with formation of (3) as the only other isolable product [some acetone is formed by air oxidation of (4)]. A similar reduction of 2-*exo*-bromo-2-*endo*-nitrobornane leads to 2-nitrobornane (95% *endo*, 5% *exo*) in 97% yield. Several observations indicate that the reduction is initiated by an electron transfer from (1) to the substrate to give an ion–radical pair (a), which collapses to the observed products.[1]

[1] R. J. Kill and D. A. Widdowson, *J.C.S. Chem. Comm.*, 755 (1976).

Benzyldimethyldodecylammonium bromide, $[(CH_3)_2(C_6H_5CH_2)(n\text{-}C_{12}H_{25})N]^+$ Br^- (1). Mol. wt. 384.33.

Phase-transfer catalysis. 1,4-Dihydropyridines (2) can be N-alkylated with this catalyst in 60–85% yield in a two phase system of 50% aqueous NaOH and an organic solvent (toluene, benzene, methylene chloride). Alkylation is also

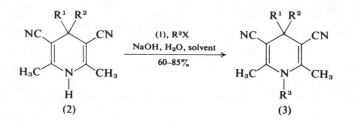

(2)   (3)

successful, but in somewhat lower yields, without this catalyst using KOH in DMSO.[1]

[1] J. Paleček and J. Kuthan, *Synthesis*, 550 (1976).

3-Benzyl-5-(2-hydroxyethyl)-4-methyl-1,3-thiazolium chloride (1), 6, 38–39, 289; 3-ethyl-5-(2-hydroxyethyl)-4-methyl-1,3-thiazolium chloride (2).

Review.[1] In the classical benzoin condensation cyanide ion is used as the catalyst. However, in the case of aliphatic aldehydes, thiazolium salts such as (1) and (2) are usually superior catalysts. The salts catalyze 1,4-addition of aldehydes to α,β-unsaturated ketones, esters, and nitriles. Examples:

$$CH_3CHO + CH_2{=}CHCOCH_3 \xrightarrow[61\%]{(1)} CH_3COCH_2CH_2COCH_3$$

$$C_6H_5CHO + CH_2{=}CHCOCH_3 \xrightarrow[65\%]{(2)} C_6H_5COCH_2CH_2COCH_3$$

$$OHC(CH_2)_2CHO + 2\ CH_2{=}CHCOCH_3 \xrightarrow[43\%]{(1)} CH_3CO(CH_2)_2CO(CH_2)_2CO(CH_2)_2COCH_3$$

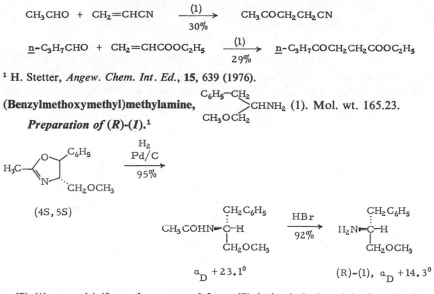

$$CH_3CHO + CH_2{=}CHCN \xrightarrow[30\%]{(1)} CH_3COCH_2CH_2CN$$

$$\underline{n}\text{-}C_3H_7CHO + CH_2{=}CHCOOC_2H_5 \xrightarrow[29\%]{(1)} \underline{n}\text{-}C_3H_7COCH_2CH_2COOC_2H_5$$

[1] H. Stetter, *Angew. Chem. Int. Ed.*, **15**, 639 (1976).

(Benzylmethoxymethyl)methylamine, $\begin{matrix}C_6H_5{-}CH_2\\ \\ CH_3OCH_2\end{matrix}\rangle CHNH_2$ (1). Mol. wt. 165.23.

Preparation of (R)-(1).[1]

(4S, 5S)

$$CH_3COHN{-}\overset{CH_2C_6H_5}{\underset{CH_2OCH_3}{\overset{|}{C}}}{-}H \xrightarrow[92\%]{HBr} H_2N{-}\overset{CH_2C_6H_5}{\underset{CH_2OCH_3}{\overset{|}{C}}}{-}H$$

$\alpha_D +23.1^0$ \qquad\qquad (R)-(1), $\alpha_D +14.3^0$

(S)-(1), $\alpha_D -14.4°$, can be prepared from (S)-(−)-ethyl phenylalaninate hydro-chloride by sodium borohydride reduction, followed by treatment with KH and CH_3I.

   Enantioselective alkylation of cyclohexanone.[1] Méa-Jacheet and Horeau[2] and Yamada *et al.*[3] have reported enantioselective alkylation of cyclohexanones via imines formed from chiral amines. Meyers *et al.*[1] have since reported efficient enantioselective alkylations of imines from (R)-(1). Thus the imine (2) is alkylated to give (3); the enantiomeric excess of the ketone is in the range of 82–95%.

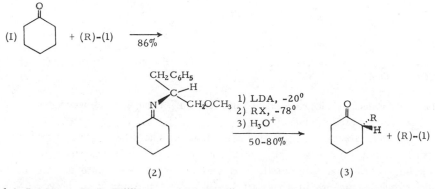

(2) \qquad\qquad\qquad\qquad (3)

[1] A. I. Meyers, D. R. Williams, and M. Druelinger, *Am. Soc.*, **98**, 3032 (1976).
[2] D. Méa-Jacheet and A. Horeau, *Bull. soc.*, 4571 (1968).
[3] M. Kitamoto, K. Hiroi, S. Terashima, and S. Yamada, *Chem. Pharm. Bull. Japan*, **22**, 459 (1974).

3-Benzyl-4-methylthiazolium chloride, 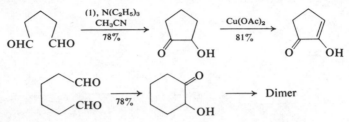 Cl^- (1). Mol. wt. 225.75.

This salt is prepared by reaction of 4-methyl-1,3-thiazole with benzyl chloride.

Cyclic α-ketols. The benzoin-like condensation has been extended to a synthesis of cyclic α-ketols by condensation of pentane- and hexanedials with this thiazolium salt. Oxidation of the products results in 2-hydroxy-2-enones.[1]

Examples:

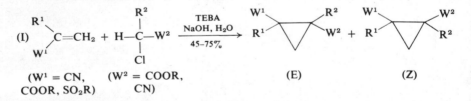

[1] R. C. Cookson and R. M. Lane, *J.C.S. Chem. Comm.*, 804 (1976).

Benzyltriethylammonium chloride (TEBA) **1,** 53; **3,** 19; **4,** 27–31; **5,** 26–28; **6,** 41–43.

The salt can be prepared in high purity by refluxing benzyl chloride and triethylamine in absolute ethanol for 64 hours. The salt is precipitated from the cooled solution with ether and purified by precipitation with ether from acetone solution.[1]

Cyclopropanes. Cyclopropanes are formed by condensation of α-halo-carbanions with electrophilic alkenes under phase-transfer conditions (equation

$$(I)\quad \begin{array}{c} R^1 \\ \diagdown \\ W^1 \end{array}\!\!C{=}CH_2 \ + \ H{-}\!\!\begin{array}{c} R^2 \\ | \\ C \\ | \\ Cl \end{array}\!\!{-}W^2 \quad \xrightarrow[45-75\%]{\substack{TEBA \\ NaOH, H_2O}} \quad \begin{array}{c} W^1 \triangle R^2 \\ R^1 \quad\quad W^2 \end{array} \ + \ \begin{array}{c} W^1 \triangle W^2 \\ R^1 \quad\quad R^2 \end{array}$$

$$(W^1 = CN, \quad (W^2 = COOR, \qquad\qquad (E) \qquad\qquad (Z)$$
$$COOR, SO_2R) \quad CN)$$

I). Two isomeric cyclopropanes are formed, but in the five published examples the (E)-isomer predominates.[2]

Dichlorocarbene. Japanese chemists[3] have modified the Makosza procedure by addition of benzene to make an emulsion. In this way hydrolysis in the aqueous alkaline solution is suppressed. This carbene reacts with epoxides to form dichlorocyclopropanes in yields of 15–40%. This reaction was shown to be stereospecific (equations I and II).

Ring expansion. Some five-membered nitrogen heterocycles have been expanded to six-membered heterocycles by use of dichlorocarbene generated by

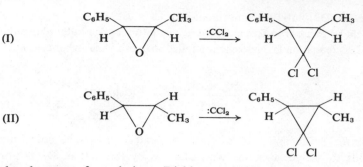

(I)

(II)

the phase-transfer technique. Dichloromethyl derivatives are usually formed as minor products.[4]

Examples:

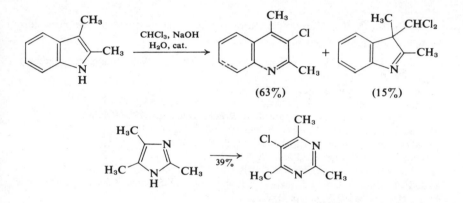

(63%)   (15%)

39%

Diazo transfer. Diazo transfer from tosyl azide to aryl amines can be effected by phase-transfer catalysis.[5]

$$ArNH_2 + TsN_3 \xrightarrow[\substack{NaOH, H_2O, C_6H_6 \\ 20-94\%}]{Cat.} ArN_3$$

Alkylation of nitriles. An example (equation I) of a general method for alkylation of active methylene groups by phase-transfer catalysis has been described in detail.[6]

(I) $C_6H_5CH_2CN + C_2H_5Br \xrightarrow[\substack{NaOH, H_2O \\ 78-84\%}]{[(C_2H_5)_3\overset{+}{N}CH_2C_6H_5]\,C\bar{l}} C_6H_5\underset{\underset{C_2H_5}{|}}{C}HCN$

Halogenation. Di-*t*-butyl phosphorohalidates (X = Cl, Br) can be prepared in *ca.* 90% yield by phase-transfer catalyzed halogenation of di-*t*-butyl phosphite with CCl_4 or CBr_4 (equation I). This reaction failed with other dialkyl phosphites.[7]

(I) $[(CH_3)_3CO]_2P{\overset{O}{\underset{H}{\diagup}}}$ + CX_4 $\xrightarrow[\sim 90\%]{\overset{\text{Cat.}}{CH_2Cl_2, \ NaOH, \ H_2O}}$ $[(CH_3)_3CO]_2P{\overset{O}{\underset{X}{\diagup}}}$ + CHX_3

                                           X = Cl, Br

α-Vinylnitriles.[8] The anion of a nitrile generated with solid potassium hydroxide adds to acetylenic bonds with catalysis by this salt in DMSO or HMPT to form α-vinylnitriles in 60–95% yields.

$$\underset{\underset{R}{|}}{ArCHCN} + HC\equiv CR' \xrightarrow[60\text{-}95\%]{\overset{\text{KOH, Cat.}}{DMSO}} \underset{\underset{R}{|}}{\overset{\overset{CH=CHR'}{|}}{ArCCN}}$$

Isonitriles (**5**, 27). The preparation of *t*-butyl isocyanide[9] by the Hofmann carbylamine reaction[10] has now been published.

[1] F. A. Souto-Bachiller, S. Masamune, C. J. Talkowski, and W. A. Sheppard (checkers), *Org. Syn.*, **55**, 97 (1976).
[2] A. Jończyk and M. Mąkosza, *Synthesis*, 387 (1976).
[3] I. Tabushi, Y. Kuroda, and Z. Yoshida, *Tetrahedron*, **32**, 997 (1976).
[4] F. DeAngelis, A. Gambacorta, and R. Nicoletti, *Synthesis*, 798 (1976).
[5] M. Nakajima and J.-P. Anselme, *Tetrahedron Letters*, 4421 (1976).
[6] M. Mąkosza and A. Jończyk, *Org. Syn.*, **55**, 91 (1976).
[7] T. Gajda and A. Zwierzak, *Synthesis*, 243 (1976).
[8] M. Mąkosza, J. Czyewski, and M. Jawdosiuk, *Org. Syn.*, **55**, 99 (1976).
[9] P. T. Hoffmann, G. Gokel, D. Marquarding, and I. Ugi, *Isonitrile Chemistry*, Academic Press, New York, 1971, Chap. II.
[10] G. W. Gokel, R. P. Widera, and W. P. Weber, *Org. Syn.*, **55**, 96 (1976).

Benzyltrimethylammonium hydroxide (Triton B), 1, 1252–1254; 5, 29.

Dehydrohalogenation. This base in benzene, toluene, or methylene chloride is a satisfactory reagent for dehydrohalogenation of haloalkenes to form alkynes.[1] One advantage over KOH or NaOH is that an aprotic solvent can be used.

Examples:

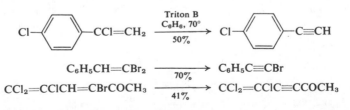

[1] A. Gorgues and A. Le Coq, *Bull. soc.*, 125 (1976).

Biacetyl (2,3-Butanedione), $CH_3COCOCH_3$. Mol. wt. 86.09, b.p. 88°. Suppliers: Aldrich, Eastman, others.

Epoxidation. Dye-sensitized photooxidation of aliphatic olefins results in formation of allylic hydroperoxides from reaction with singlet oxygen. Photo-oxidation with α-diketone sensitizers follows a different course: epoxides are the main products. Of several α-diketones examined, epoxides are obtained in highest yields with biacetyl, which is oxidized at the same time mainly to acetic acid. In some cases yields of epoxides in biacetyl photooxidation are 90% or more. Both *cis*- and *trans*-alkenes yield mainly *trans*-epoxides. Both epoxidation and ketone destruction are much slower with benzil. The nature of this epoxidation reaction is discussed, although many points of ambiguity remain. This reaction is of practical importance.[1]

[1] N. Shimizu and P. D. Bartlett, *Am. Soc.*, **98**, 4193 (1976).

Birch reduction, 1, 54–56; **2**, 27–29; **3**, 19–20; **4**, 31–33; **5**, 30.

α-Alkylcyclohexenones. These useful intermediates in organic synthesis can be prepared easily, albeit in modest yield, from *o*-methoxybenzoic acids by the method shown in equation (I).[1]

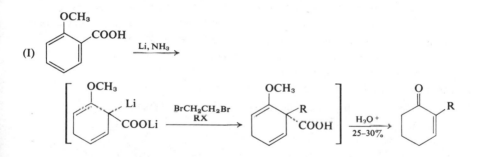

[1] D. F. Taber, *J. Org.*, **41**, 2649 (1976); see also M. D. Bach, J. W. Epstein, Y. Herzberg-Minzly, and H. J. E. Loewenthal, *ibid.*, **34**, 126 (1969).

Bis(acetonitrile)dichloropalladium(II), $(CH_3CN)_2PdCl_2$. Mol. wt. 259.42. Supplier: ROC/RIC.

Indoles. Hegedus, Sjöberg *et al.*[1] have reported a palladium-assisted amination of alkenes by secondary amines to produce tertiary amines. This reaction has

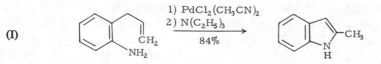

been extended to a more useful intramolecular amination to obtain indoles.[2] Thus addition of *o*-allylaniline to a solution of PdCl$_2$(CH$_3$CN)$_2$ in THF gives a yellow precipitate; upon addition of triethylamine the solid dissolves and Pd(0) is deposited with formation of 2-methylindole in 84% yield. This reaction is compatible with substituents on the benzene ring and on the 2- and 3-positions of the allyl side chain. The ring closure occurs at the most substituted terminus of the double bond: methallylaniline → 2,2-dimethylindoline.

A further application is the synthesis of 3-methylisocoumarin (equation II).

(II)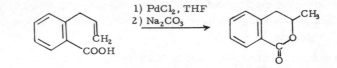

¹ B. Åkermark, J. E. Bäckvall, L. S. Hegedus, K. Zetterberg, K. Siirala-Hansén, and K. Sjöberg, *J. Organometal. Chem.*, **72**, 127 (1974).
² L. S. Hegedus, G. F. Allen, and E. L. Waterman, *Am. Soc.*, **98**, 2674 (1976).

1,4-Bis(chloromethoxy)butane; 1-chloro-4-chloromethoxybutane, 6, 104–105.

Warning. These chloromethyl ethers should be handled with care; they may be less dangerous than bis(chloromethyl) ether because of decreased volatility.

Chloromethylation. Olah and co-workers¹ have reported further studies on chloromethylation of arenes with these reagents. The reaction shows a wide range of selectivities with variation in conditions.

¹ G. A. Olah, D. A. Beal, and J. A. Olah, *J. Org.*, **41**, 1627 (1976).

Bis(1,3-diphenyl-2-imidazolidinylidene),   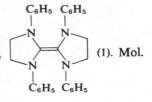   (1). Mol.

wt. 444.55, m.p. 285° dec.

Preparation: **2,** 430.

*Phenolic aldehydes.*¹ The reagent reacts with phenols in refluxing DMF to form cyclic aminals which are hydrolyzed quantitatively by mineral acids to an aldehyde and the salt of N,N'-diphenylethylenediamine. The reaction occurs preferably *para* to the phenolic OH group; if this position is blocked or hindered, reaction occurs at the *ortho* position.

Examples:

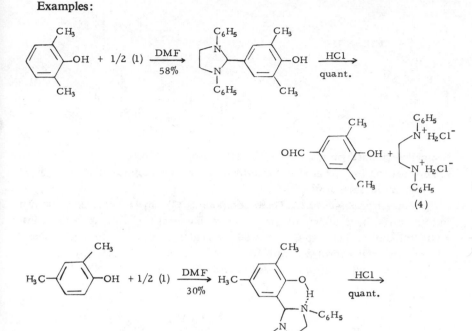

(4)

[1] J. Hocker, H. Gieseche, and R. Merten, *Angew. Chem. Int. Ed.*, **15**, 169 (1976).

Bis(ethenyl) sulfone, $(CH_2{=}CH)_2SO_2$. Mol. wt. 118.15, b.p. 90–92°/8 mm.; vesicant, mild lachrymator.

The sulfone is prepared conveniently by dehydration of di-β-hydroxyethyl sulfone with H_3PO_4 at 280° or by pyrolysis of di-β-acetoxyethyl sulfone at 500°.[1]

Homo-Diels-Alder reaction.[2] The reaction of benzyne with (1) gives, in addition to diphenylene, (2), (3), (4), and (5) as major products. Products (3)–(5) are evidently derived from (2). This product is considered to be formed via a [2 + 2 + 2]cycloaddition to give (a), followed by elimination of SO_2. Naphthalene (4) is a dehydrogenation product of (2); (3) is also derived from (2) by an ene reaction and is a precursor of (5). The combined yield of (2) and (4) is 26.8%; the yield of (3) is 11.2% and the yield of (5) is 0.5%.

[1] C. G. Overberger, D. L. Schoene, P. M. Kamath, and I. Tashlick, *J. Org.*, **19**, 1486 (1954).
[2] E. E. Nunn, *Tetrahedron Letters*, 4199 (1976).

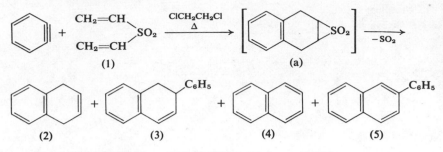

(1)   (a)

(2)   (3)   (4)   (5)

Bis(methylthio)methyllithium, $(CH_3S)_2CHLi$. Mol. wt. 114.16.

This reagent is prepared by treatment of bis(methylthio)methane with *n*-butyllithium in THF at 0°.

3-Formylcycloalkenones.[1] These compounds (3) can be prepared as shown in equation (I) by treatment of 3-ethoxycycloalkenones (1)[2] with bis(methylthio)-methyllithium in THF at 0° followed by acid hydrolysis to form (2). These adducts are then hydrolyzed with mercuric oxide in THF/H₂O. This synthesis has been used with cyclopentenones and cyclohexenones.

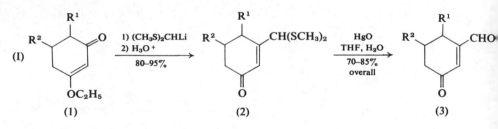

(1)   (2)   (3)

[1] M. L. Quesada and R. H. Schlessinger, *Syn. Comm.*, **6**, 555 (1976).
[2] Preparation: G. Stork and R. L. Danheiser, *J. Org.*, **38**, 1775 (1973).

Bismuth chloride, $BiCl_3$. Mol. wt. 315.34, m.p. 230–232°. Suppliers: Alfa, ROC/RIC.

Rearrangement of anilides. N-Acylanilines rearrange when heated with this catalyst at 160–245° for 2–9 minutes (equation I).[1]

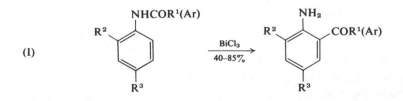

(I)

[1] A. Basha, S. S. Ahmed, and T. A. Farooqui, *Tetrahedron Letters*, 3217 (1976).

Bis(phenylthio)methane, $(C_6H_5S)_2CH_2$. Mol. wt. 232.36, m.p. 39.5–40.5°.

The reagent is prepared by the reaction of sodium thiophenoxide in ethanol with diiodomethane (80% yield).[1]

Ketone synthesis.[2] The anion of (1) can be alkylated in high yield to give (2), but a second alkylation of (2) is difficult. However, the anion of (2), if prepared in the presence of TMEDA, reacts readily with an aldehyde or ketone to form an alcohol (3). When (3) is dissolved in TFA at 20°, diphenyl disulfide separates and a ketone (4) is obtained from the filtrate (equation I). One advantage over the dithiane route is that the hydrolysis step is easier.

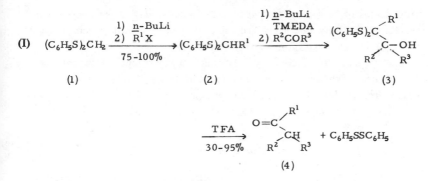

The same paper reports the use of the related reagent (5), prepared as shown in equation (II), for synthesis of ketones (equation III). The limitation of the synthesis is that ketones do not react satisfactorily.

(II)

$$R^1CH_2X \xrightarrow{P(C_6H_5)_3} (C_6H_5)_3\overset{+}{P}CH_2R^1 \underset{X^-}{\xrightarrow[H_2O]{OH^-}}$$

$$(X = Cl, Br)$$

$$(C_6H_5)_2\overset{\overset{O}{\|}}{P}CH_2R^1 \xrightarrow[2)\ (C_6H_5S)_2]{1)\ n\text{-}BuLi,\ TMEDA} (C_6H_5)_2\overset{\overset{O}{\|}}{P}\overset{\overset{R^1}{|}}{\underset{\underset{H}{|}}{C}}SC_6H_5$$

(5)

(III) (5) $\xrightarrow[75-80\%]{\overset{n\text{-}BuLi}{R^2CHO}} R^2CH=C\overset{R^1}{\underset{SC_6H_5}{\diagup}} \xrightarrow[80-90\%]{TFA} O=C\overset{R^1}{\underset{CH_2R^2}{\diagup}}$

α-Phenylthio ketones (7) can be prepared by use of an aldehyde rather than a ketone in equation (I). The intermediates (2) form adducts (6) with aldehydes, which on brief treatment with *p*-toluenesulfonic acid in benzene give α-phenylthio ketones (7), with the original carbonyl group of the aldehyde being regenerated.

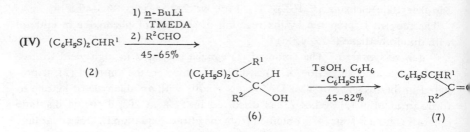

(IV) $(C_6H_5S)_2CHR^1$

(2)

(6)

(7)

The route seems to be general for synthesis of unsymmetrical ketones bearing a sulfenyl group on only one CH_2 group, a type that is not generally readily available by direct sulfenylation.[3]

Secondary alcohols. Secondary alcohols can be prepared in satisfactory yield by reaction of the anion of this reagent with trialkylboranes in the presence of excess mercuric chloride to induce migration of two alkyl groups. Oxidation (alkaline H_2O_2) of the intermediate borane compound leads to the alcohol.[4]

$$R_3B + LiCH(SC_6H_5)_2 \longrightarrow Li^+[R_3BCH(SC_6H_5)_2]^- \xrightarrow{-LiSC_6H_5}$$

$$R_2BCHR(SC_6H_5) \xrightarrow{HgCl_2} RBClCHR_2 \xrightarrow[75-90\%]{H_2O_2,\ NaOH} R_2CHOH$$

Yields of alcohols are appreciably lower when 2-lithio-1,3-dithiane is used in this sequence in place of bis(phenylthio)methyllithium.

[1] E. J. Corey and D. Seebach, *J. Org.*, **31**, 4097 (1966).
[2] P. Blatcher, J. I. Grayson, and S. Warren, *J.C.S. Chem. Comm.*, 547 (1976).
[3] P. Blatcher and S. Warren, *ibid.*, 1055 (1976).
[4] R. J. Hughes, A. Pelter, K. Smith, E. Negishi, and T. Yoshida, *Tetrahedron Letters*, 87 (1976).

Bis(tri-*n*-butyltin) oxide, $[(C_4H_9)_3Sn]_2O$. Mol. wt. 596.08, b.p. 180°/2 mm. Suppliers: Aldrich, Alfa, ROC/RIC.

Tributylstannyl ethers. These derivatives of carbohydrates can be prepared by azeotropic dehydration of the tin oxide and the sugar in refluxing toluene (Dean–Stark trap). Ethers containing a free hydroxyl group are stable to moisture for several days, but aprotic ethers are rapidly hydrolyzed when exposed to air.[1]

Oxidation of alcohols. Mukaiyama et al.[2] have reported that oxidation of benzylic, allylic, and secondary alcohols by this oxide (1) in combination with bromine gives the corresponding carbonyl compounds. More recently, normal primary alcohols have been found to give at the most only traces of aldehydes.[3] This mild procedure is therefore useful for selective oxidation of dihydroxy compounds to ω-hydroxy ketones.

Examples:

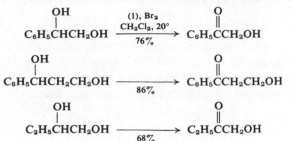

[1] A. J. Crowe and P. J. Smith, *J. Organometal. Chem.*, **110**, C57 (1976).
[2] K. Saigo, A. Morikawa, and T. Mukaiyama, *Bull. Chem. Soc. Japan*, **49**, 1656 (1976).
[3] Y. Ueno and M. Okawara, *Tetrahedron Letters*, 4597 (1976).

1,3-Bis(trimethylsilyloxy)-1,3-butadiene, (1). Mol. wt., 230.45, b.p. 58° (4 mm.).

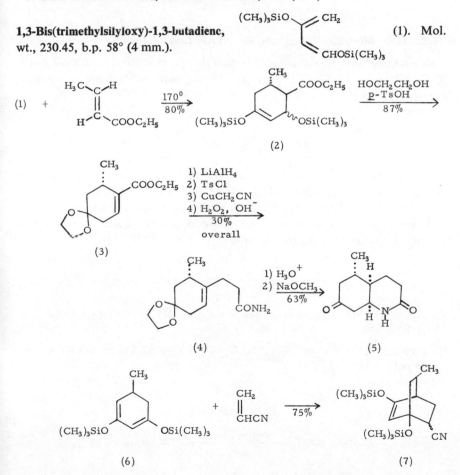

This diene is prepared by reaction of sodioacetoacetaldehyde and chloro-trimethylsilane.

Diels-Alder reaction.[1] This diene reacts with ethyl crotonate to give a single adduct (2), which was converted as shown into the keto lactam (5). The lactam (5) was also synthesized by an initial Diels-Alder reaction of the related cyclic diene (6). This work was carried out in connection with a total synthesis of a neurotoxin, pumiliotoxin (8), isolated from neotropical frogs.

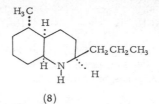

(8)

[1] T. Ibuka, Y. Mori, and Y. Inubushi, *Tetrahedron Letters*, 3169 (1976).

N,O-Bis(trimethylsilyl) sulfamate (Trimethylsilyl trimethylsilylamidosulfonate), $(CH_3)_3SiNHSO_3Si(CH_3)_3$ (1). Mol. wt. 241.46, m.p. 50°, soluble in common organic solvents.

Preparation:

$$[(CH_3)_3Si]_2NH + NH_2SO_3NH_4 \xrightarrow[78\%]{-2NH_3} (1)$$

Silylation.[1] This is an efficient silyl donor; in fact, it is superior to N,O-bis(trimethylsilyl)acetamide (**1**, 61; **2**, 30; **3**, 23–24) or to N,O-bis(trimethylsilyl) carbamate, $(CH_3)_3SiNHCO_2Si(CH_3)_3$.[2] Yields of silylated alcohols, phenols, and carboxylic acids are 92–98%. The by-product, sulfamic acid, is insoluble in the medium and removable by filtration.

$$R\!-\!OH + (1) \xrightarrow[92-98\%]{CH_2Cl_2} (CH_3)_3SiOR + H_2NSO_3H$$

[1] B. E. Cooper and S. Westall, *J. Organometal. Chem.*, **118**, 135 (1976).
[2] L. Birkofer and P. Sommer, *ibid.*, **99**, Cl (1975).

Bis(trimethylsilyl)thioketene, $[(CH_3)_3Si]_2C\!=\!C\!=\!S$. Mol. wt. 157.36, b.p. 45°/1 mm., stable even at 140°.

Preparation:

$$(CH_3)_3SiC\!\equiv\!CH \xrightarrow[90\%]{\begin{array}{l}1)\ \underline{n}\text{-BuLi}\\2)\ S\\3)\ (CH_3)_3SiCl\end{array}} [(CH_3)_3Si]_2C\!=\!C\!=\!S$$

(1)

Reactions:

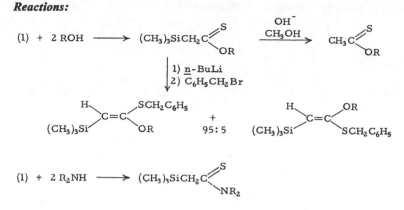

(1) + 2 ROH ⟶ $(CH_3)_3SiCH_2C\overset{S}{\underset{OR}{\diagup}}$ $\xrightarrow[CH_3OH]{OH^-}$ $CH_3C\overset{S}{\underset{OR}{\diagup}}$

$\downarrow$ 1) $\underline{n}$-BuLi
2) $C_6H_5CH_2Br$

$\overset{H}{\underset{(CH_3)_3Si}{}}C=C\overset{SCH_2C_6H_5}{\underset{OR}{}}$ + $\overset{H}{\underset{(CH_3)_3Si}{}}C=C\overset{OR}{\underset{SCH_2C_6H_5}{}}$

95 : 5

(1) + 2 R_2NH ⟶ $(CH_3)_3SiCH_2C\overset{S}{\underset{NR_2}{\diagup}}$

¹ S. J. Harris and D. R. M. Walton, *J.C.S. Chem. Comm.*, 1008 (1976).

9-Borabicyclo[3.3.1]nonane (9-BBN), 2, 31; 3, 24–29; 4, 41; 5, 46–47, 6, 62–64.

Hydroboration. Brown *et al.*[1] have examined the relative reactivities of various alkenes toward hydroboration with 9-BBN. This reagent has lesser steric requirements than disiamylborane, but is more sensitive to electron availability of the double bond.

Selective reductions.[2] The reducing properties of 9-BBN in THF have been reviewed in detail. Aldehydes and ketones are reduced rapidly and in high yield with some bias, in the case of unsymmetrical ketones, for attack from the less hindered side. α,β-Unsaturated aldehydes and ketones are reduced selectively to allylic alcohols. Anthraquinone (1) is reduced cleanly to 9,10-dihydro-9,10-anthracenediol (2). Carboxylic acids, acid chlorides, and esters are reduced efficiently. Epoxides are reduced only slowly by 9-BBN, but are reduced readily

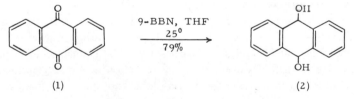

$\xrightarrow[79\%]{\substack{9-BBN, THF \\ 25^0}}$

(1)   (2)

in the presence of added lithium borohydride. Alkyl halides are inert to 9-BBN. Tertiary amides are reduced to alcohols. This reaction follows a course different from that of other organoboranes as shown in equation (I).

(I) $RCONR'_2$

$\xrightarrow{BH_3, THF} RCH_2NR'_2$

$\xrightarrow{Sia_2BH, THF} RCHO$

$\xrightarrow{9-BBN, THF} RCH_2OH$

γ,δ-*Unsaturated ketones*.[3] A B-alkenyl-9-BBN, prepared by hydroboration of alkynes with 9-BBN, undergoes 1,4-addition to methyl vinyl ketone (THF, reflux). Hydrolysis provides γ,δ-unsaturated ketones.

Examples:

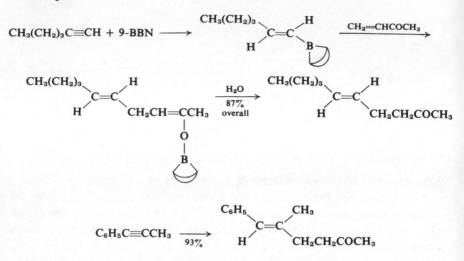

This reaction can be used with other cisoid α,β-enones, but transoid α,β-enones (*e.g.*, cyclohexenone-2) give complex mixtures containing none of the desired product.

[1] H. C. Brown, R. Liotta, and C. J. Scouten, *Am. Soc.*, **98**, 5297 (1976).
[2] H. C. Brown, S. Krishnamurthy, and N. M. Yoon, *J. Org.*, **41**, 1778 (1976).
[3] P. Jacob, III, and H. C. Brown, *Am. Soc.*, **98**, 7832 (1976).

9-Borabicyclo[3.3.1]nonane ate complexes, 6, 62–63.

Reduction of carbonyl groups. Japanese chemists[1] have reported some novel stereo-, chemo-, and regioselective reductions of carbonyl groups with the lithium di-*n*-butyl ate complex of 9-borabicyclo[3.3.1]nonane (1). Controlled stereoselective reduction is exemplified by reduction of 4-methylcyclohexanone (equation I). Either *cis*- or *trans*-4-methylcyclohexanol can be obtained depending on the presence of an additive.

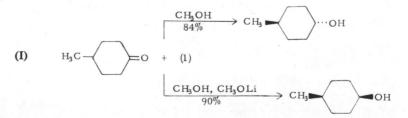

Chemoselectivity is evidenced by selective reduction of aldehydes in the presence of ketones (equation II).

(II) $CH_3(CH_2)_5CHO + CH_3(CH_2)_4COCH_3 \xrightarrow[57\%]{(1)}$

$$CH_3(CH_2)_5CH_2OH + CH_3(CH_2)_4CHOHCH_3$$
$$(95\%) \qquad\qquad\qquad (5\%)$$

Examples of regioselectivity are the reduction of a mixture of 2- and 4-heptanone (equation III) and the reduction of 4-(p-benzoylphenyl)-2-butanone (equation IV).

(III) $CH_3(CH_2)_4COCH_3 + CH_3(CH_2)_2CO(CH_2)_2CH_3 \xrightarrow[56\%]{(1)}$

$$CH_3(CH_2)_4CHOHCH_3 + CH_3(CH_2)_2CHOH(CH_2)_2CH_3$$
$$(91\%) \qquad\qquad\qquad (9\%)$$

(IV) C_6H_5CO —⟨⟩— $CH_2CH_2COCH_3 \xrightarrow[77\%]{(1)} C_6H_5CO$—⟨⟩—$CH_2CH_2CHOHCH_3$

[1] Y. Yamamato, H. Toi, A. Sonoda, S.-I. Murahashi, *Am. Soc.*, **98**, 1965 (1976).

Boron trifluoride, 1, 68–69; **3,** 32–33; **5,** 51–52.

Rearrangement of a 10-epieudesmane derivative to a nootkatane derivative. Treatment of the cyclopropyl ketone (1) with BF_3 at 20° leads to (2) in 46% yield (GLC) together with four minor, unidentified products. The reaction

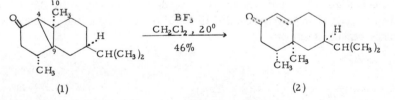

(1)   (2)

involves cleavage of the 4,9-bond and migration of the C_{10}-methyl group. The rearrangement is of interest as a possible model for biosynthesis of nootkatanes.[1]

[1] D. Caine and S. L. Graham, *Tetrahedron Letters*, 2521 (1976).

Boron trifluoride etherate, 1, 70–72, **2,** 35–36; **3,** 33; **4,** 44–45; **5,** 52–55; **6,** 65–70.

Cleavage of tosylhydrazones. Carbonyl compounds can be regenerated in high yield (80–95%) from tosylhydrazones by exchange with aqueous acetone in the presence of distilled BF_3 etherate (20°, 6–18 hours).[1] The paper reports 25 successful experiments.

Cyclization with rearrangement. In connection with studies directed toward a synthesis of anthracycline antibiotics, Sih and co-workers[2] treated the tetra-hydronaphthyl ester of 3-methoxyphthalic acid (1), prepared as shown, with BF_3 etherate at 90°. The product (2) is formed by a Fries rearrangement followed by cyclodehydration.

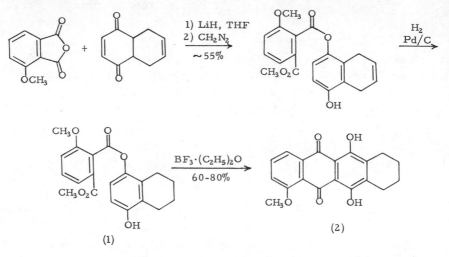

Regeneration of 3β-hydroxy-Δ⁵-steroids from 3,5-cyclo derivatives. Japanese chemists[3] have carried out this transformation with acetic acid in the presence of BF_3 etherate. *t*-Butyldimethylsilyl ether groups are cleaved at the same time. This newer method appears to be superior to the use of zinc acetate–acetic acid, aluminum oxide–acetic acid, or acetic anhydride–boron trifluoride etherate.[4]

Lactonization.[5] ω-Hydroxycarboxylic acids can be cyclized by BF_3 etherate in the presence of polystyrene beads.[6] In the absence of the polystyrene yields are lower, especially when n > 13.

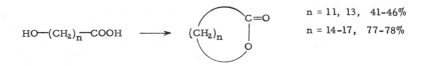

[1] C. E. Sacks and P. L. Fuchs, *Synthesis*, 456 (1976).
[2] D. G. Miller, S. Trenbeath, and C. J. Sih, *Tetrahedron Letters*, 1637 (1976).
[3] H. Hosoda, K. Yamashita, N. Chino, and T. Nambara, *Chem. Pharm. Bull. Japan*, **24**, 1860 (1976).
[4] C. R. Narayanan, S. R. Prakash, and B. A. Nagasampagi, *Chem. Ind.*, 966 (1974).
[5] L. T. Scott and J. O. Naples, *Synthesis*, 738 (1976).
[6] Rohm and Haas XE 305 polystyrene beads crosslinked with 3–5% divinylbenzene were used.

Boron trifluoride etherate–Ethanethiol.

Demethylation of ethers. Primary and secondary alkyl methyl ethers and aryl methyl ethers are cleaved to alcohols by BF_3 etherate in several thiols (ethanethiol is particularly effective). The reaction occurs at room temperature, but requires 1–12 days. Retention of stereochemistry is observed.[1]

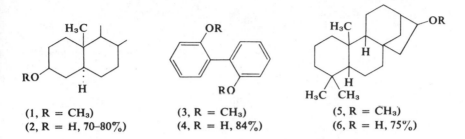

(1, R = CH₃)
(2, R = H, 70–80%)

(3, R = CH₃)
(4, R = H, 84%)

(5, R = CH₃)
(6, R = H, 75%)

[1] M. Node, H. Hori, and E. Fujita, *J.C.S. Perkin I*, 2237 (1976).

Boronic acid resins, (P)—B(OH)₂. *Cf.* **Phenylboronic acid, 1,** 833–834; **2,** 317; **3,** 221–222; **5,** 513–514.

Separation of cis- and trans-1,2- and 1,3-diols. Polystyrylboronic acid resins react selectively with *cis*-1,2- and 1,3-diols to form derivatives that can be cleaved to the diol by acetone–water to regenerate the polymer. The regenerated resins can be reused efficiently.[1]

[1] E. Seymour and J. M. J. Fréchet, *Tetrahedron Letters*, 1149, 3669 (1976).

Bromine, 3, 34; **4,** 46–47; **5,** 55–57; **6,** 70–73.

Photobromination.[1] When a mixture of chlorobenzene and bromine is irradiated with visible light, a slow reaction occurs to form *meta*-bromochlorobenzene in yields as high as 80%. In ionic brominations, a chlorine substituent is *ortho–para* directing. Some bromobenzene is also formed, the first report of radical replacement of chlorine by bromine. This unexpected reaction was investigated further with the following results. Chlorine is the most effective *meta*-directing halogen. NBS is more reactive than bromine itself in *meta* brominations. 1,2,3-Trihalogenobenzenes predominate in photobrominations of 1,3-dihalobenzenes rather than 1,3,5-trihalobenzenes. Photobromination of *p*-dihalobenzenes is extremely slow.

Free-radical bromination of ketones (**6,** 71). Details are available for regioselective bromination of ketones at the more highly substituted α-position by photocatalytic bromination in CCl₄ at 20° in the presence of 1,2-epoxycyclohexane. The epoxide is a scavenger for HBr and thus promotes a free-radical bromination and inhibits bromination through the enol.[2]

Examples:

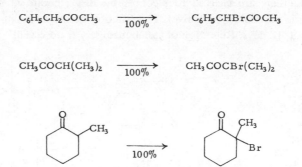

$C_6H_5CH_2COCH_3 \xrightarrow[100\%]{} C_6H_5CHBrCOCH_3$

$CH_3COCH(CH_3)_2 \xrightarrow[100\%]{} CH_3COCBr(CH_3)_2$

Bromomethyl ketones. Direct bromination of methyl ketones as a route to bromomethyl ketones is usually unsatisfactory because bromine mainly attacks the more substituted carbon. One expedient is reaction of ketals of the ketone with phenyltrimethylammonium perbromide (**1, 855**) or with bromine in methanol.[3] Actually methyl ketones can be brominated directly with bromine in methanol, but the regiospecificity depends on the structure of the ketone. Thus 1-bromo-3-methyl-2-butanone (**2**) can be obtained directly from (**1**) in good yield as shown in equation (I). It should be noted that dimethyl ketals are formed as well and must be hydrolyzed during work-up.[4]

(I) $(CH_3)_2CHCOCH_3 \xrightarrow[95\% \text{ crude}]{\begin{array}{l}1)\ Br_2,\ CH_3OH\\ 2)\ H_2O\end{array}} (CH_3)_2CHCOCH_2Br$

(1)   (2)

α-Bromo carbonyl compounds. Bromination of trimethylsilyl enol ethers either with bromine in carbon tetrachloride or pentane or with NBS in carbon tetrachloride or THF provides an excellent method for regiospecific introduction of bromine at the α-position of carbonyl compounds. At present it is not possible to predict which method of bromination is preferable.[5]

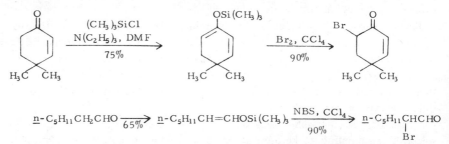

$n\text{-}C_5H_{11}CH_2CHO \xrightarrow[65\%]{} n\text{-}C_5H_{11}CH=CHOSi(CH_3)_3 \xrightarrow[90\%]{NBS,\ CCl_4} n\text{-}C_5H_{11}\underset{\underset{Br}{|}}{C}HCHO$

Alkyl bromides. Methyl or phenyl alkyl selenides (**1**) react with bromine, NBS, or bromine and triethylamine to form alkyl bromides (**2**). The third method generally gives the highest yield. Yields are poor when $R^1 = R^2 = H$. In this case, the alkyl iodide can be obtained in 65–80% yield with CH_3I and NaI in DMF (2, 324).[6]

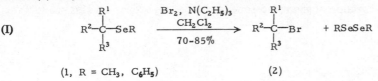

$$(I) \qquad R^2\text{-}\underset{\underset{R^3}{|}}{\overset{\overset{R^1}{|}}{C}}\text{-SeR} \xrightarrow[\substack{CH_2Cl_2 \\ 70\text{-}85\%}]{Br_2,\ N(C_2H_5)_3} R^2\text{-}\underset{\underset{R^3}{|}}{\overset{\overset{R^1}{|}}{C}}\text{-Br} \quad + \text{ RSeSeR}$$

$$(1,\ R = CH_3,\ C_6H_5) \qquad\qquad (2)$$

This reaction permits the homologation of alkyl halides in about 60% overall yield (equation II). Alkyl bromides or chlorides are obtained by treatment of

$$(II) \quad R^3X + Li\underset{\underset{R^2}{|}}{\overset{\overset{R^1}{|}}{C}}\text{-SeR} \longrightarrow R^3\text{-}\underset{\underset{R^2}{|}}{\overset{\overset{R^1}{|}}{C}}\text{-SeR} \xrightarrow[55\text{-}95\%]{\substack{1)\ O_3 \\ 2)\ HX,\ N(C_2H_5)_3}} R^3\text{-}\underset{\underset{R^2}{|}}{\overset{\overset{R^1}{|}}{C}}\text{-X}$$

$$(X = Br,\ I)$$

selenoxides with gaseous hydrobromic acid or hydrochloric acid in methylene chloride containing triethylamine. Yields, however, are low when $R^1 = R^2 = H$.[7]

The exact role of the tertiary amine in these transformations is not clear, but the reaction proceeds more readily and in higher yield when the amine is present.

[1] P. Gouverneur and J. P. Soumillion, *Tetrahedron Letters*, 133 (1976).
[2] V. Calò, L. Lopez, and G. Pesce, *J.C.S. Perkin I*, 501 (1977).
[3] M. Gaudry and A. Marquet, *Tetrahedron*, **26**, 5611, 5617 (1970).
[4] *Idem, Org. Syn.*, **55**, 24 (1976).
[5] L. Blanco, P. Amice, and J. M. Conia, *Synthesis*, 194 (1976).
[6] M. Sevrin, W. Dumont, L. Hevesi, and A. Krief, *Tetrahedron Letters*, 2647 (1976).
[7] L. Hevesi, M. Sevrin, and A. Krief, *ibid.*, 2651 (1976).

Bromine–Hexamethylphosphoric triamide, Br_2–HMPT.

Oxidation of alcohols.[1] Bromine or chlorine in HMPT and in the presence of a base (NaH_2PO_4) oxidizes primary and secondary alcohols to carbonyl compounds in high yield. Since secondary alcohols are oxidized more readily than primary ones, selective oxidations are possible (last example).

Examples:

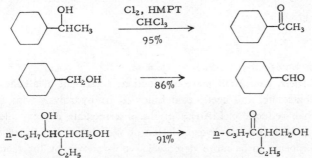

[1] M. Al Neirabeyeth, J.-C. Ziegler, and B. Gross, *Synthesis*, 811 (1976).

Bromine–Silver tetrafluoroborate.

Bromocyclization. Wolinsky and Faulkner[1] have published a synthesis of 10-bromo-α-chamigrene (5), the key step of which involves a bromonium ion initiated cyclization of geranylacetone (1) to the vinyl ether (2). This product was

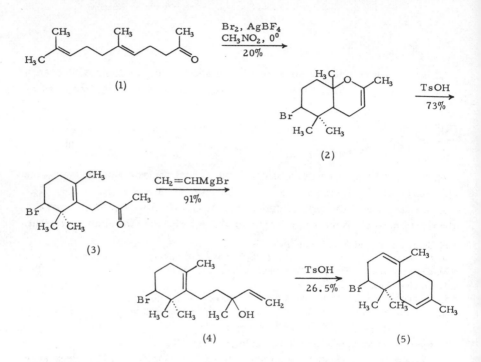

converted into (5) by conventional methods. This synthesis may have biosynthetic significance since several brominated sesquiterpenes have been isolated as a natural product.

[1] L. E. Wolinsky and D. J. Faulkner, *J. Org.*, **41**, 597 (1976).

N-Bromoacetamide–Silver acetate–Acetic acid.

cis-1,2-Diols. cis-1,2-Diols are obtained by treatment of alkenes at 0° with NBA, silver acetate, and acetic acid followed by hydrolysis with dilute acetic acid and then reduction (LiAlH₄) of the intermediate acetoxy alcohols. The reaction is related to the *cis*-hydroxylation of Woodward and Brutcher (1, 1002–1003), but appears to be more stereoselective as shown by the reaction of (1). In addition, yields are generally higher.[1]

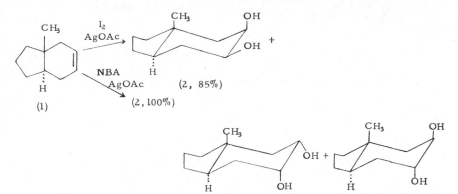

[1] D. Jasserand, J. P. Girard, J. C. Rossi, and R. Granger, *Tetrahedron Letters*, 1581 (1976).

2-(2-Bromoethyl)-1,3-dioxane, $\langle\!\!\langle$—CH_2CH_2Br (1). Mol. wt. 195.06, b.p. 67–70° (2.8mm.).

The dioxane is prepared from 1,3-propanediol, acrolein, and hydrogen bromide.[1]

γ-*Keto aldehydes.*[2] The Grignard reagent of this acetal reacts with acid chlorides in refluxing THF to form an intermediate that is hydrolyzed by aqueous oxalic acid to a γ-keto aldehyde (example I). The Grignard derivative of 2-

(I) $\langle\!\!\langle$—CH_2CH_2MgBr + $Cl\overset{\overset{O}{\|}}{C}(CH_2)_5CH_3$ $\xrightarrow[92\%]{THF,\ \Delta}$

$\langle\!\!\langle$—$CH_2CH_2\overset{\overset{O}{\|}}{C}(CH_2)_5CH_3$ $\xrightarrow[89\%]{(HOOC)_2,\ H_2O}$ $H\overset{\overset{O}{\|}}{C}CH_2CH_2\overset{\overset{O}{\|}}{C}(CH_2)_5CH_3$

(2-bromoethyl)-1,3-dioxolane has also been used, but this reagent is less stable than (1).

[1] D. C. Kriesel and O. Gisvold, *J. Pharm. Sci.*, **60**, 1250 (1971).
[2] J. C. Stowell, *J. Org.*, **41**, 560 (1976).

N-Bromosuccinimide, 1, 78–80; **2,** 40–42; **3,** 34–36; **4,** 49–53; **5,** 65–66; **6,** 74–76.

Allylic oxidation (**3,** 35). NBS is somewhat superior to the Ratcliffe reagent (CrO_3–Py, **4,** 96) for oxidation of (1) to α- and β-levantenolide, (2) and (3), diterpenes of the Nicotiana species. The reaction was carried out in dioxane–water in the presence of calcium carbonate. An intermediate furane may account for the formation of the anomalous β-isomer (3).[1]

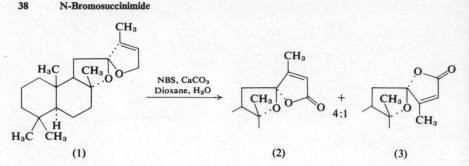

(1)   (2)   (3)

Oxidative esterification. Japanese chemists[2] have reported a novel oxidation discovered in the carbohydrate field. Reaction of the tri-*n*-butylstannyl alkoxide (1)[3] with 1 equiv. of NBS in CCl_4 at 20° gives the ester (2) and succinimide (3) in high yield. The reaction involves oxidation of (1) to an aldehyde, which then reacts with (1) to form the ester (2).

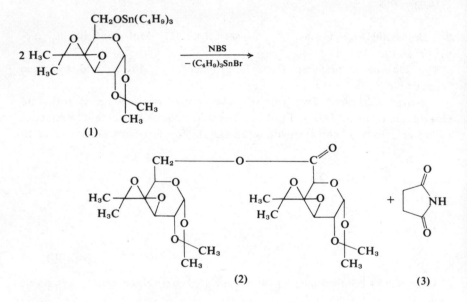

This reaction was shown to be general. In the absence of an added aldehyde, primary alkoxides give dimeric esters, RCH_2OCOR, and secondary alkoxides are converted into ketones. If an aldehyde (RCHO) is added to either reaction, esters are formed, R^1CH_2OCOR from a primary alkoxide and $R^1R^2CHOCOR$ from a secondary alkoxide.

Examples:

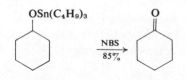

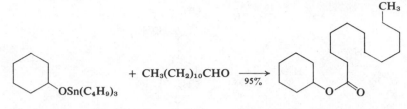

$$+ CH_3(CH_2)_{10}CHO \xrightarrow{95\%}$$

The method is related to that of Saigo *et al.*[4] in which an alcohol is converted into the triethyltin alkoxide by reaction with triethyltin methoxide; an aldehyde or ketone is formed by oxidation of the alkoxide with bromine in the presence of triethyltin methoxide (**6**, 613).

$$\underset{R^2}{\overset{R^1}{>}}CHOH + (C_2H_5)_3SnOCH_3 \xrightarrow{80-90\%} \underset{R^2}{\overset{R^1}{>}}CHOSn(C_2H_5)_3 \xrightarrow[80-95\%]{\overset{(C_2H_5)_3SnOCH_3}{Br_2}}$$

$$\underset{R^2}{\overset{R^1}{>}}C=O + (C_2H_5)_3SnBr + CH_3OH$$

Cyclization of methyl farnesate. Reaction of methyl *trans,trans*-farnesate (1) with NBS and cupric acetate in *t*-butyl alcohol–HOAc yields the cyclic product (2) in 12% yield. The product was transformed into snyderol (3), a sesquiterpene in *Laurencia snyderae*.[5]

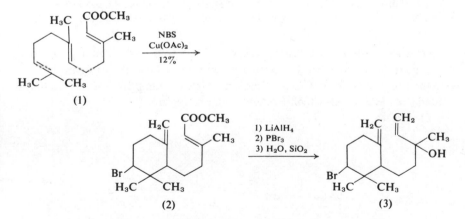

Halogenation of a C_{16}-macrolide. Reaction of the pyrrolidine enamine of the antibiotic leucomycin A_3 (1) with NBS yields, after hydrolysis, the 17-bromo derivative (62%). A similar reaction is observed with NCS (66% yield). In contrast, reaction with perchloryl fluoride in limited amounts results in fluorination at C_{17} and C_{18}. The C_{17}-monofluoride was made indirectly by reaction first with NBS and then with perchloryl fluoride; the bromofluoro

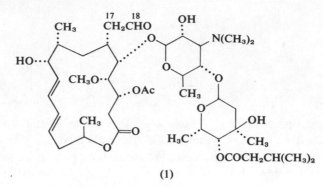

(1)

product was then reduced selectively with triphenylphosphine to give the 17-monofluoro derivative.[6]

α-Bromo acids (**5**, 65). The NBS route to α-bromo acids has been published.[7]

[1] A. G. González, C. G. Francisco, R. Freire, R. Hernández, J. A. Salazar, and E. Suárez, *Tetrahedron Letters*, 2725 (1976).

[2] T. Ogawa and M. Matsui, *Am. Soc.*, **98**, 1629 (1976).

[3] The alkoxides are prepared by reaction of the alcohol with bis(tri-*n*-butyltin) oxide, [(C_4H_9)$_3$Sn]$_2$O, in refluxing toluene, with azeotropic removal of water [A. G. Davies, *Synthesis*, 56 (1969)].

[4] K. Saigo, A. Morikawa, and T. Mukaiyama, *Chem. Letters*, 145 (1975).

[5] A. G. González, J. D. Martin, C. Pérez, and M. A. Ramírez, *Tetrahedron Letters*, 137 (1976).

[6] N. N. Girota, A. A. Patchett, and N. L. Wendler, *Tetrahedron*, **32**, 991 (1976).

[7] D. N. Harpp, L. Q. Bao, C. Coyle, J. G. Gleason, and S. Horovitch, *Org. Syn.*, **55**, 27 (1976).

Bromotrichloromethane, CCl_3Br. Mol. wt. 198.28, b.p. 104°. Supplier: Aldrich.

Benzylic bromination. Under irradiation alkyl aromatics can be brominated in the benzylic position with this reagent.[1] Yields are improved if the reaction is conducted at 30° rather than 60°, and under nitrogen.[2]

Examples:

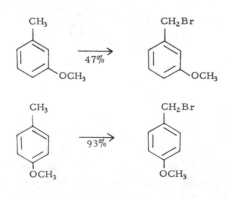

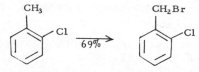

[1] E. S. Huyser, *Am. Soc.*, **82**, 391 (1960).
[2] S. W. Baldwin and T. H. O'Neill, *Syn. Comm.*, **6**, 109 (1976).

5-Bromo-2,2,5-trimethyl-1,3-dioxane-4,6-dione, 5, 66–68.

Transfer alkylation (**5**, 67–68). The complete paper is available.[1]

[1] B. M. Trost and L. S. Melvin, Jr., *Am. Soc.*, **98**, 1204 (1976).

t-**Butoxybis(dimethylamino)methane** $[(CH_3)_2N]_2CHOC(CH_3)_3$ (**1**), **5**, 71–73.

α-Diketones.[1] Ketones can be converted into α-diketones in high yield by reaction with this derivative of DMF to form an enamino ketone followed by cleavage of the C=C bond by singlet oxygen. In practice isolation of the intermediate is not necessary. Yields of α-diones by this method are usually superior to those obtained with SeO_2 (Riley oxidation, **1**, 993).

Examples:

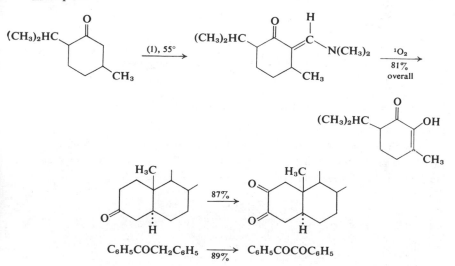

[1] H. H. Wasserman and J. L. Ives, *Am. Soc.*, **98**, 7868 (1976).

t-**Butyl azidoformate, 1**, 84–85; **2**, 44–45; **3**, 36; **4**, 54; **6**, 77–78.

Caution. Fenlon (Eastman Kodak)[1] and Aldrich[2] warn that this substance is extremely shock sensitive and unstable between 80 and 135°. O-*t*-Butyl S-phenyl thiocarbonate[3] is recommended as a substitute.

A serious detonation of undetermined origin during the preparation of the compound has since been reported.[4]

[1] W. J. Fenlon, *Chem. Eng. News*, May 24, 3 (1976).
[2] Aldrich Chem. Co., *ibid.*, Sept. 20, 5 (1976).
[3] $(CH_3)_3COCOSC_6H_5$, supplier: Eastman.
[4] P. Feyen, *Angew. Chem. Int. Ed.*, **16**, 115 (1977).

t-Butyl bis(trimethylsilyl)lithioacetate, $LiC[Si(CH_3)_3]_2COOC(CH_3)_3$ (1). Mol. wt. 266.75.

Preparation:

$$LiCHSi(CH_3)_3COOC(CH_3)_3 \xrightarrow[-78°]{(CH_3)_3SiCl} \underset{(70\%)}{HC[Si(CH_3)_3]_2COOC(CH_3)_3} \xrightarrow[THF, -78°]{LDA} (1)$$

α-*Trimethylsilyl vinyl esters*.[1] The reagent reacts with aldehydes, but not ketones, to form these compounds in 70–90% yield (GLPC), equation (I).

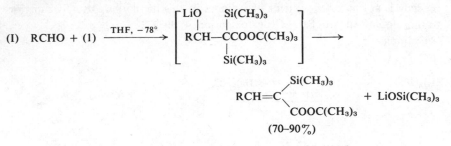

$$(I) \quad RCHO + (1) \xrightarrow{THF, -78°} \left[\begin{array}{c} LiO \quad Si(CH_3)_3 \\ | \qquad | \\ RCH—CCOOC(CH_3)_3 \\ | \\ Si(CH_3)_3 \end{array} \right] \longrightarrow$$

$$RCH=C\underset{COOC(CH_3)_3}{\overset{Si(CH_3)_3}{\diagup}} \quad + LiOSi(CH_3)_3$$

$$(70–90\%)$$

[1] S. L. Hartzell and M. W. Rathke, *Tetrahedron Letters*, 2737 (1976).

t-Butyl chromate, 1, 86–87; **2,** 48; **3,** 37; **4,** 54–55; **5,** 73–74.

Allylic oxidation. *t*-Butyl chromate was found superior to selenium dioxide or $CrO_3 \cdot Py_2$ for oxidation of the trienone (1) to (2). This reaction was used in a transformation of α-santonin (4) into yomogin (5).[1]

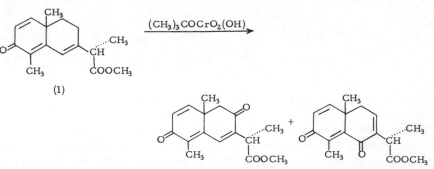

(2, 28-39%)   (3, 10-20%)

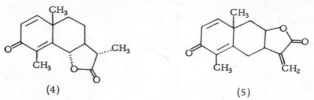

(4)   (5)

[1] K. Yamakawa, K. Nishitani, and A. Yamamoto, *Chem. Letters*, 177 (1976).

_t_-Butylcyanoketene, 4, 55; **6,** 78.

Preparation: Detailed instructions for the preparation have been published.[1] It is stable to self-condensation in benzene solution at 20°, but it has not been isolated as such.

[1] W. Weyler, Jr., W. G. Duncan, M. B. Liewen, and H. W. Moore, *Org. Syn.*, **55,** 32 (1976).

_t_-Butyl hydroperoxide, 1, 88–89; **2,** 49–50; **3,** 37–38; **5,** 75–77; **6,** 81–82.

1,3-Transposition of allylic alcohols. This reaction can be accomplished in three steps, the first of which is Sharpless epoxidation (**5,** 75–76). The resulting epoxide is then converted into the epoxy mesylate. The last step is reduction with either sodium and liquid ammonia or sodium naphthalenide.[1]

Examples:

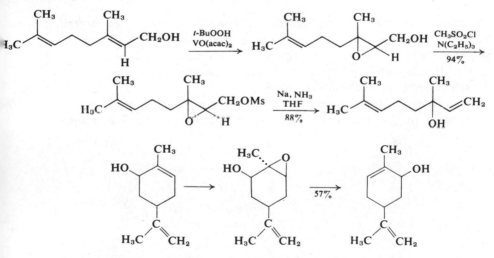

Epoxidation of a latent enedione. The last step in a total synthesis of the antibiotic flavipucine (2), isolated from *Aspergillus flavipes*, involves treatment of (1) with *t*-butyl hydroperoxide and potassium *t*-butoxide in *t*-butanol (0–25°). Presumably the enedione (a) is formed and then epoxidized. The product consists of (±)-flavipucine and (±)-isoflavipucine, separable by fractional crystallization.[2]

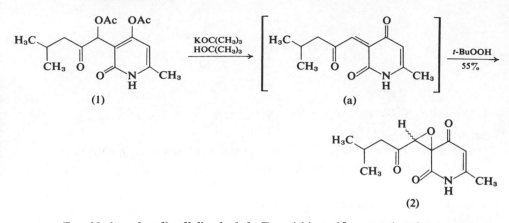

Epoxidation of cyclic allylic alcohols. Teranishi *et al.*[3] report that the stereo-

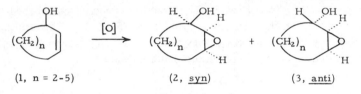

(1, n = 2-5)   (2, <u>syn</u>)   (3, <u>anti</u>)

chemical course of epoxidation of allylic alcohols (1) depends on the reagent. *t*-Butyl hydroperoxide oxidation catalyzed by VO(acac)$_2$ leads in all cases to almost exclusive formation of *syn*-epoxides. This course is also observed for (1, n = 2) with *m*-chloroperbenzoic acid. But as n increases to 4 or 5, the *anti*-epoxide is formed almost exclusively (99.8%).

Epoxidation of the cyclic dienol (4) with *m*-chloroperbenzoic acid proceeds as expected to give the *anti*-2,3-epoxy-4-enol (5); epoxidation with *t*-butyl

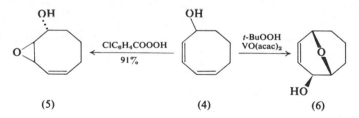

(5)   (4)   (6)

hydroperoxide and the vanadium catalyst gives the oxabicyclic compound (6) (*cf.* 5, 77–78). The latter reaction involves two steps, normal epoxidation followed by a transannular rearrangement.[4]

Stereospecific chloroepoxide synthesis. Treatment of the allylic alcohol (1) with *t*-butyl hypochlorite (2 equiv.) in CCl$_4$ at 20° leads to the chloroepoxide (2) as the major product. A corresponding product was not formed from the epimeric

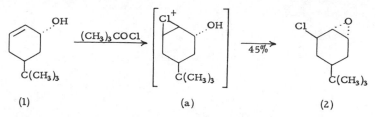

(1)   (a)   (2)

alcohol (pseudo-equatorial). Ganem presents evidence for an ionic pathway involving the hydroxychloronium ion (a).[5]

[1] A. Yasuda, H. Yamamoto, and H. Nozaki, *Tetrahedron Letters*, 2621 (1976).
[2] N. N. Girotra, Z. S. Zelawski, and N. L. Wendler, *J.C.S. Chem. Comm.*, 566 (1976).
[3] T. Itoh, K. Kaneda, and S. Teranishi, *ibid.*, 421 (1976).
[4] T. Itoh, K. Jitsukawa, K. Kaneda, and S. Teranishi, *Tetrahedron Letters*, 3157 (1976).
[5] B. Ganem, *Am. Soc.*, **98**, 858 (1976).

***n*-Butyllithium**, **1**, 95–96; **2**, 51–53; **4**, 60–63; **5**, 78; **6**, 85–91.

Carbodiimides. Carbodiimides can be prepared in 30–60% yield by reaction of thioureas with 2 equiv. of *n*-butyllithium followed by thermal decomposition. The yield is improved to 65–85% by addition of carbon disulfide.[1]

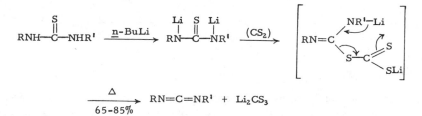

$$\xrightarrow[65-85\%]{\Delta} RN=C=NR' + Li_2CS_3$$

Lithiation of epoxides. Styrene oxide is metallated exclusively at the carbon atom α to the phenyl group by *n*-butyllithium, but only to the extent of ~50% (equation I). However, β-trimethylsilylstyrene oxide is metallated in high yield under the same conditions (equation II).

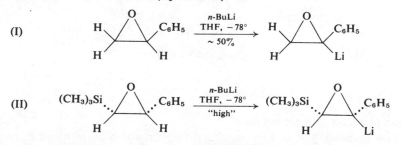

Epoxyalkylsilanes such as epoxyethyltriphenylsilane are metallated at the carbon α to silicon (equation III).[2]

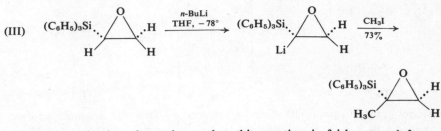

(III)

Later investigations have shown that this reaction is fairly general for α-heterosubstituted epoxides, particularly those containing Si, P, or S. *n*- or *t*-Butyllithium (sometimes with TMEDA) or LDA can be used as the base; useful solvents include THF, hexane, ether, and combinations thereof.[3]

α-Lithiocarboxylates. These useful intermediates are generally prepared with LDA. In the case of dihydro-1,3-thiazole-4-carboxylic acids (1), this method of metallation is not successful, but two protons can be substituted by use of *n*-butyllithium in THF. The reaction was investigated as a method for direct oxygenation of luciferin analogs.[4]

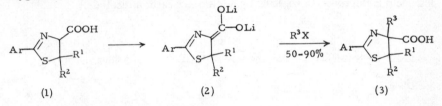

(1)   (2)   (3)

Intramolecular cyclization. The epoxide (1), derived from *trans*-nerolidol in 71% yield, is cyclized by *n*-butyllithium in the presence of DABCO to the two isomeric hydroxy thioethers (2) and (3). This reaction is a convenient route to 10-membered sesquiterpenes of the germacrane type and may be involved in

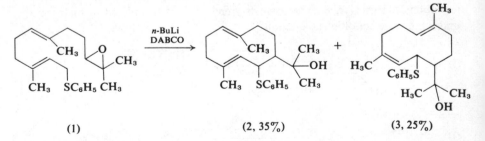

(1)   (2, 35%)   (3, 25%)

biogenetic pathways to these terpenes.[5]

Sulfone synthesis. The reaction of phenyl tosylate with *n*-butyllithium gives *n*-butyl *p*-tolyl sulfone in 85% yield. This synthesis is general for sulfonate esters, and any organolithium compound can be used.[6]

$$CH_3C_6H_4SO_2OC_6H_5 + C_4H_9Li \xrightarrow{\text{Ether}} CH_3C_6H_4SO_2C_4H_9 + C_6H_5OLi$$
$$(85\%)$$

[1] S. Sakai, T. Fujinami, N. Otani, and T. Aizawa, *Chem. Letters*, 811 (1976).
[2] J. J. Eisch and J. E. Galle, *Am. Soc.*, **98**, 4646 (1976).
[3] *Idem, J. Organometal. Chem.*, **121**, C10 (1976).
[4] W. Adam and V. Ehrig, *Synthesis*, 817 (1976).
[5] M. Kodama, Y. Matsuki, and S. Ito, *Tetrahedron Letters*, 1121 (1976).
[6] W. H. Baarschers, *Canad. J. Chem.*, **54**, 3056 (1976).

_t_-Butyllithium, 1, 96–97; 5, 79–80.

Vinyllithium derivatives.[1] Terminal alkenyllithiums (2) are not generally available by Br/Li exchange of vinyl bromides (1) with alkyllithiums because of dehydrobromination to alkynes. However, they are accessible by reaction of (1) with 2 equiv. of *t*-butyllithium at $-20°$ in THF–ether–pentane.[2] Subsequent

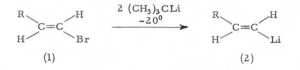

(1)   (2)

reactions of (2) with electrophiles can be carried out at $20°$ if desired.
Examples:

$$CH_2{=}CHLi + C_6H_5CHO \xrightarrow[76\%]{-78 \to 20°} CH_2{=}CHCH(OH)C_6H_5$$

$$CH_2{=}CHLi + \underline{n}\text{-}C_8H_{17}I \xrightarrow[52\%]{-90 \to 20°} CH_2{=}CHC_8H_{17}\text{-}\underline{n}$$

$$CH_2{=}CHLi + C_6H_5SSC_6H_5 \xrightarrow{74\%} CH_2{=}CHSC_6H_5$$

[1] H. Neumann and D. Seebach, *Tetrahedron Letters*, 4839 (1976).
[2] G. Köbrich and H. Trapp, *Ber.*, **99**, 680 (1966).

_n_-Butyllithium–Tetramethylethylenediamine, 2, 403; 3, 284; 4, 485–489; 5, 678–680.

Vinyllithium reagents. Bond et al.[1] have modified the Shapiro-Heath olefin synthesis (2, 418–419; 4, 511; 5, 678–680; 6, 598–599) by use of TMEDA as solvent. In this way, the intermediate vinyl carbanions can be trapped by a variety of electrophilic reagents. At least 3 equiv. of *n*-butyllithium must be used, otherwise the olefin becomes a major product.

Examples:

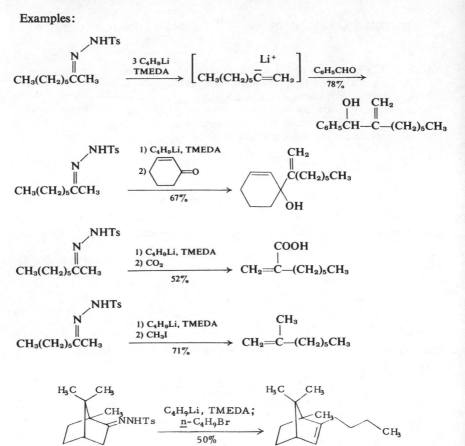

One attractive feature is the regioselectivity in formation of the vinyl carbanion. Thus 2-octanone is converted almost entirely into the 1-octenyl anion. The products containing the unit $\diagdown$C=CH₂ can be oxidized to $\diagdown$C=O by NaIO₄–KMnO₄. One drawback is that epoxides do not react cleanly in this process. Another is that aldehyde tosylhydrazones cannot be used.

[1] J. E. Stemke, A. R. Chamberlin, and F. T. Bond, *Tetrahedron Letters*, 2947 (1976).

t-**Butyl nitrite**, (CH₃)₃CONO. Mol. wt. 103.12. This nitrite can be prepared according to the general procedure of Noyes.[1]

   gem-*Dihalides*.[2] Unbranched primary amines undergo oxidative deamination when treated with an alkyl nitrite and an anhydrous copper(II) halide in acetonitrile (1 hour, 65°). gem-Dihalides are obtained in fair to high yield. *t*-Butyl nitrite is somewhat superior to isopentyl nitrite.

(I) $RCH_2NH_2 + 2\,CuX_2 + (CH_3)_3CONO \xrightarrow[45-80\%]{CH_3CN}$
         (X = Cl, Br)

$$RCHX_2 + 2\,CuX + (CH_3)_3COH + N_2$$

[1] W. A. Noyes, *Org. Syn. Coll. Vol.*, **2**, 108 (1943).
[2] M. P. Doyle and B. Sigfried, *J.C.S. Chem. Comm.*, 433 (1976).

t-Butyl perbenzoate, **1**, 98–101; **2**, 54–55; **5**, 66.

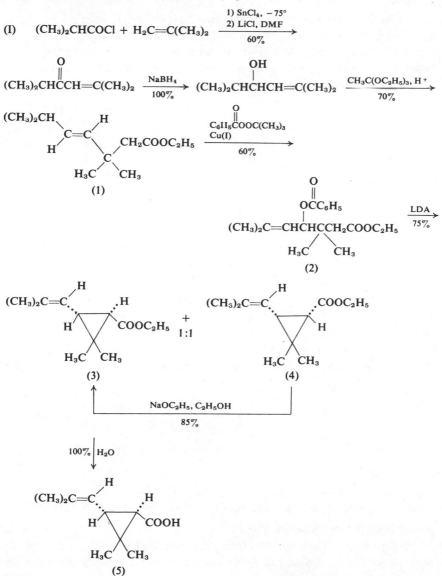

(I) $(CH_3)_2CHCOCl + H_2C{=}C(CH_3)_2 \xrightarrow[60\%]{\substack{1)\ SnCl_4,\ -75°\\ 2)\ LiCl,\ DMF}}$

Allylic oxidation.[1] A recent, simple synthesis of (±)-chrysanthemic acid (5, the (+) acid is the most active constituent of the insecticide pyrethrine) is outlined in scheme I. The overall yield is 16%. A key step involves an allylic oxidation of (1) with rearrangement to give the benzoate (2); this product can be cyclized by a 1,3-elimination to give a mixture of (3) and (4), both of which are convertible into (±)-chrysanthemic acid (5). Various homologs can be prepared in the same way.

Oxidation of camphene. When camphene (1) is allowed to react in acetonitrile at 80° with *t*-butyl perbenzoate and catalytic amounts of cuprous chloride and cupric benzoate, mainly alcohols are obtained after saponification (23% yield) rather than ethers (8% yield). The two principal alcohols are (2) and (3); the former is identical to nojigiku, isolated originally from a Japanese chrysanthemum.

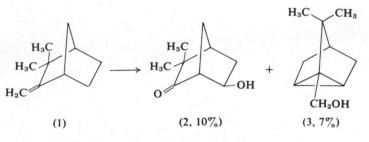

| (1) | (2, 10%) | (3, 7%) |

When optically active camphene is used, the alcohol (2) is also active.[2]

[1] J. Ficini and J. d'Angelo, *Tetrahedron Letters*, 2441 (1976).
[2] M. Julia, D. Mansuy, and P. Detraz, *ibid.*, 2141 (1976).

n-**Butyl N-(*p*-toluenesulfonyl)iminoacetate,** $C_4H_9OOCCH{=}NO_2SC_6H_4CH_3$ (1). Mol. wt. 283.4, oil. This imine is prepared by the reaction of *n*-butyl glyoxylate with N-sulfinyl-*p*-toluenesulfonamide (AlCl$_3$, 67% yield).[1]

Ene reactions.[2] This imine forms ene adducts with alkenes (equation I). The products can be transformed by standard procedures into γ,δ-unsaturated

(I)

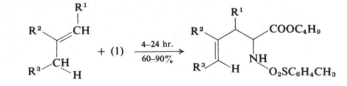

amino acids.

[1] R. Albrecht and G. Kresze, *Ber.*, **98**, 1431 (1965).
[2] O. Achmatowicz, Jr., and M. Pietraszkiewicz, *J.C.S. Chem. Comm.*, 484 (1976).

t-**Butyl trimethylsilylacetate, 5,** 371–372.

β-*Keto esters.*[1] The anion (1) reacts with acetylimidazole to give, after acidic work-up, *t*-butyl acetoacetate in 94% yield (equation I), with elimination of the

(I) $\underset{(1)}{\text{LiCHCO}_2\text{C(CH}_3)_3} \overset{\text{Si(CH}_3)_3}{\mid} + \text{CH}_3\text{CON} \overset{}{\underset{\text{N}}{\bigcirc}} \xrightarrow[-78 \to 25°]{\text{THF}} \xrightarrow{\text{HCl}} \underset{(94\%)}{\text{CH}_3\text{COCH}_2\text{COOC(CH}_3)_3}$

trimethylsilyl group. The reaction with simple acylating reagents is complicated, but proceeds well with various acylimidazoles (70–95% yield).

An interesting variation is reaction of (1) with 4-bromobutanoylimidazole (equation II).

(II) $\text{(1)} + \text{Br(CH}_2)_3\text{CON} \overset{}{\underset{\text{N}}{\bigcirc}} \longrightarrow \left[\text{Br(CH}_2)_3\overset{\overset{\text{OLi}}{\mid}}{\text{C}}{=}\text{CHCOOC(CH}_3)_3 \right] \xrightarrow[75\%]{}$

$$\overset{}{\underset{\text{O}}{\bigcirc}}{=}\text{CHCOOC(CH}_3)_3$$

[1] S. L. Hartzell and M. W. Rathke, *Tetrahedron Letters*, 2757 (1976).

C

Cadmium(0)–Solvent slurries.

An active form of cadmium metal and zinc metal slurried in several solvents can be prepared by codeposition of metal vapors (atoms) and excess solvent on the walls of a metal atom reactor at 77°K. The slurries are sometimes so fine that they can be handled with a syringe.

Dialkylcadmium(zinc) compounds. These slurries react with alkyl halides to form R_2Cd and R_2Zn compounds. The yields depend on the solvent, being highest in diglyme and dioxane.[1]

[1] T. O. Murdock and K. J. Klabunde, *J. Org.*, **41**, 1076 (1976).

Calcium ethoxide, $Ca(OC_2H_5)_2$.

Reaction of epoxides with isocyanates. Calcium ethoxide is the most efficient catalyst for the reaction of phenyl isocyanate (1) with the epoxide (2) to form the 2-oxazolidinone (3).[1]

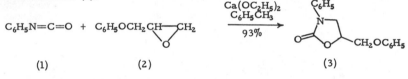

      (1)   (2)   (3)

[1] D. Braun and J. Weinert, *Ann.*, 221 (1976).

Carbon, activated.

Racemization of R-(−)- and S-(+)-binaphthyl. Solutions of optically active binaphthyl are completely racemized within minutes at room temperature by small amounts of carbon blacks or decolorizing carbons. The activity of the carbons increases with the surface area. Graphite is inactive; Norit is highly active.[1]

[1] R. E. Pincock, W. M. Johnson, and J. Haywood-Farmer, *Canad. J. Chem.*, **54**, 548 (1976).

Carbon dioxide, 3, 40–41; 5, 93–94; 6, 94–95.

2-Oxazolidinones. Under iodine catalysis, carbon dioxide reacts with aziridines to form 2-oxazolidinones (equation I). In the case of aziridine itself the yield is somewhat low (21.5%) because of polymerization.[1]

(I)   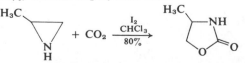

[1] K. Soga, S. Hosoda, H. Nakamura, and S. Ikeda, *J.C.S. Chem. Comm.*, 617 (1976).

Carbon monoxide, 2, 60, 204; **3,** 41–43; **5,** 96.

Carbonylation of vinylmercurials. Vinylmercuric chlorides, available by hydroboration–mercuration of alkynes,[1] react with carbon monoxide (1 atm.) and an alcohol in the presence of Li_2PdCl_4 to form α,β-unsaturated esters in high yield (equation I).[2]

(I)

$$\underset{H}{\overset{R}{>}}C=C\underset{HgCl}{\overset{H}{<}} + CO + R'OH \xrightarrow[85-100\%]{Li_2PdCl_4}$$

$$\underset{H}{\overset{R}{>}}C=C\underset{COOR'}{\overset{H}{<}} + HgCl_2 + 2\ LiCl + Pd + HCl$$

This reaction has now been extended to a synthesis of $\Delta^{\alpha,\beta}$-butenolides from propargylic alcohols (equation II). Apparently it is possible to avoid isolation of

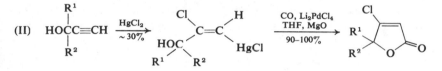

the intermediate vinylmercurial by reaction of the propargylic alcohol with $HgCl_2$, LiCl and $PdCl_2$ in THF under 1 atm. of CO. The chlorobutenolides are alkylated normally with Gilman reagents.[3]

[1] R. C. Larock and H. C. Brown, *J. Organometal. Chem.,* **36,** 1 (1972); R. C. Larock, S. K. Gupta, and H. C. Brown, *Am. Soc.,* **94,** 4371 (1972).
[2] R. C. Larock, *J. Org.,* **40,** 3237 (1975).
[3] R. C. Larock and B. Riefling, *Tetrahedron Letters,* 4661 (1976).

Carbon suboxide, 6, 96.

This reagent has been prepared by thermolysis of bis(trimethylsilyl)malonate in the presence of P_2O_5 at 160°; yield 54%.[1]

[1] L. Birkofer and P. Sommer, *Ber.,* **109,** 1701 (1976).

Carbonylhydridotris(triphenylphosphine)rhodium(I), $HRhCO[P(C_6H_5)_3]_3$, **5,** 331. Mol. wt. 918.79, m.p. 138° dec., light yellow. Suppliers: Alfa, ROC/RIC, Strem.

Hydroformylation.[1] This rhodium compound is an efficient homogeneous catalyst for hydroformylation of 1-hexene at 1 atm. and 45°. With added excess triphenylphosphine the ratio of straight-chain to branched aldehyde is 99:1.

$$CH_3(CH_2)_3CH=CH_2 + CO + H_2 \xrightarrow{cat.}$$

$$CH_3(CH_2)_3CH_2CH_2CHO + CH_3(CH_2)_3\underset{CHO}{\overset{|}{C}HCH_3}$$
$$99:1$$

Polymer-attached reagent, [Ⓟ—P(C₆H₅)₂]₃RhH(CO). Hydroformylation of 1-pentene with this catalyst favors formation of the linear product (1) (equation I).

(I) $CH_3(CH_2)_2CH=CH_2$ $\xrightarrow[\text{Cat.}]{H_2, CO, C_6H_6}$ $CH_3(CH_2)_4CHO$ + $CH_3(CH_2)_2\overset{\overset{\displaystyle CH_3}{|}}{C}HCHO$

(1)   12:1   (2)

The ratio of (1) to (2) is 3.3:1 when the homogeneous catalyst itself is used. Some other unusual effects of temperature and pressure were also noted, but these differences are not clearly understood at present.[2]

[1] W. Strohmeier and A. Kühn, *J. Organometal. Chem.*, **110**, 265 (1976).
[2] C. U. Pittman, Jr., and R. M. Hanes, *Am. Soc.*, **98**, 5402 (1976).

Catecholborane (1,3,2-Benzodioxaborole), **4**, 25, 69–70; **5**, 100–101; **6**, 33–34, 98.

Hydroboration. This process has been reviewed (66 references).[1]

Deoxygenation (**6**, 98). Tosylhydrazones of α,β-unsaturated carbonyl compounds are reduced by 1 equiv. of this hydride (1) to methylene derivatives, with migration of the double bond. The reaction is believed to proceed through a diazene intermediate.[2]

Examples:

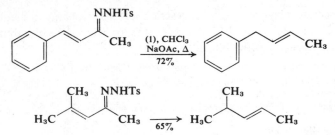

The reduction of the tosylhydrazone of Δ⁴-cholesten-3-one with catecholborane provides the best known route to Δ³-5β-cholestene (equation I).[3]

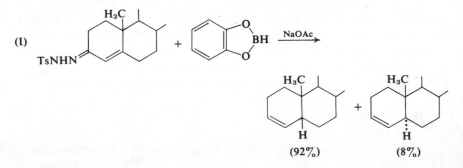

[1] C. F. Lane and G. W. Kabalka, *Tetrahedron*, 981 (1976).
[2] G. W. Kabalka, D. T. C. Yang, and J. D. Baker, Jr. *J. Org.*, **41**, 574 (1976).
[3] D. T. C. Yang and G. W. Kabalka, *Org. Prep. Proc. Int.*, **9**, 85 (1977).

Ceric ammonium nitrate (CAN), **1,**120–121; **2,** 63–65; **3,** 44–45; **4,** 71–74; **5,** 101–102; **6,** 99.

p-Benzoquinones. 1,4-Dimethoxybenzene and derivatives are oxidized by CAN in aqueous acetonitrile to *p*-benzoquinones in yields of 60–95%. This oxidation is not useful for preparation of *o*-quinones.[1]

Oxidation of camphorquinone. Camphor itself is not affected by treatment with CAN in CH_3OH at 20°, but the nonenolizable camphorquinone (1) is

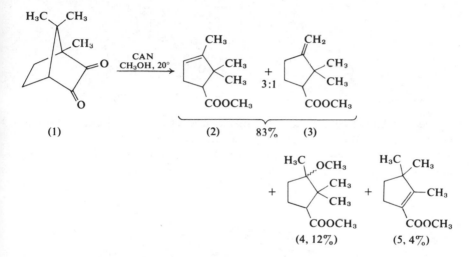

oxidized easily, mainly to an inseparable mixture of (2) and (3). The paper presents a plausible mechanism.[2]

Oxidation of 2,2'-dipyrromethanes. These pyrrole derivatives are oxidized to 2,2'-dipyrroketones by CAN in aqueous acetic acid buffered with sodium acetate. An example is formulated in equation (I). This reaction had been conducted with lead tetraacetate, with lead dioxide, and with sulfuryl chloride and bromine.[3]

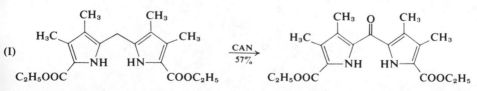

Cyclobutadiene (**4,** 72). The reaction of cyclobutadiene in a cycloaddition reaction with *p*-benzoquinone has now been published. The diene can also be liberated from the iron tricarbonyl complex with lead tetraacetate in pyridine.[4]

[1] P. Jacob, III, P. S. Callery, A. T. Shulgin, and N. Castagnoli, Jr., *J. Org.*, **41**, 3627 (1976).
[2] B. Danieli and G. Palmisano, *Chem. Ind.*, 565 (1976).
[3] J. B. Paine, III, and D. Dolphin, *Canad. J. Chem.*, **54**, 411 (1976).
[4] L. Brenner, J. S. McKennis, and R. Pettit, *Org. Syn.*, **55**, 43 (1976).

Ceric ammonium sulfate, CAS [Ammonium cerium (IV) sulfate], $Ce(NH_4)_4(SO_4)_4 \cdot 2 H_2O$. Mol. wt. 632.56. Supplier: Alfa.

Baeyer-Villiger oxidation of polycyclic ketones.[1] Oxidation of 1,3-bishomo-cubanone (1) with ceric ammonium nitrate or sulfate results in formation of the lactone (2) in 78% yield. This product (2) is a minor product of the oxidation of

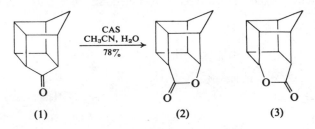

(1)   (2)   (3)

(1) with *m*-chlorperbenzoic acid, the major product being the isomeric lactone (3); the ratio of (3) to (2) is 5:1 in the peracid oxidation.

Oxidation of 1,4-bishomocubanone (4) gives the only possible lactone (5), and this is also the main product of peracid oxidation of (4).

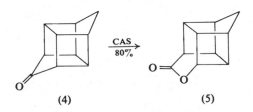

(4)   (5)

Oxidation of (6) with CAS is also regioselective and gives (7) in high yield. Only a dilactone could be isolated from oxidation of (6) with peracid.

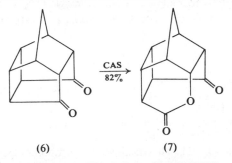

(6)   (7)

[1] G. Mehta, P. N. Pandey, and T.-L. Ho, *J. Org.*, **41**, 953 (1976).

Cesium carbonate, Cs_2CO_3. Mol. wt. 325.82. Suppliers: Alfa, ROC/RIC. Soluble in acetone, ethanol, ether, THF.

Selective cleavage of aryl esters. Resorcinol dibenzoate is selectively cleaved to the monobenzoate in >95% yield by Cs_2CO_3 (50% excess) in refluxing DME (24 hours). $CsHCO_3$ can also be used, but in this case highest yields are obtained with 1 equiv. of the salt. The cesium salts are definitely superior to the corresponding potassium salts for this cleavage, possibly because cesium salts are more soluble in organic solvents than potassium salts. The phenanthrene-2,4-diacetate (1) can be converted in quantitative yield into the monoacetate (2) by a 10% excess of Cs_2CO_3 in THF or DME.[1]

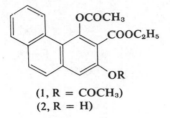

(1, R = $COCH_3$)
(2, R = H)

[1] H. E. Zaugg, *J. Org.*, **41**, 3419 (1976).

Cesium fluoride, CsF. Mol. wt. 151.90, m.p. 682°. Suppliers: Alfa, ROC/RIC.

Transesterification of phosphate triesters.[1] Phenyl groups in phosphate triesters are replaced quantitatively by alkyl groups when dissolved in alcohols containing excess cesium fluoride (equation I) or tetra-*n*-butylammonium fluoride (**4**, 477–478; **5**, 645).

(I) $(C_6H_5O)_3PO + 3 ROH \xrightarrow[\text{Quant.}]{\text{CsF}} (RO)_3PO + 3 C_6H_5OH$

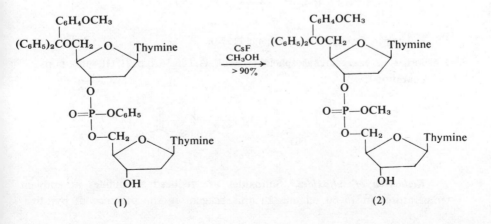

This transesterification is of interest in nucleotide chemistry; for example, (1) can be converted into (2) in $>90\%$ yield.

[1] K. K. Ogilvie and S. L. Beaucage, *J.C.S. Chem. Comm.*, 443 (1976).

Chloramine-T, 4, 75, 445–446; 5, 104.

Caution: An explosion of anhydrous chloramine-T has been reported.[1]

2-Alkylidene-1,3-dithianes.[2] These useful ketene thioacetals can be prepared as formulated in equation (I). The preparation involves N-tosylsulfilimines, which are cleaved readily to 2-alkylidene-1,3-dithianes.

(I)

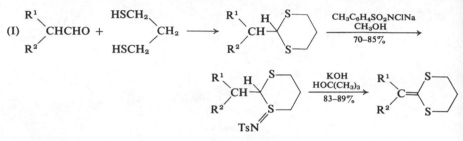

[1] I. L. Klundt, *Chem. Eng. News*, Dec. 5, 56 (1977).
[2] H. Yoshida, T. Ogata, and S. Inokawa, *Synthesis*, 552 (1976).

Chlorine, 5, 105–106; 6, 101–102.

Sulfonyl chlorides. Alkyl benzyl sulfides are converted into alkanesulfonyl chlorides by reaction with chlorine in aqueous acetic acid (30°). The benzylic group is selectively cleaved.[1]

Examples:

$$C_6H_5CH_2SCH_3 \xrightarrow[75\%]{\underset{HOAc}{Cl_2}} CH_3SO_2Cl \left[+ \underset{\sim 80\%}{C_6H_5CH_2Cl} \right]$$

$$C_6H_5CH_2S(CH_2)_4CH_3 \xrightarrow[95\%]{} CH_3(CH_2)_4SO_2Cl$$

[1] R. F. Langler, *Canad. J. Chem.*, **54**, 498 (1976).

2-Chloro-1,3,2-benzodioxaphosphole (1). Mol. wt. 155.56, b.p. 91°/18 mm., m.p. 30°.
   Preparation[1]:

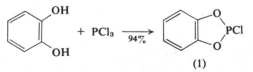

Reduction of sulfoxides.[2] Sulfoxides are reduced to sulfides at ambient temperatures in 15–60 minutes by this reagent in the presence of pyridine.

Benzene is used as solvent. Yields are in the range of 70–100%. *Cf.* **3-Phenoxy-1,3,2-benzodioxaphosphole, 5,** 512.

[1] P. C. Crofts, J. H. H. Markes, and H. N. Rydon, *J. Chem. Soc.,* 4250 (1958).
[2] D. W. Chasar and T. M. Pratt, *Synthesis,* 262 (1976).

***p*-Chlorobenzyl alcohol,** $ClC_6H_4CH_2OH$. Mol. wt. 142.59, m.p. 70–72°. Suppliers: Aldrich, Fluka.

Protection of carboxyl groups.[1] *p*-Chlorobenzyl esters are about twice as stable as benzyl esters to hydrolysis by trifluoroacetic acid in methylene chloride at 45°. They can be removed by liquid HF at 0°. These esters are recommended for protection of the β- and γ-carboxyl group of aspartic acid and glutamic acid in the synthesis of moderate-sized peptides.

p-Nitrobenzyl esters are much more stable to HBr than *p*-chlorobenzyl esters. They are recommended for protection of the terminal carboxyl group in solid-phase peptide synthesis. A tri- and a hexapeptide have been prepared by this approach.

[1] R. L. Prestidge, D. R. K. Harding, and W. S. Hancock, *J. Org.,* **41,** 2579 (1976).

Chlorodicarbonylrhodium(I) dimer, 6, 108.

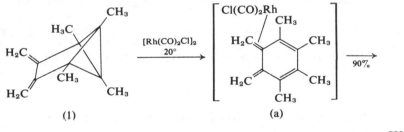

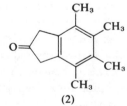

(2)

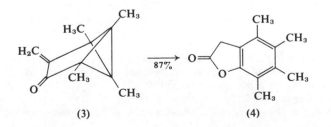

Carbon monoxide insertion reactions.[1] The reaction of the diene (1) with this transition metal complex at 20° results in rearrangement and insertion of CO to form (2) in high yield. A transient species has been identified tentatively as (a) by IR and NMR. A similar reaction of (3) results in (4). The reaction can be catalytic with respect to the rhodium complex if carried out under a CO pressure of 40 atm.

The simpler substrate *o*-benzoquinone methide (5) affords (6), but only in 12% yield.

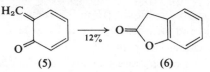

(5)   (6)

[1] R. F. Heldeweg and H. Hogeveen, *Am. Soc.*, **98**, 6040 (1976).

Chlorodifluoromethane, HCF_2Cl. Mol. wt. 86.48, b.p. −40.8°. Supplier: du Pont (Fluorocarbon 22).

1,1-Difluoro-1-alkenes. Alkylidene ylides can function as bases to generate difluorocarbene from chlorodifluoromethane, and also as a trap for the carbene (equation I). Yields of olefins are in the range of 80–100%. One advantage is

$$2(C_6H_5)_3\overset{+}{P}-\overset{-}{C}R^1R^2 + HCF_2Cl \longrightarrow (C_6H_5)_3\overset{+}{P}CHR^1R^2 + (C_6H_5)_3P + F_2C{=}CR^1R^2$$

that triphenylphosphine, rather than the oxide, is formed and can be recycled to prepare the ylide.[1]

[1] G. A. Wheaton and D. J. Burton, *Tetrahedron Letters*, 895 (1976).

Chlorohydridotris(triphenylphosphine)ruthenium(II), $HRuCl[P(C_6H_5)_3]_3$. Mol. wt. 924.41, red violet, extremely air-sensitive.

This homogeneous catalyst is prepared *in situ* in $C_2H_5OH-C_6H_6$ or C_2H_5OH by hydrogenation of $RuCl_2[P(C_6H_5)_3]_3$. It is a very active catalyst for hydrogenation of 1-hexene and ethyl acrylate.[1]

This ruthenium complex is one of the most efficient of known homogeneous catalysts for hydrogenation,[2,3] but has not found wide use because of the difficult preparation. Recently Geoffrey and Bradley[4] found that it can be prepared efficiently by photo-induced decarbonylation of the relatively stable RuHCl(CO)–$[P(C_6H_5)_3]_3$[5] in benzene. The solutions can be used directly or the complex can be isolated in slightly impure form.

[1] W. Strohmeier and G. Buckow, *J. Organometal. Chem.*, **110**, C17 (1976).
[2] P. S. Hallman, B. R. McGarvey, and G. Wilkinson, *J. Chem. Soc.* (*A*), 3143 (1968).
[3] J. E. Lyons, *J.C.S. Chem. Comm.*, 412 (1975).
[4] G. L. Geoffrey and M. G. Bradley, *ibid.*, 20 (1976).
[5] N. Ahmad, J. J. Levison, S. D. Robinson, and M. F. Uttley, *Inorg. Syn.*, **15**, 45 (1974).

2-Chloro-3-methylbenzo-1,3-thiazolium trifluoromethanesulfonate (1). Mol. wt. 272.27, m.p. 130°.

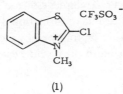

(1)

The reagent is prepared from 2-chlorobenzothiazole (Aldrich) and methyl trifluoromethanesulfonate.

Thiol esters and amides. Reagent (1) reacts exothermally with a carboxylic acid in the presence of triethylamine to form 3-methylbenzothiazolidone and the acid chloride. Addition of a thiol or an amine to this mixture results in formation of a thiol ester or an amide in generally high yield (equation I).[1]

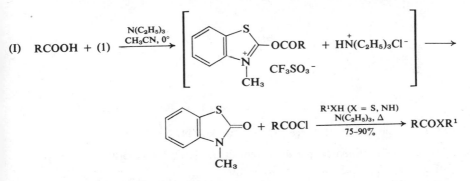

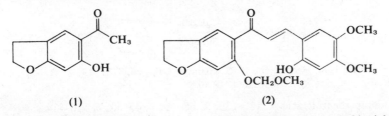

[1] F. Souto-Bachiller, G. S. Bates, and S. Masamune, *J.C.S. Chem. Comm.*, 719 (1976).

Chloromethyl methyl ether, 1, 132–135; 5, 120.

Protection of phenolic hydroxyl groups. Rall *et al.*[1] encountered difficulties in methoxymethylation of the phenol (1) by standard procedures (reaction with sodium ethoxide, ethanol, and chloromethyl methyl ether). They then found

(1)   (2)

that the methoxymethyl ether of (1) could be obtained in about 80% yield by reaction of the potassium salt of (1) in acetonitrile containing 18-crown-6 with chloromethyl methyl ether. This ether was used for the synthesis of the 2′-hydroxychalcone (2) in 95% yield.

Hydracrylic and glycidic esters. Treatment of the methoxymethyl ester (1) with LDA in THF at $-78 \rightarrow 20°$ results in formation of the methyl hydracrylate (2). The reaction involves fragmentation followed by recombination (equation I). By a similar reaction (3) is converted to the glycidic ester (4).[2]

$$(I) \quad (CH_3)_2CH\overset{\overset{O}{\|}}{-}C-OCH_2OCH_3 \xrightarrow{\ LDA\ } \left[(CH_3)_2C{=}C{=}O + HCHO + CH_3O^- \right.$$

(1)

$$\longrightarrow (CH_3)_2C\overset{CH_2O^-}{\underset{COOCH_3}{\big<}} \left] \xrightarrow[69\%]{H^+} (CH_3)_2C\overset{CH_2OH}{\underset{COOCH_3}{\big<}}$$

(2)

$$(II) \quad \overset{H_3C}{\underset{Br}{>}}CH\overset{\overset{O}{\|}}{C}-OCH_2OCH_3 \xrightarrow[\substack{2)\ H^+ \\ 61\%}]{1)\ LDA} H_2C\overset{\overset{O}{\|}}{-}C\overset{COOCH_3}{\underset{CH_3}{\big<}}$$

(3)   (4)

[1] G. J. H. Rall, M. E. Oberholzer, D. Ferreira, and D. G. Roux, *Tetrahedron Letters*, 1033 (1976).
[2] A. G. Schultz and M. H. Berger, *J. Org.*, **41**, 585 (1976).

Chloromethyltriphenylphosphonium chloride, $[(C_6H_5)_3P^+CH_2Cl]Cl^-$, (1). Mol. wt. 347.23, m.p. 241–245° dec. Supplier: Aldrich.

Preparation:

$$(C_6H_5)_3P + CH_2O + HCl \xrightarrow[60-70\%]{} [(C_6H_5)_3\overset{+}{P}CH_2OH]Cl^- \xrightarrow[70-80\%]{SO_2Cl_2}$$

$$(1) + SO_2 + HCl$$

The reagent is a useful Wittig reagent for synthesis of vinylic chlorides.[1]

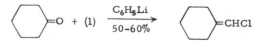

[1] D. Seyferth, J. K. Heeran, and S. O. Grim, *J. Org.*, **26**, 4783 (1961); D. Seyferth and J. K. Heeren, *Org. Syn.*, submitted (1976).

***m*-Chloroperbenzoic acid, 1,** 135–139; **2,** 68–69; **3,** 45–50; **6,** 110–114.

trans-2,3-Epoxycyclohexanol. The oxidation of 2-cyclohexene-1-ol with perbenzoic acid gives the *cis*-epoxide stereoselectively.[1] Oxidation of the acetate with perbenzoic acid gives a mixture of about equal amounts of the *cis*- and *trans*-epoxide. If the hydroxyl group is first silylated, then oxidation with *m*-chloroperbenzoic acid gives the *trans*-epoxy ether as the main product.[2]

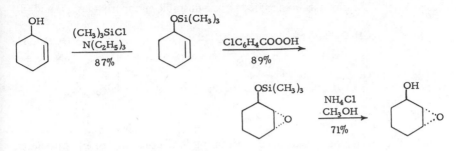

Epoxidation of 1,3-cyclohexadienes. In research directed toward the synthesis of the natural 1,3-diepoxide crotepoxide (4), White and co-workers[3] investigated the epoxidation of the 1,3-cyclohexadiene (1). Under usual conditions (25°), only two isomeric monoepoxides were obtained with *m*-chloroperbenzoic acid. However, under forcing conditions in the presence of an antioxidant (4, 85–86), two isomeric diepoxides, (2) and (3), were formed. The minor product (3) was converted into the natural product (4).

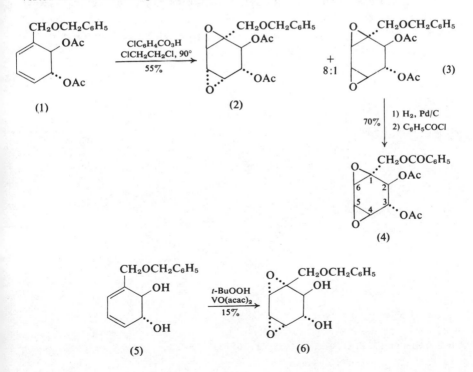

Epoxidation of (5) with *t*-butyl hydroperoxide with VO(acac)$_2$ as catalyst (5, 75–76) led stereospecifically to the *cis*-dioxide (6), but in low yield (15%).

Preparation of a diepoxide by rearrangement of an endoperoxide was also investigated. Photooxidation of (5) led to two unstable endoperoxides, (7) and (8),

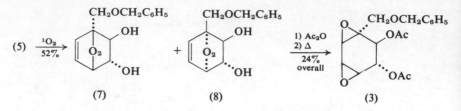

(7)   (8)   (3)

which, after acetylation, rearranged on heating in the presence of an antioxidant to the desired *cis*-diepoxide (3).

[1] H. B. Henbest and R. A. L. Wilson, *J. Chem. Soc.*, 1958 (1957).
[2] C. G. Chavdarian and C. H. Heathcock, *Syn. Comm.*, **6**, 277 (1976).
[3] M. R. Demuth, P. E. Garrett, and J. D. White, *Am. Soc.*, **98**, 634 (1976).

5-Chloro-1-phenyl-1-H-tetrazole, 2, 319–320.

Removal of an aromatic methylenedioxy group. This reaction can be carried out in three steps. Treatment with boron trichloride is known to cleave methylenedioxy groups preferentially (**4**, 43). The catechol group is then removed by etherification with 5-chloro-1-phenyl-1-*H*-tetrazole and hydrogenolysis of the ether. An example is formulated.[1]

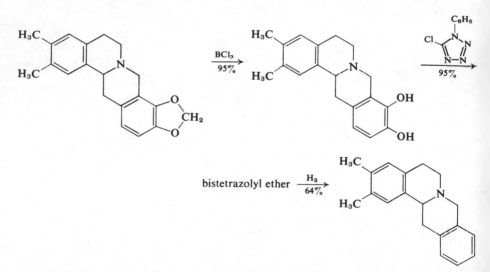

[1] S. Teitel and J. P. O'Brien, *J. Org.*, **41**, 1657 (1976).

N-Chlorosuccinimide, 1, 139; **2,** 69–70; **5,** 127–129; **6,** 115–118.

Aldehydes from primary halides and from terminal olefins.[1] α-Chlorination of phenyl sulfides prepared from these starting materials followed by hydrolysis (one flask operation) results in aldehydes. Two typical procedures are outlined.

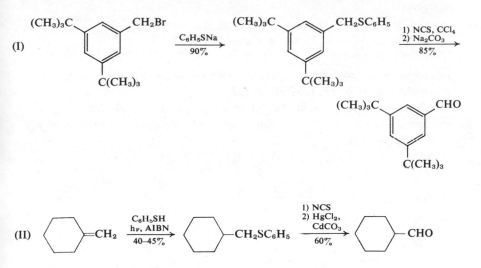

(I)

(II)

[1] L. A. Paquette, W. D. Klobucar, and R. A. Snow, *Syn. Comm.*, **6,** 575 (1976).

Chlorosulfonyl isocyanate (CSI), 1, 117–118; **2,** 70; **3,** 51–53; **4,** 90–94; **5,** 132–136; **6,** 122.

Reaction with barrelene.[1] The reaction of CSI with barrelene (1) leads to the expected lactam (2); when heated in DMF, (2) is converted to the γ-chloro nitrile (3).[2] This substance undergoes 1,3-elimination of hydrogen chloride when

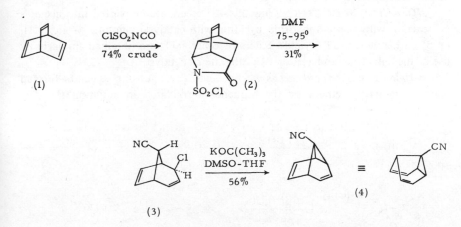

(1)   (2)   (3)   (4)

treated with base to form 1-cyanosemibullvalene (4). NMR and X-ray analysis show that (4) does not undergo Cope rearrangement.

Similar results were obtained with benzo and dibenzo derivatives of barrelene.

Review.[3] Recent synthetic uses of this isocyanate have been reviewed (86 references).

[1] L. A. Paquette and W. E. Volz, *Am. Soc.*, **98**, 2910 (1976).
[2] E. J. Moriconi and C. C. Jalandoni, *J. Org.*, **35**, 3796 (1970).
[3] J. K. Rasmussen and A. Hassner, *Chem. Revs.*, **76**, 389 (1976).

1-Chloro-N,N,2-trimethylpropenylamine **(N,N-Dimethyl-1-chloro-2-methyl-1-propenylamine)**, **4**, 94–95; **5**, 136–138; **6**, 122–123.

Aminoalkenylation. Enamines of this type can be metallated at C_1 with replacement of Cl, and the resulting compounds behave as equivalents of acyl anions (Umpolung).[1]

Example:

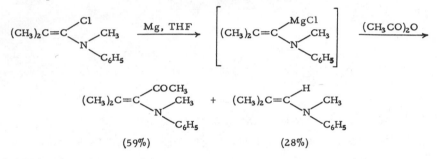

(59%)   (28%)

[1] C. Wiaux-Zamar, J.-P. Dejonghe, L. Ghosez, J. F. Normant, and J. Villieras, *Angew. Chem., Int. Ed.*, **15**, 371 (1976).

Chlorotrimethylsilane (Trimethylchlorosilane), **1**, 1232; **2**, 535–438; **3**, 310–312; **4**, 537–539; **5**, 709–713; **6**, 626–628.

ROSi(CH₃)₃ → ROCOR′. Trimethylsilyl ethers are converted into acetates by reaction with acetic anhydride–pyridine with catalytic amounts of 48% HF or BF₃ etherate (*ca.* 85% yield). Esters of other acids are obtained similarly by use of the anhydride and HF or BF₃ etherate as catalyst.[1]

α-Halo-α,β-unsaturated carbonyl compounds. These substances can be formed from trimethylsilyl ethers by the sequence formulated in equation (I). The

(I)

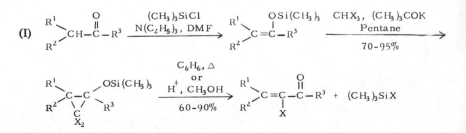

method was used for preparation of eight bromo ketones and one chloro ketone.[2]

O-Trimethylsilyl hemithioacetals and hemithioketals.[3] These substances can

$$(I) \qquad R^1COR^2 + RSH + (CH_3)_3SiCl \xrightarrow[75-90\%]{Py} R^1-\overset{\overset{\displaystyle OSi(CH_3)_3}{|}}{\underset{\underset{\displaystyle SR}{|}}{C}}-R^2$$

$$(1)$$

be obtained in satisfactory yield by reaction of carbonyl compounds with a thiol and chlorotrimethylsilane in the presence of a base (pyridine) [equation (I)].

The products react with an alkyllithium in THF or ether predominantly with cleavage of the Si—O bond (equation II); however, in HMPT or TMEDA substitution at carbon occurs (equation III). Apparently the silyl group is responsible for the unusual cleavage by an alkyllithium.

$$(II) \qquad (1) + R^3Li \xrightarrow[ether]{THF\ or} R^1\overset{\overset{\displaystyle O}{\|}}{C}R^2 + RSLi + R^3Si(CH_3)_3$$

$$(III) \qquad (1) + R^3Li \xrightarrow[TMEDA]{HMPT\ or} R^1-\overset{\overset{\displaystyle OSi(CH_3)_3}{|}}{\underset{\underset{\displaystyle R^3}{|}}{C}}-R^2 \quad + RSLi$$

Dithioacetals and dithioketals.[4] Reaction of carbonyl compounds with a thiol and chlorotrimethylsilane in the absence of pyridine results in dithioacetals and dithioketals (equation IV).

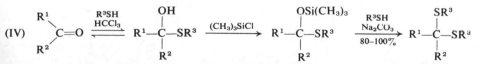

$$(IV) \quad \overset{R^1}{\underset{R^2}{>}}C=O \underset{HCCl_3}{\overset{R^3SH}{\rightleftharpoons}} R^1-\overset{\overset{OH}{|}}{\underset{\underset{R^2}{|}}{C}}-SR^3 \xrightarrow{(CH_3)_3SiCl} R^1-\overset{\overset{OSi(CH_3)_3}{|}}{\underset{\underset{R^2}{|}}{C}}-SR^3 \xrightarrow[80-100\%]{\overset{R^3SH}{Na_2CO_3}} R^1-\overset{\overset{SR^3}{|}}{\underset{\underset{R^2}{|}}{C}}-SR^3$$

[1] D. J. Voaden and R. M. Waters, *Org. Prep. Proc. Int.*, **8**, 227 (1976).
[2] P. Amice, L. Blanco, and J. M. Conia, *Synthesis*, 196 (1976).
[3] T. H. Chan and B. S. Ong, *Tetrahedron Letters*, 319 (1976).
[4] B. S. Ong and T. H. Chan, *Syn. Comm.*, **7**, 283 (1977).

Chlorotris(methyldiphenylphosphine)rhodium(I), $[CH_3P(C_6H_5)_2]_3RhCl$. Mol. wt. 738.32.

The complex has been prepared from μ-dichlorotetraethylenedirhodium and methyldiphenylphosphine.[1] It has been used successfully for decarbonylation of two aldehydo sugars that were resistant to chlorotris(triphenylphosphine)-rhodium(I),[2] possibly because of steric factors.

[1] K. C. Dewhirst, W. Keim, and C. A. Reilly, *Inorg. Chem.*, **7**, 546 (1968).
[2] D. I. Ward, W. A. Szarek, and J. K. N. Jones, *Chem. Ind.*, 162 (1976).

Chlorotris(triphenylphosphine)rhodium(I). 1, 1252; **2,** 248–253; **3,** 325–329; **4,** 559–562; **5,** 736–740; **6,** 652–653.

Homogeneous hydrogenation catalysts. Birch and Williamson[1] have reviewed these catalysts (344 references). In addition to Wilkinson's catalyst (probably the most widely used), [RhCl(py)$_2$(DMF)$_2$BH$_4$]Cl, IrCl(CO)[P(C$_6$H$_5$)$_3$]$_2$, K$_3$[Co(CN)$_5$], Co$_2$(CO)$_8$, Henbest's catalyst, and soluble Ziegler catalysts are covered in detail. The authors conclude that at the present time many of the catalysts are not particularly useful for synthesis, probably because most of the investigations have been directed at mode of action and modifications of the catalysts.

[1] A. J. Birch and D. H. Williamson, *Org. React.*, **24,** 1 (1976).

Chromic acid, 1, 142–144; **2,** 70–72; **3,** 54; **4,** 95–96; **5,** 138–140; **6,** 123–124.

γ-Butyrolactones. Cyclobutanones tetrasubstituted at C$_2$, C$_3$ or at C$_2$, C$_4$ are oxidized in generally satisfactory yields to γ-butyrolactones by potassium dichromate in dilute sulfuric acid (**6,** 124).[1]

Examples:

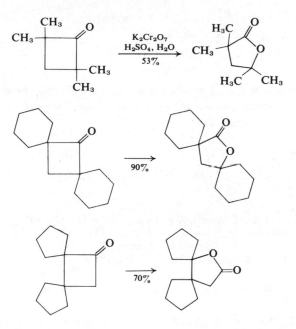

This oxidation is not applicable to larger cyclic ketones; 2,2,5,5-tetramethyl-cyclopentanone is not oxidized.

1,4-Ketols and 1,4-diketones. Tertiary cyclobutanols are oxidatively cleaved by Jones reagent to 1,4-ketols or 1,4-diketones.[2]

Examples:

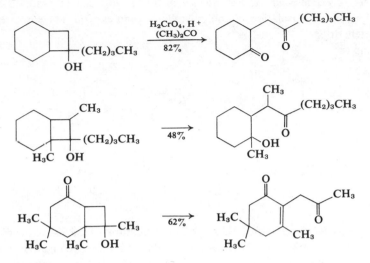

Jones reagent. Oxidation of the alcohol (1) with Jones reagent does not give the expected ketone, but the δ-lactone (2). The *p*-methoxy group was shown to be essential for this oxidation of a methyl group.[3] The expected ketone is obtained under Oppenauer conditions.

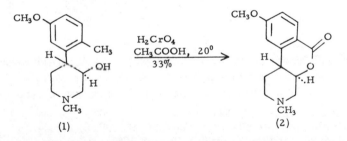

(1)   (2)

Polymer-supported reagent. Italian chemists[4] have prepared a supported form of this oxidant by reaction of the Cl⁻ form of an anion exchange resin (Amberlyst A-26, Amberlyst A-29, Amberlite IRA 400, or Amberlite 904) in H_2O with chromium trioxide to obtain a CrO_4H^- form of the resin. This polymeric reagent oxidizes primary and secondary alcohols in high yield (usually 85–95%). The chloride form of the resin is regenerated by wash with NaOH and HCl solutions.

[1] R. Jeanne-Carlier and F. Bourelle-Wargnier, *Bull. soc.*, 297 (1976).

[2] H.-J. Liu, *Canad. J. Chem.*, **54**, 3113 (1976).

[3] R. A. Jones, J. F. Saville, and S. Turner, *J.C.S. Chem. Comm.*, 231 (1976).

[4] G. Cainelli, G. Cardillo, M. Orena, and S. Sandri, *Am. Soc.*, **98**, 6737 (1976).

Chromic anhydride, 1, 144–147; **2,** 72–75; **3,** 54–57; **4,** 96–97; **5,** 140–141.

CrO₃·pyridine complex. Ratcliffe's reagent (**4,** 96)[1] can be used for allylic oxidations without isolation of the crystalline complex if all reagents are kept completely dry.[2]

Examples:

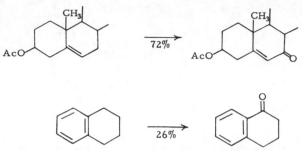

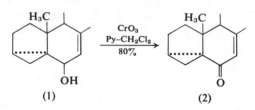

CrO₃–pyridine–methylene chloride. The oxidation of 3α,5α-cyclocholest-7-ene-6-ol (1) to the ketone (2) is unsatisfactory with either manganese dioxide (no reaction below 40°) or CrO₃ in pyridine (dehydration to the $\Delta^{6,8(14)}$-diene).

However, satisfactory yields are obtained with CrO₃ in pyridine–methylene chloride (1:3.6). The reaction was used in one step in the synthesis of Δ^{5}-14β-cholestene-3β-ol, of interest as a possible inhibitor of cholesterol synthesis.[3]

[1] R. W. Ratcliffe, *Org. Syn.,* **55,** 84 (1976).
[2] D. S. Fullerton and C.-M. Chen, *Syn. Comm.,* **6,** 217 (1976).
[3] M. Anastasia, A. Scala, and G. Galli, *J. Org.,* **41,** 1064 (1976).

Chromic anhydride–Acetic anhydride.

Oxidation of macrocyclic lactones. Macrocyclic lactones (1) can be oxidized to monoketo lactones (2) by chromium trioxide and acetic anhydride in acetic acid. The reaction shows some regioselectivity. Thus when n = 15, attack is

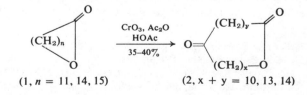

(1, *n* = 11, 14, 15)   (2, x + y = 10, 13, 14)

mainly at C_9 (x = 8, y = 6). Presumably the preferred conformation of (1) determines the possible reaction sites.[1] This selectivity may also be involved in biosynthesis of macrolide antibiotics from macrocyclic lactones.

[1] G. K. Eigendorf, C.-L. Ma, and T. Money, *J.C.S. Chem. Comm.*, 561 (1976).

Chromic anhydride–Hexamethylphosphoric triamide.

Oxidation of alcohols.[1] Alcohols can be oxidized by a complex prepared by addition of CrO_3 (20 mmole) to dry HMPT (6 ml.).[2] The complex oxidizes saturated primary alcohols in about 80% yield; lower yields are obtained from secondary alcohols. Highest yields are obtained from α,β-unsaturated primary and secondary alcohols. Optimum yields require only a 2:1 molar ratio of oxidant to alcohol.

Examples:

$$C_6H_5\overset{\overset{O}{\|}}{\underset{\underset{O}{\|}}{S}}CH_2\overset{\overset{CH_3}{|}}{CH}=CHCH_2OH \xrightarrow[100\%]{} C_6H_5\overset{\overset{O}{\|}}{\underset{\underset{O}{\|}}{S}}CH_2\overset{\overset{CH_3}{|}}{CH}=CHCHO$$

$$C_6H_5CH_2OH \xrightarrow[85\%]{} C_6H_5CHO$$

$$n\text{-}C_9H_{19}CHOHCH_3 \xrightarrow[48\%]{} n\text{-}C_9H_{19}COCH_3$$

$$C_6H_5CHOHC_6H_5 \xrightarrow[100\%]{} C_6H_5COC_6H_5$$

[1] G. Cardillo, M. Orena, and S. Sandri, *Synthesis*, 394 (1976).
[2] *Caution:* A violent decomposition can result if crushed CrO_3 is added to HMPT; add CrO_3 in small portions with stirring at 20°.

Chromium hexacarbonyl (Hexacarbonylchromium), 5, 142–143; 6, 125–126.

Annelation of Cr(CO)₃ complexes.[1] The reaction of either *endo-* or *exo-*2-methyl-1-indanone–Cr(CO)₃ complex (1) with methyl vinyl ketone gives (2) as the major product. The minor product undergoes cyclization, as expected, to (4). The major product (2), however, on cyclization gives only minor amounts of the expected enone (5). The major products are the two isomers (6) and (7). Evidently the complexed benzene ring activates the α-methylene protons.

The same annelation sequence was observed with the Cr(CO)₃ complex of 2-methyl-1-tetralone. It was also applied to optically active indanone and tetralone complexes and resulted in optically active products.

[1] G. Jaouen and A. Meyer, *Tetrahedron Letters*, 3547 (1976).

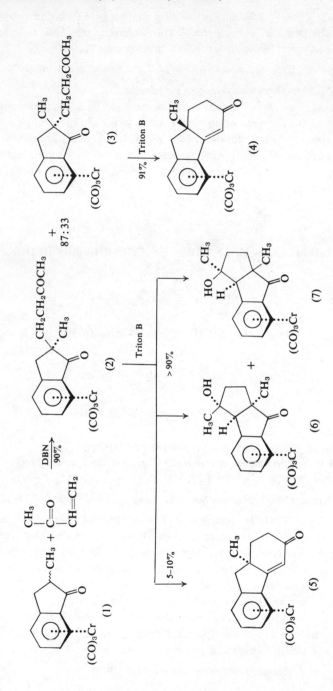

Chromous chloride, 1, 149–150; **2,** 76–77; **3,** 60–61; **4,** 144.

Deoxygenation of amine oxides. Chromous chloride is useful for deoxygenation of aromatic amine oxides (yields ~ 60–90%).[1]

[1] Y. Akita, K. Misu, T. Watanabe, and A. Ohta, *Chem. Pharm. Bull. Japan,* **24,** 1839 (1976).

Copper halide nitrosyls, $(CuX_2 \cdot NO)_2$, X = Cl, Br.

The violet-black complexes are formed by passing nitric oxide into a mixture of CuX_2 in acetonitrile.

gem-Dihalides. Primary amines coordinated with CuX_2 are oxidized by these complexes to *gem*-dihalides (equation I) together with products of the type RCN, RCH_2X, and RCH_2OH.[1]

$$\text{(I)} \qquad RCH_2NH_2 \cdot CuX_2 + (CuX_2 \cdot NO)_2 \xrightarrow[15-60\%]{\substack{CH_3CN \\ 25°}} RCHX_2$$

[1] M. P. Doyle, B. Siegfried, and J. J. Hammond, *Am. Soc.,* **98,** 1627 (1976).

Copper, 1, 157–158; **2,** 82–84; **3,** 63–65; **4,** 102–103; **5,** 146–148.

Isomerization of benzvalene (**1**) *and benzobenzvalene* (**4**). The two hydrocarbons react in different ways to silver or zinc and to copper, nickel, platinum, or gold (equations I and II). More information concerning the mechanism of these rearrangements has been obtained by isomerization of 1-deuterobenzo-

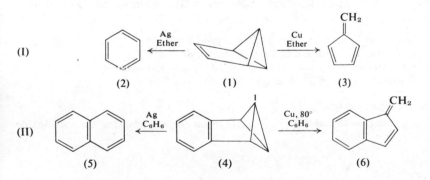

benzvalene. Isomerization with silver gives only β-deuteronaphthalene. Isomerization with copper, on the other hand, gives three isotopomeric benzofulvenes. Clearly, these rearrangements proceed by different mechanisms; these are discussed in the original paper.[1]

Cyclopropanes.[2] The reaction of alkenes with *gem*-dihalides and copper[3] results in cyclopropane derivatives, often in good yields. The reaction is carried out in a refluxing aromatic hydrocarbon: benzene, toluene, ethylbenzene.

Examples:

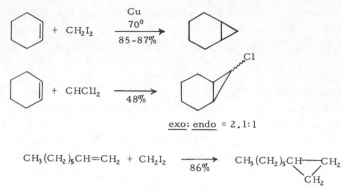

exo: endo = 2.1:1

$$CH_3(CH_2)_5CH{=}CH_2 + CH_2I_2 \xrightarrow[86\%]{} CH_3(CH_2)_5CH{-\!-}CH_2$$
$$\underset{CH_2}{\diagdown\!\diagup}$$

Ullmann reaction (5, 146).[4] Arylmercuric salts are converted into biaryls, usually in 70–90% yield, by reaction with copper catalyzed by PdCl₂ in pyridine at 115° (1–2 hours) (equation I). The use of pyridine as solvent is an important

(I)
$$2\ ArHgX + Cu \xrightarrow[\text{Py, }115^0]{PdCl_2} Ar{-}Ar + Hg + CuX_2$$

factor; yields are low in nonbasic solvents. The reaction is subject to steric hindrance by bulky *ortho*-substituents; certain acidic functional groups (OH, COOH) decrease yields; little selectivity is observed in the synthesis of unsymmetrical biaryls.

[1] U. Burger and F. Mazenod, *Tetrahedron Letters*, 2885 (1976).
[2] N. Kawabata, M. Naka, and S. Yamashita, *Am. Soc.*, **98**, 2676 (1976).
[3] Copper is treated with some iodine before the reaction is conducted.
[4] R. A. Kretchmer and R. Glowinski, *J. Org.*, **41**, 2661 (1976).

Copper(I) cyanoacetate, 5, 171.

Activated carbon dioxide. Copper(I) cyanoacetate transfers carbon dioxide

(I)
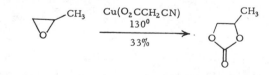

to propylene oxide to produce propylene carbonate (equation I). Metallic copper, CuCN, and CuOAc are ineffective catalysts.[1]

This CO_2 transfer is of interest because some biological oxidations are known to involve such reactions (*e.g.*, biotin-dependent carboxylations).

[1] T. Tsuda, Y. Chujo, and T. Saegusa, *J.C.S. Chem. Comm.*, 415 (1976).

Copper(I) methyltrialkylborates, $R_3B^-CH_3Cu^+$.

Alkylbenzenes.[1] These borates (1), prepared as shown from trialkylboranes, react with benzyl bromides (2) in THF–ether to form alkylbenzenes (3) and 1,2-diphenylethanes (4).

$$R_3B \xrightarrow{CH_3Li} R_3B^-CH_3Li^+ \xrightarrow{CuX} R_3B^-CH_3Cu^+ \xrightarrow[0-25°]{ArCH_2Br(2)}$$
$$(1)$$

$$ArCH_2R \quad + \; ArCH_2CH_2Ar$$
$$(3, 45-70\%) \quad (4, 15-45\%)$$

[1] N. Miyaura, M. Itoh, and A. Suzuki, *Synthesis*, 618 (1976).

Copper(I) trifluoromethanesulfonate, 6, 130–133.

1,3-Dienes substituted by phenylthio groups. Dienes of this type are valuable because the Diels-Alder adducts bear a synthetically useful functional group. Cohen *et al.*[1] have developed a general route based on the elimination of thiophenol with this Cu(I) salt and diisopropylethylamine. The preparations of two of these dienes are formulated in equations (I) and (II).

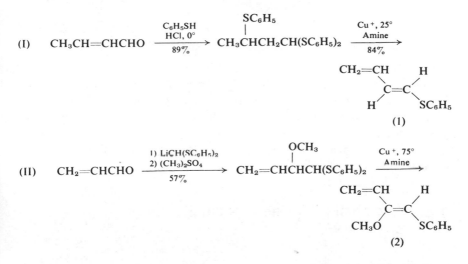

(I) $CH_3CH=CHCHO \xrightarrow[89\%]{\substack{C_6H_5SH \\ HCl, 0°}} CH_3\overset{\overset{\displaystyle SC_6H_5}{|}}{C}HCH_2CH(SC_6H_5)_2 \xrightarrow[84\%]{\substack{Cu^+, 25° \\ Amine}}$

(1)

(II) $CH_2=CHCHO \xrightarrow[57\%]{\substack{1)\ LiCH(SC_6H_5)_2 \\ 2)\ (CH_3)_2SO_4}} CH_2=CH\overset{\overset{\displaystyle OCH_3}{|}}{C}HCH(SC_6H_5)_2 \xrightarrow[]{\substack{Cu^+, 75° \\ Amine}}$

(2)

These dienes undergo Diels-Alder reactions readily; a trace of a radical inhibitor (*e.g.*, hydroquinone) is added. Typical Diels-Alder reactions are formulated in equations (III) and (IV).

(III)

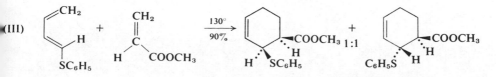

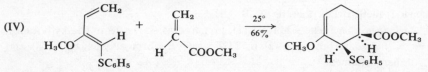

(IV)

[1] T. Cohen, A. J. Mura, Jr., D. W. Shull, E. R. Fogel, R. J. Ruffner, and J. R. Falck, *J. Org.*, **41**, 3218 (1976).

Copper(II) trifluoromethanesulfonate, 5, 152.

Iminoaziridines. Methyl diazoacetate reacts with N,N′-diisopropylcarbo-diimide in the presence of either copper(II) triflate or rhodium(II) acetate at room temperature with evolution of N_2 to form 1-isopropyl-2-methoxycarbonyl-3-isopropyliminoaziridine in 70% yield (equation I). In the absence of a catalyst, N_2 is evolved only at 150° and the yield of the aziridine is only 15%. Formation of dimethyl maleate and fumarate (dimers of the intermediate carbene) is not observed.[1]

(I) $(CH_3)_2CH-N=C=NCH(CH_3)_2$ + $N_2CHCOOCH_3$ $\xrightarrow[70\%]{cat.}$

$$(CH_3)_2CH-N-C=NCH(CH_3)_2$$

[1] A. J. Hubert, A. Feron, R. Warin, and P. Teyssie, *Tetrahedron Letters*, 1317 (1976).

Crown ethers, 4, 142–143; 5, 152–155; 6, 133–137.

Caution: Stott (Parish Chem. Co.)[1] has reported the occurrence of a severe explosion in the purification of 18-crown-6 according to the procedure of Cram, Liotta, and co-workers (5, 152), probably due to formation of *p*-dioxane. Some precautions are suggested in the report.

Review. The preparation and synthetic uses of crown ethers have been reviewed (190 references).[2]

(I) $ClCH_2CH_2OCH_2CH_2Cl$ + $HO-(CH_2CH_2O)_3CH_2CH_2OH$ $\xrightarrow[30\%]{KOH \atop THF, \Delta}$
        (excess)

(1, m.p. 38–38.5°)

(II) $ClCH_2CH_2OCH_2CH_2Cl + HO-(CH_2CH_2O)_2CH_2CH_2OH$ $\xrightarrow[\sim 38\%]{\substack{\text{Dioxane, }\Delta \\ \text{KOH}}}$
(excess)

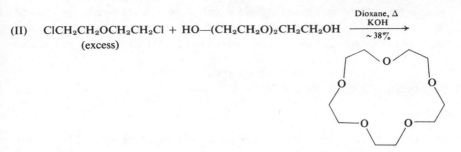

(2, b.p. 96°/0.02 torr)

Improved preparation. Reese *et al.*[3] have published new, improved syntheses of 18-crown-6 (1) and 15-crown-5 (2), both of which use the readily accessible bis[2-chloroethyl] ether (Aldrich, Eastman) [equations I and II).

Alkyl-substituted crown ethers. Italian chemists have prepared a series of crown ethers (1) substituted with a long alkyl chain and of diaza-crown ethers (2) similarly substituted. These proved to be as efficient as any of the most efficient phase-transfer catalysts known to date.

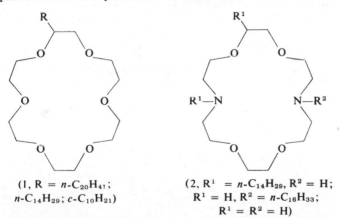

(1, R = n-C$_{20}$H$_{41}$;
n-C$_{14}$H$_{29}$; c-C$_{10}$H$_{21}$)

(2, R^1 = n-C$_{14}$H$_{29}$, R^2 = H;
R^1 = H, R^2 = n-C$_{16}$H$_{33}$;
R^1 = R^2 = H)

Cyanoformates. These compounds can be obtained from chloroformates by reaction with potassium cyanide in CH_2Cl_2 with 18-crown-6 catalysis.[5]

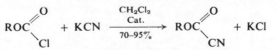

Dialkylvinylidenecyclopropanes (*cf.* 5, 460). Various crown ethers, particularly dicyclohexyl-18-crown-6, are somewhat more effective than quaternary ammonium salts as catalysts for the synthesis of these substances from 3-chloro-3-methyl-1-butyne and alkenes in the presence of aqueous KOH.[6]

Cannizzaro reaction. Gokel and co-workers[7] observed that the condensation of dimethyl sulfone and benzaldehyde under phase-transfer conditions resulted in formation of 3,5-diphenylthiodioxane-S,S-dioxide as the main product (equation I). However, when 18-crown-6 is used as catalyst, benzoic acid is also

(I) $CH_3SO_2CH_3 + 2\ C_6H_5CHO \xrightarrow[\text{NaOH, H}_2\text{O}]{\overset{+}{C_6H_5CH_2N(C_2H_5)_3}Cl^-}$ 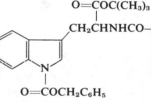 $+\ C_6H_5COOH$

(~50%)   (trace)

formed in significant amounts, evidently via a Cannizzaro reaction. The authors also examined the different effects of benzyltriethylammonium chloride and 18-crown-6 on the Cannizzaro reaction and found that the quaternary ammonium salt does not function as a catalyst in this reaction. This report is one of the few examples of differences between phase-transfer catalysis and crown ether catalysis. The differing behavior is rationalized on steric grounds.

   Tryptophan protection.[8] The trytophan nucleus has been protected by acylation with benzyl *p*-nitrophenyl carbonate[9] to form the CB derivative. The reaction is effected with KF solubilized in 18-crown-6 as base to abstract a proton from the indole ring (equation I).

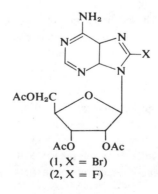

(1, X = Br)
(2, X = F)

Fluorination (5, 153–154). The very unstable and reactive 2',3',5'-tris-O-acetyl-8-fluoroadenosine (2) has been prepared for the first time by reaction of (1) with KF in the presence of 18-crown-6 with acetonitrile as solvent; yield 25%.[10] The product is of interest since 5-fluoropyrimidine nucleosides have anti-tumor activity.

[1] P. E. Stott, *Chem. Eng. News*, Sept. 6, 5 (1976).
[2] G. W. Gokel and H. D. Durst, *Synthesis*, 168 (1976).
[3] G. Johns, C. J. Ransom, and C. B. Reese, *ibid.*, 515 (1976).
[4] M. Cinquini and P. Tundo, *ibid.*, 516 (1976).
[5] M. E. Childs and W. P. Weber, *J. Org.*, 41, 3486 (1976).
[6] T. Sasaki, S. Eguchi, M. Ohno, and F. Nakata, *ibid.*, 41, 2408 (1976).
[7] G. W. Gokel, H. M. Gerdes, and N. W. Rebert, *Tetrahedron Letters*, 653 (1976).
[8] M. Chorev and Y. S. Klausner, *J.C.S. Chem. Comm.*, 596 (1976).
[9] Suppliers: Eastman, K + K.
[10] Y. Kobayashi, I. Kumadaki, A. Ohsawa, and S. Murakami, *J.C.S. Chem. Comm.*, 430 (1976).

Cupric chloride, 1, 163; **2,** 84–85; **3,** 66; **4,** 105–107; **5,** 158–160; **6,** 139–141.

Oxidative chlorination of an epoxide. Treatment of the monoepoxide of isoprene (1) with cupric chloride and lithium chloride in chloroform–ethyl acetate gives (2) as the major product. The yield is lower if lithium chloride is omitted.

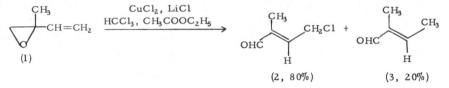

These products are also formed, albeit in lower yields, by treatment of (1) with *t*-butyl hypochlorite.[1]

[1] G. Eletti-Bianchi, F. Centini, and L. Re, *J. Org.*, 41, 1648 (1976).

Cupric perchlorate, $Cu(ClO_4)_2$. Mol. wt. 262.43. Suppliers: Alfa, ROC/RIC.

Resolution of amino acids. The dibasic amino acids DL-aspartic acid and DL-glutamic acid (A) have been resolved by formation of a copper(II) complex with a basic L-amino acid (arginine, lysine, or ornithine, B). When DL-A and L-B are mixed with $Cu(ClO_4)_2$ in the ratio 2:1:1, a blue complex Cu(D-A)(L-B) crystallizes. The copper(II) is eliminated from the complex with H_2S, and D-A and L-B are separated by chromatography on an ion exchange resin. D-Aspartic acid and D-glutamic acid have been obtained in this way in optical yields of 40–90%. The preferential formation of the copper complex with D-A and L-B ligands is probably due to electrostatic interactions.[1]

[1] T. Sakurai, O. Yamauchi, and A. Nakahara, *J.C.S. Chem. Comm.*, 553 (1976).

Cuprous bromide, 1, 165–166; **2,** 90–91; **3,** 67; **4,** 108; **5,** 163–164; **6,** 143–144.

Allenic ethers. Under catalysis with CuBr, Grignard reagents react with propionaldehyde diethyl acetal to form allenic ethers, hydrolyzable by acid to *trans*-α, β-unsaturated aldehydes (equation I).[1]

$$\text{(I)} \quad RMgBr + HC\equiv CCH(OC_2H_5)_2 \xrightarrow[\sim 80\%]{\underset{\text{Ether}}{CuBr}} RCH=C=CHOCH_3 \xrightarrow{H_3O^+} RCH=CHCHO$$

Cleavage of cyclic ethers.[2] Tetrahydrofuranes are cleaved by alkyllithiums or lithium dialkyl cuprates in the presence of catalytic amounts of CuBr or CuI

(I)   RLi + [cyclic ether with R'] $\xrightarrow[\sim 65\%]{\substack{1)\ CuBr,\ 0° \\ 2)\ H_3O^+}}$ [R–chain–R'–OH]

(equation I). Tetrahydropyrane is cleaved under these conditions, but in much lower yields ($\sim 10\%$). Oxetane is cleaved to primary alcohols in 40–80% yields.

(II)   RLi + [oxetane] $\xrightarrow[40–80\%]{\substack{1)\ CuBr \\ 2)\ H_3O^+}}$ $RCH_2CH_2CH_2OH$

[1] G. Tadema, P. Vermeer, J. Meijer, and L. Brandsma, *Rec. trav.*, **95**, 66 (1976).
[2] J. Millon and G. Linstrumelle, *Tetrahedron Letters*, 1095 (1976).

Cuprous bromide–Lithium trimethoxyaluminum hydride, $LiCuHBr \cdot Al(OCH_3)_3$.

cis-1-Alkenyl sulfides. This complex is equivalent to "CuH." It is prepared in THF at $-60°$. The complex reduces 1-alkynyl sulfides to *cis*-1-alkenyl sulfides in yields $> 90\%$ (equation I). Reduction with lithium aluminum hydride in THF

(I) $R^1-C\equiv C-SR^2$ $\xrightarrow[>90\%]{\substack{1)\ "CuH" \\ 2)\ H_2O}}$ [alkene with R^1, SR^2 cis, H, H]

(II) $R^1-C\equiv C-SR^2$ $\xrightarrow[\sim 100\%]{\substack{1)\ LiAlH_4 \\ 2)\ H_2O}}$ [alkene with H, SR^2; R^1, H]

leads to almost quantitative yields of *trans*-1-alkenyl sulfides (equation II).[1]

[1] P. Vermeer, J. Meijer, C. Eylander, and L. Brandsma, *Rec. trav.* **95**, 25 (1976).

Cuprous chloride, 1, 166–169; **2,** 91–92; **3,** 67–69; **4,** 109–110; **5,** 164–165; **6,** 145–146.

Catalyst for oxygenations (**5,** 165). Cuprous chloride or cuprous acetate can be used in catalytic amounts for oxygenation of various nitrogen compounds if it is first treated with oxygen to form a cupric salt.[1]

Examples:

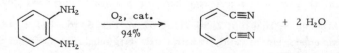

$$\text{[benzene-1,2-diamine]} \xrightarrow[94\%]{O_2,\ cat.} \text{[diene-dinitrile]} + 2\ H_2O$$

$$C_6H_5\underset{\underset{H_2N}{\overset{\|}{N}}}{\overset{\|}{C}}-\underset{\underset{NH_2}{\overset{\|}{N}}}{\overset{\|}{C}}-C_6H_5 \quad \xrightarrow[78\%]{O_2, \text{ cat.}} \quad C_6H_5C{\equiv}CC_6H_5 \;+\; 2\,N_2 \;+\; 2\,H_2O$$

$$C_6H_5CONHNH_2 \quad \xrightarrow[82\%]{O_2, \text{ cat.}} \quad C_6H_5COOH \;+\; N_2 \;+\; H_2O$$

Oxidation of phenol. Tsuji et al.[2] have found that catechol is cleaved to monomethyl muconate in high yield by oxygen catalyzed by cuprous chloride in Py–CH$_3$OH. Even more interesting from the biochemical standpoint is that phenol itself can be oxidized in the same way (equation I).[3] The reaction is slow;

(I)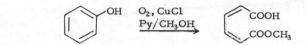

the acid is obtained after 60 hours in 44% yield. Presumably the slow step is oxygenation of phenol to catechol.

Oxidation of o-phenylenediamines (**5**, 165). The detailed paper on copper-catalyzed oxidation of o-phenylenediamines to *cis,cis*-muconitriles has been published.[4] A possible mechanism is discussed.

Octamethylcyclododeca-1,3,7,9-tetrayne (**3**). Deprotonation of the THP ether (**1**) with *n*-BuLi in THF followed by addition of a small amount of CuCl gives (**3**) in 5% yield. The intermediate (**2**) has been isolated. Physical properties of (**3**) are unusual and distinctly different from those of linear diacetylenes.[5]

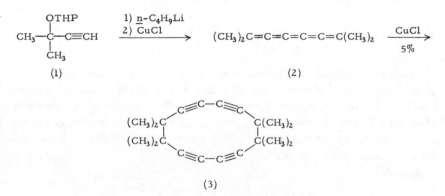

(3)

[1] J. Tsuji, H. Takayanagi, and Y. Toshida, *Chem. Letters*, 147 (1976).
[2] J. Tsuji, H. Takayanagi, and I. Sakai, *Tetrahedron Letters*, 1245 (1975).
[3] J. Tsuji and H. Takayanagi, *ibid.*, 1365 (1976).
[4] T. Kajimoto, H. Takahashi, and J. Tsuji, *J. Org.*, **41**, 1389 (1976).
[5] L. T. Scott and G. J. De Cicco, *Tetrahedron Letters*, 2663 (1976).

Cuprous iodide, 1, 169; **2**, 92; **3**, 69–71; **5**, 167–168; **6**, 147.

Alkenes. Alkenes can be prepared by alkylation of vinylic Grignard reagents with iodides and tosylates in the presence of this salt.[1] The purity of the salt is critical (*see* Linstrumelle et al.[2]).

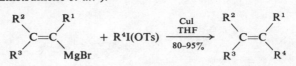

(I)

Vinylsilanes.[3] The reaction of a vinylborane such as (1), prepared as shown, in sequence with methyllithium, cuprous iodide, and triethyl phosphite yields an α-silylated vinylcopper (a) that is more reactive than vinylcopper itself.[4] It can be alkylated readily after addition of HMPT. This sequence is stereoselective and results mainly in (Z)-vinylsilanes (2).

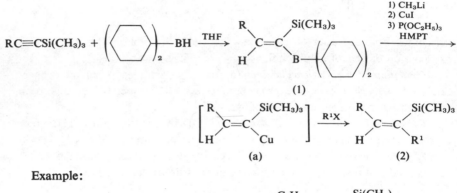

Example:

$$n\text{-}C_6H_{13}\text{—}C{\equiv}CSi(CH_3)_3 \xrightarrow[88\%]{CH_3CH_2I}$$

Copper dienolates of α,β-unsaturated acids.[5] Lithium dienolates of α,β-unsaturated acids (and esters) are alkylated mainly at the α- rather than the γ-carbon. In contrast, copper dienolates are alkylated mainly at the γ-carbon. These compounds are prepared by prior formation of the dilithium dianion (LDA) followed by addition of CuI (−78°). The example formulates the preferential γ-alkylation of crotonic acid.

Example:

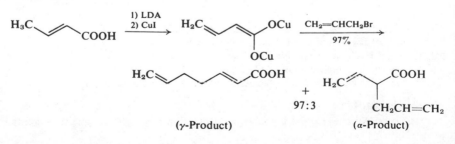

(γ-Product)   (α-Product)

The γ-alkylation has been used for a synthesis of several isoprenoids, for example, of *dl*-lanceol from *dl*-limonene (equation I).

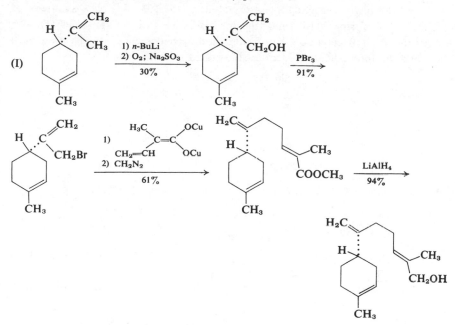

(I)

β-*Ketophosphonates.*[6] These useful Wittig-Horner reagents are easily prepared by reaction of an α-lithiophosphonate (a) with CuI or CuBr to form an organocopper derivative (b), which reacts readily with an acid chloride to form a β-ketophosphonate (2).

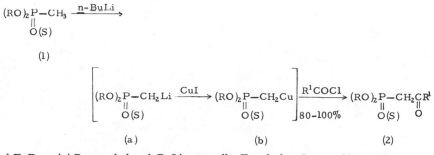

(a)   (b)   (2)

[1] F. Derguini-Boumachal and G. Linstrumelle, *Tetrahedron Letters*, 3225 (1976).
[2] G. Linstrumelle, J. K. Krieger, and G. M. Whitesides, *Org. Syn.*, **55**, 103 (1976).
[3] K. Uchida, K. Utimoto, and H. Nozaki, *J. Org.*, **41**, 2941 (1976).
[4] J. F. Normant, G. Cahiez, C. Chuit, and J. Villieras. *J. Organometal. Chem.*, **77**, 269 (1974).
[5] J. A. Katzenellenbogen and A. L. Crumrine, *Am. Soc.*, **98**, 4925 (1976).
[6] P. Savignac and F. Mathey, *Tetrahedron Letters*, 2829 (1976).

Cyclopentadienylcobalt dicarbonyl, 5, 172–173; **6,** 153–154.

Acetylene cooligomerization. Vollhardt has extended his synthesis of the strained 4,5-bis(trimethylsilyl)benzocyclobutene (**6,** 153–154) to a synthesis of 2,3,6,7-tetrakis(trimethylsilyl)naphthalene (3) by reaction of the ether (1) with bis(trimethylsilyl)acetylene (2) using this homogeneous catalyst. The product is useful as an intermediate to various substituted naphthalenes.[1]

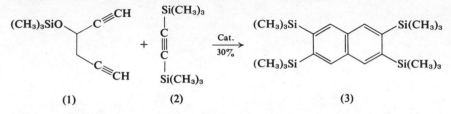

(1)   (2)   (3)

[1] R. L. Funk and K. P. C. Vollhardt, *J.C.S. Chem. Comm.*, 833 (1976).

Cyclopropyltrimethylsilane, ▷⟨H,Si(CH₃)₃⟩ (1). Mol. wt. 111.24, b.p. 95°.

This reagent is prepared by cyclopropanation of trimethylvinylsilane with diiodomethane and zinc–copper (70% yield).

Cyclopropyl ketones. These ketones can be prepared by reaction of (1) with acid chlorides (aluminum chloride, CH_2Cl_2).[1]

$$(1) + RCOCl \xrightarrow[40-75\%]{AlCl_3} \text{▷⟨H, COR⟩}$$

[1] M. Grignon-Dubois, J. Dunoguès, and R. Calas, *Synthesis*, 737 (1976).

D

trans,trans-1,4-Diacetoxybutadiene, **1**, 183–185; **3**, 73.

   Diels-Alder reactions. Holbert and Ganem[1] have presented chemical evidence consistent with an all-*cis*-arrangement in the product (1) of cycloaddition of the diene with methyl acrylate. The paper reports some other related Diels-Alder reactions.

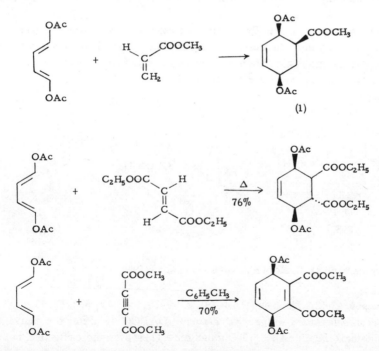

(1)

[1] G. W. Holbert and B. Ganem, *J. Org.*, **41**, 1655 (1976).

Diallylcopperlithium, 5, 175.

Conjugate addition–alkylation. This cuprate has been used as an acetic acid synthon in a new short synthesis of methyl *dl*-jasmonate (4).[1]

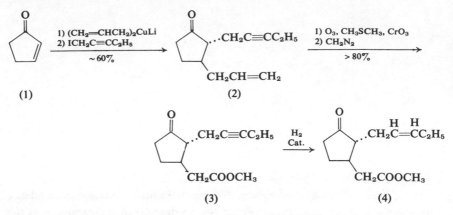

(1)   (2)

(3)   (4)

[1] A. E. Greene and P. Crabbé, *Tetrahedron Letters,* 4867 (1976).

1,5-Diazabicyclo[4.3.0]nonene-5 (DBN), 1, 189–190; **2,** 98–99; **4,** 116–119; **5,** 176; **6,** 157.

Esterification of thermally unstable carboxylic acids. Bicyclic β-keto acids can be converted into ethyl esters by diethyl sulfate at 20° or lower with DBN as base. Methyl esters are obtained in the same way. Typical esters and the yields obtained by this procedure are given.[1]

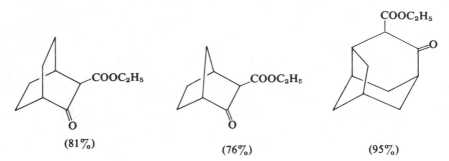

(81%)   (76%)   (95%)

[1] J. Fairhurst and D. C. Horwell, *Syn. Comm.,* **6,** 89 (1976).

1,4-Diazabicyclo[2.2.2]octane (DABCO), 2, 99–101; **4,** 119; **5,** 176–177.

Decarboalkylation of geminal diesters. DABCO is useful for cleavage of these diesters (equation I).   3-Quinuclidinol (1) is even more effective, but also more expensive.[1]

(I)

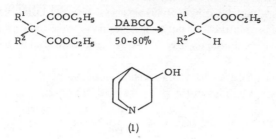

(1)

[1] D. H. Miles and B.-S. Huang, *J. Org.*, **41**, 208 (1976).

1,5-Diazabicyclo[5.4.0]undecene-5 (DBU), **2**, 101; **4**, 16–18; **5**, 177–178; **6**, 158.

Synthesis of arenes.[1] Barnett and Needham[2] have reported that the sodium salt of 1,4-dihydrobenzoic acid (1) is converted by bromination into the halo-β-lactone (2). Ganem notes that a nonnucleophilic base such as DBU effects

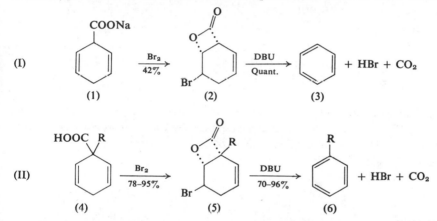

dehydrobromination of (2) to give presumably a β-lactone that loses carbon dioxide spontaneously to form benzene. Similar halo-β-lactones can be prepared from

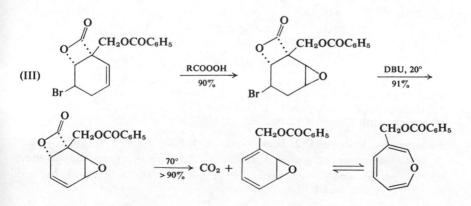

1-, 2-, or 6-substituted 1,4-dihydrobenzoic acids. Usually DBU is the most effective base for the aromatization, but triethylamine is sometimes more satisfactory.

The lactones are suitable as precursors to arene oxides as shown in equation (III).

Dialkylation of CH₂-active compounds. Activated CH_2 compounds can be dialkylated by alkyl halides in the presence of DBU (2 equiv.).

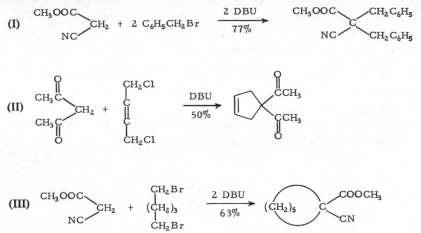

Anhydronucleosides. Treatment of the 5'-O-(methanesulfonyl)uridine (1) with DBU in CH_2Cl_2 results in conversion to the 2,5'-anhydronucleoside (2) in 85% yield. DBU is more efficient than DBN for this conversion.[4]

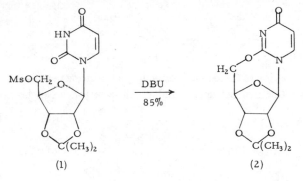

[1] G. W. Holbert, L. B. Weiss, and B. Ganem, *Tetrahedron Letters*, 4435 (1976).
[2] W. E. Barnett and L. L. Needham, *J. Org.*, **40**, 2843 (1975) and references cited therein.
[3] H. Oediger and F. Möller, *Ann.*, 348 (1976).
[4] J. A. Secrist, III, *Carbohydrate Res.*, **42**, 379 (1975).

Diazomethane, 1, 191–195; **2,** 102–104; **3,** 74; **4,** 120–122; **5,** 179–182; **6,** 158.

α,β-Unsaturated diazomethyl ketones.[1] The direct preparation of these ketones by the Arndt-Eistert reaction is not generally satisfactory because of preferred reaction of diazomethane with the C=C bond. They can be prepared satisfactorily by protection of the double bond by addition of HBr. The method is illustrated for preparation of allenyl diazomethyl ketone (equation I).

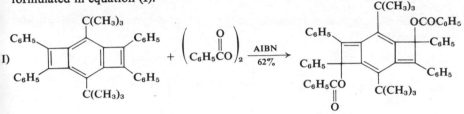

(I) $CH_2=C=CHCOOH$ $\xrightarrow[88\%]{HBr}$ $CH_2=CBrCH_2COOH$ $\xrightarrow[85\%]{\substack{1)\ (COCl)_2 \\ 2)\ CH_2N_2}}$

$CH_2=CBrCH_2COCHN_2$ $\xrightarrow[59\%]{DBN}$ $CH_2=C=CHCCHN_2$

[1] N. R. Rosenquist and O. L. Chapman, *J. Org.*, **41**, 3326 (1976).

Dibenzoyl peroxide, 1, 196–198; **4,** 122–123; **5,** 182–183; **6,** 160–161.

Addition reaction. Japanese chemists[1] have reported the 1,8-addition formulated in equation (I).

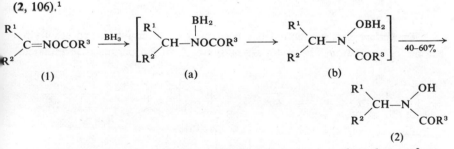

I)

[1] F. Toda, K. Tanaka, and T. Yoshiska, *Chem. Letters*, 657 (1976).

Diborane, 1, 199–207; **2,** 106–108; **3,** 76–77; **4,** 124–126; **5,** 184–186; **6,** 161–162.

Hydroxamic acids. Oximino esters (1) are reduced by 1 equiv. of BH_3 in THF at $-78°$ to hydroxamic acids (2). The reaction involves an O to N acyl shift, (a) → (b). Ketoximes are reduced completely to amines by excess diborane (**2,** 106).[1]

$$\underset{R^2}{\overset{R^1}{\diagdown}}C=NOCOR^3 \xrightarrow{BH_3} \left[\underset{R^2}{\overset{R^1}{\diagdown}}\underset{}{\overset{BH_2}{CH-NOCOR^3}} \longrightarrow \underset{R^2}{\overset{R^1}{\diagdown}}CH-N\underset{COR^3}{\overset{OBH_2}{\diagup}}\right] \xrightarrow{40-60\%}$$

(1)   (a)   (b)

$$\underset{R^2}{\overset{R^1}{\diagdown}}CH-N\underset{COR^3}{\overset{OH}{\diagup}}$$

(2)

Complex borohydrides are not suitable for this reduction, since they preferentially reduce the ester carbonyl.

Review. Pelter[2] has reviewed reduction with diborane and some reagents based on diborane.

[1] B. Ganem, *Tetrahedron Letters*, 1951 (1976).
[2] A. Pelter, *Chem. Ind.*, 888 (1976).

Dibromomethyllithium, 1, 223–224, **2,** 119; **3,** 89; **4,** 138–129; **5,** 398, **6,** 162–163.

Ring expansion (**6,** 162–163). Dibromomethyllithium, generated from methylene bromide and lithium tetramethylpiperidide, has been used in a new synthesis of *dl*-muscone (3) from 2-methylcyclotetradecanone (1). The direction of

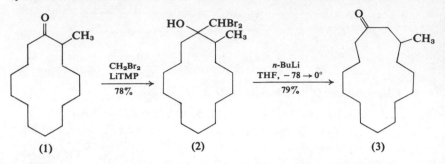

insertion of methylene is dependent on the reaction temperature, the solvent, and ring size.[1]

[1] H. Taguchi, H. Yamamoto, and H. Nozaki, *Tetrahedron Letters*, 2617 (1976).

1,3-Dibromo-2-pentene, (1). Mol. wt. 227.95, b.p. 88°/3 mm.
    Preparation:

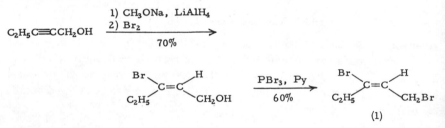

Annelation. This compound has been used as an equivalent to ethyl vinyl ketone for annelation. The use in a new synthesis of β-cyperone is formulated.[1]

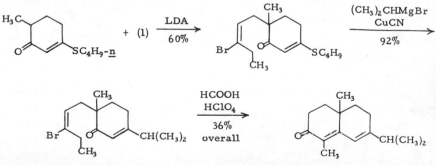

[1] R. B. Gammill and T. A. Bryson, *Syn. Comm.*, **6,** 209 (1976).

2,2-Dibromo-1,3-propanediol, $Br_2C(CH_2OH)_2$. Mol. wt. 233.91, m.p. 110–111°.
Supplier: Fluka.
    Preparation:

$$(CH_3O)_2CHCH_2CH(OCH_3)_2 \xrightarrow[\text{2) OH}^-]{\text{1) HCl, Br}_2} Na^+Br\bar{C}(CHO)_2 \xrightarrow[37\%]{\substack{\text{1) Br}_2 \\ \text{2) LiAlH}_4}} Br_2C(CH_2OH)_2$$
$$(1)$$

Protection of $\diagdown C{=}O$.[1] Under acid catalysis, (1) converts carbonyl compounds into 5,5-dibromo-1,3-dioxanes (2). The carbonyl group is regenerated in
high yield with zinc–silver couple (**4**, 436).

$$(2)$$

[1] E. J. Corey, E. J. Trybulski, and J. W. Suggs, *Tetrahedron Letters*, 4577 (1976).

Di-*t*-butyl dicarbonate, 4, 128. Supplier: Fluka. Do not heat above 80°.
    Protection of amino acids. Amino acids can be converted in high yield into
BOC-derivatives by reaction under mild conditions with this reagent in aqueous
dioxane, THF, or *t*-butanol in the presence of sodium hydroxide, which is
necessary for conversion of the acid into a salt. The reaction can also be carried
out in DMF with triethylamine or Triton B in equally high yield.[1]

[1] L. Moroder, A. Hallett, E. Wünsch, O. Keller, and G. Wersin, *Z. Physiol. Chem.*, **357**,
1651 (1976).

Di-*n*-butylboryl trifluoromethanesulfonate, $(n\text{-}C_4H_9)_2BOSO_2CF_3$ (1). Mol. wt.
274.11, b.p. 37°/0.12 mm.
    This triflate is prepared from equimolar amounts of tri-*n*-butylborane and
triflic acid.
    Aldol condensations.[1] This triflate, in combination with a base (diisopropyl-
ethylamine), is useful for cross-aldol reaction of ketones with aldehydes.
    Examples:

The method has been used for a synthesis of γ-ionone (3) from γ-cyclocitral (1).[2]

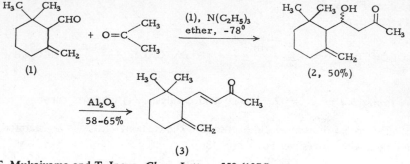

(1)   (2, 50%)

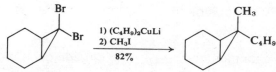

(3)

[1] T. Mukaiyama and T. Inoue, *Chem. Letters*, 559 (1976).
[2] T. Mukaiyama, K. Saigo, and O. Takazawa, *ibid.*, 1033 (1976).

Di-*n*-butylcopperlithium (Lithium di-*n*-butylcuprate), 2, 152; 5, 187–188.

Dialkylation of gem-*dihalides.* The reaction of *gem*-dihalides with dialkyl-copperlithiums has been used to effect dialkylation on the same carbon atom, particularly dimethylation. Japanese chemists[1] have developed another means of performing this reaction. *gem*-Dihalocyclopropanes (1) are treated with di-*n*-butylcopperlithium (excess) at $-40°$ (*ca.* 1 hour) and then at $-20°$ with an alkyl iodide. The products are (2) and (3), the ratio depending on the reaction

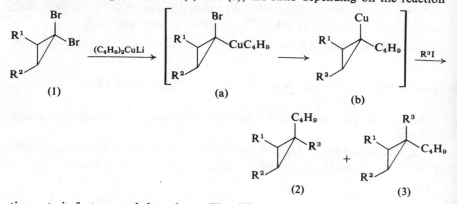

time, steric factors, and the solvent. The different behavior of di-*n*-butylcopper-lithium as compared with dimethylcopperlithium is a consequence of the difference in reactivity of the two cuprates. Note that the reaction of 7,7-dibromo-monorcarane results in exclusive formation of the *endo*-methyl isomer (equation I)

This new dialkylation reaction was used in a simple synthesis of *dl*-sirenin (6).

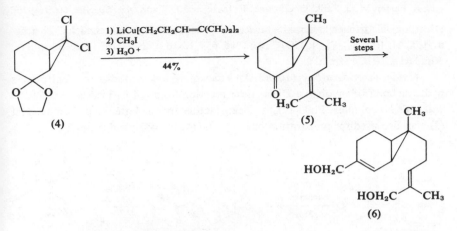

(4)   (5)   (6)

Coupling with enol diphenylphosphate esters. Ketones can be converted into olefins by conversion into the enol diphenylphosphate esters followed by treatment of a dialkylcopperlithium. Dimethylcopperlithium does not undergo this reaction, nor do hindered enol phosphates.[2]

Examples:

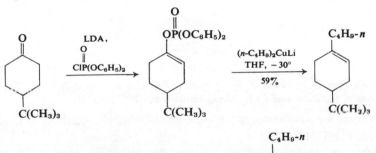

$$CH_3(CH_2)_4COCH_3 \xrightarrow[74\%]{} CH_3(CH_2)_4\overset{\overset{\displaystyle C_4H_9\text{-}n}{|}}{C}=CH_2$$

[1] K. Kitatani, T. Hiyama, and H. Nozaki, *Am. Soc.*, **98**, 2362 (1976).
[2] L. Blaszczak, J. Winkler, and S. Okuhn, *Tetrahedron Letters*, 4405 (1976).

Di-*t*-butylperoxyoxalate, 6, 166–167.

Cyclic peroxides.[1] The complete paper on the formation of cyclic peroxides from unsaturated hydroperoxides using this radical source has been published.

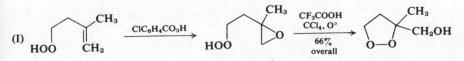

(I)

The paper includes another route via oxirane hydroperoxides (equation I).

[1] N. A. Porter, M. O. Funk, D. Gilmore, R. Isaac, and J. Nixon, *Am. Soc.*, **98**, 6000 (1976).

Dicarbonylbis(triphenylphosphine)nickel(0) [Bis(triphenylphosphine)nickel dicarbonyl], **1**, 61, $[(C_6H_5)_3P]_2Ni(CO)_2$. Mol. wt. 639.32, m.p. 213–218°, dec. Suppliers: Alfa, ROC/RIC, Strem.

Oxidative bisdecarboxylation.[1] An efficient recent synthesis of Nenitzescu's hydrocarbon (3)[2] involves a Diels-Alder reaction followed by bisdecarboxylation with this Ni complex. $Pb(OAc)_4$ is unsatisfactory for the latter step. The product (3) was also used for preparation of barrelene (4) in 24% yield from (1).

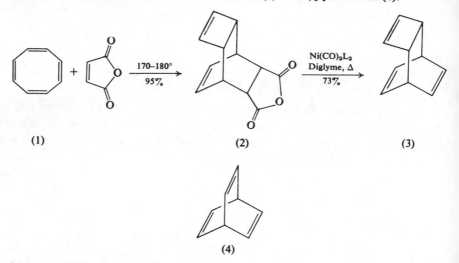

(1)   (2)   (3)

(4)

[1] W. G. Dauben, G. T. Rivers, R. J. Twieg, and W. T. Zimmerman, *J. Org.*, **41**, 887 (1976).
[2] M. Avram, E. Sliam, and C. D. Nenitzescu, *Ann.*, **636**, 184 (1960).

Dicarbonylcyclopentadienylcobalt (Cyclopentadienylcobalt dicarbonyl), $C_5H_5Co(CO)_2$, **5**, 172–173; **6**, 153–154.

Acetylene cooligomerizations. Interesting polycyclic structures can be obtained in one step by cobalt-catalyzed cooligomerization of 1,5-hexadiynes and bis(trimethylsilyl)acetylene.[1]

Examples:

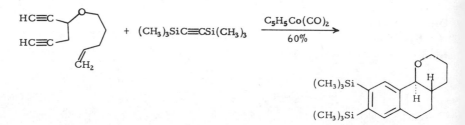

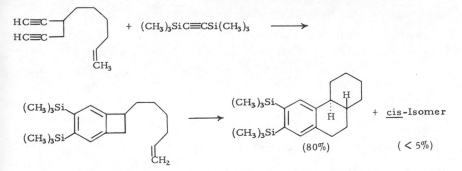

The reaction proceeds through intermediate benzocyclobutenes, as shown in the second example. The new ring junction has almost exclusively the *trans*-stereochemistry.

[1] R. L. Funk and K. P. C. Vollhardt, *Am. Soc.*, **98**, 6755 (1976).

Dichlorobis(benzonitrile)palladium(II), $PdCl_2(C_6H_5CN)_2$. Mol. wt. 383.55, yellow-orange, air-stable. Suppliers: ROC/RIC, Strem.

Preparation.[1] This complex is readily soluble in common organic solvents. Since benzonitrile is bound very loosely, the complex is used as a solubilized form of $PdCl_2$.

Oligomerization. Maitlis[2] has reviewed the trimerization of acetylenes in aprotic solvents to benzenoids catalyzed by this complex.

[1] M. S. Kharasch, R. C. Seyler, and F. R. Mayo, *Am. Soc.*, **60**, 882 (1938).
[2] P. M. Maitlis, *Accts. Chem. Res.*, **9**, 93 (1976).

Dichlorobis(triphenylphosphine)palladium(II), 6, 59.

Alkenyl–alkenyl cross-coupling. Baba and Negishi[1] have prepared a catalyst from this Pd(II) complex and 2 equiv. of diisobutylaluminum hydride that promotes this coupling reaction. Tetrakis(triphenylphosphine)palladium(0) is inactive, as is material prepared *in situ* from palladium chloride, triphenylphosphine and $HAl(i-C_4H_9)_2$. A nickel catalyst prepared from $Ni(acac)_2$, $P(C_6H_5)_3$, and diisobutylaluminum hydride is somewhat less efficient. The coupling involves (E)-alkenylalanes (4, 158, 159) and alkenyl halides. The products are (E,E)- and (E,Z)-dienes.

Examples:

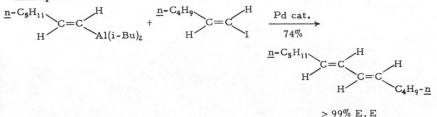

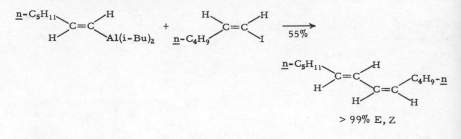

> 99% E, Z

[1] S. Baba and E. Negishi, *Am. Soc.*, **98**, 6729 (1976).

Dichloroborane diethyl etherate, 4, 130; 5, 191.

Hydroboration. The complete paper on hydroboration of alkenes and alkynes with this reagent is available.[1]

[1] H. C. Brown and N. Ravindran, *Am. Soc.*, **98**, 1798 (1976).

2,3-Dichloro-5,6-dicyano-1,4-benzoquinone (DDQ), 1, 215–219; **2,** 112–117; **3,** 83–84; **4,** 130–134; **5,** 193–194; **6,** 168–170.

Regeneration from the hydroquinone (DDHQ), 1, 215. The hydroquinone

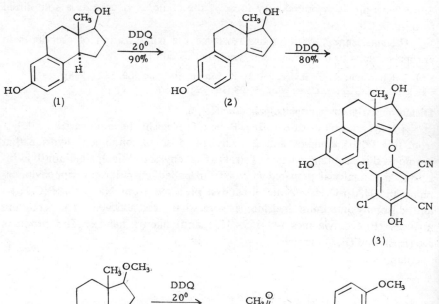

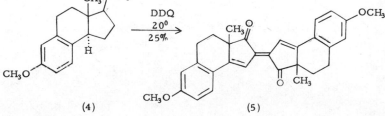

is also oxidized to the quinone in high yield by a mixture of perchloric acid, sulfuric acid, and nitric acid in the ratio 1:4:2.5 in glacial acetic acid.[1]

Oxidative coupling. Oxidation of (1) with 1 equiv. of DDQ results in the styrene (2), which can be oxidized by DDQ further to the coupled adduct (3). Oxidation of the dimethyl ether (4) of (1) with DDQ results mainly in formation of the coupled dimer (5).[2]

A stable 1,5-naphthoquinone. 3,7-Di-*t*-butyl-1,5-naphthalenediol (1) is oxidized to the 1,5-quinone (2) in high yield by DDQ. The reverse reaction can

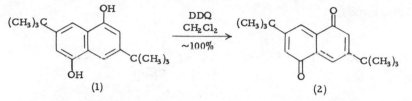

be carried out with zinc and HCl. Water and air must be excluded during the oxidation to obtain satisfactory results.[3]

[1] K. H. Kim and G. L. Grunewald, *Org. Prep. Proc. Int.*, **8**, 141 (1976).
[2] A. B. Turner, *Chem. Ind.*, 1030 (1976).
[3] H. L. K. Schmand, H. Kratzin, and P. Boldt, *Ann.*, 1560 (1976).

(Dichloromethylene)triphenylphosphorane, $(C_6H_5)_3P\!=\!CCl_2$, **1**, 222–223. Mol. wt. 345.21, m.p. 115–122° dec., stable to storage for several days, yellow.

This ylide has now been isolated by Appel *et al.*[1] by dehydrochlorination of (dichloromethyl)triphenylphosphonium chloride (1) with an excess of bis(triphenylphosphoranylidene)methane (2).[2] The choice of solvent is crucial for successful isolation of the ylide (3). Only aprotic solvents can be used.

$$\left[(C_6H_5)_3P\!-\!CHCl_2\right]Cl^- + (C_6H_5)_3P\!=\!C\!=\!P(C_6H_5)_3 \xrightarrow[55\%]{\substack{C_6H_5Cl\\C_6H_5CH_3}} (C_6H_5)_3P\!=\!CCl_2$$

(1)   (2)   (3)

For *in situ* preparation, substitution of bromotrichloromethane for carbon tetrachloride (**1**, 223) results in more rapid formation of this Wittig reagent and also significantly improves the yields of 3,3-dichloroacrylonitriles formed by reaction with acyl nitriles.[3]

$$2(C_6H_5)_3P + Cl_3CBr \xrightarrow{0°, 30 \text{ min.}} (C_6H_5)_3P\!=\!CCl_2 + (C_6H_5)_3PBrCl$$

$$RCOCN + (C_6H_5)_3P\!=\!CCl_2 \xrightarrow[85-90\%]{} Cl_2C\!=\!CRCN$$

[1] R. Appel, F. Knoll, and H. Veltmann, *Angew. Chem., Int. Ed.*, **15**, 315 (1976).
[2] F. Ramirez, N. B. Desai, B. Hansen, and N. McKelvie, *Am. Soc.*, **83**, 3539 (1961).
[3] B. A. Clement and R. L. Soulen, *J. Org.*, **41**, 556 (1976).

α,α-Dichloromethyl methyl ether, 2, 120; 5, 200–203.

Bicyclo[3.3.1]nonane-9-one (5, 201–202). Details for the preparation of this ketone are available (83% yield). The procedure includes a preparation of the hindered base used, that is, lithium triethylcarboxide.[1]

[1] B. A. Carlson and H. C. Brown, *Org. Syn.*, submitted (1976).

(Z)-1,2-Dichloro-4-phenylthio-2-butene, (1). Mol. wt. 223.16, b.p. 79–80°/ 0.01 torr.

Preparation:

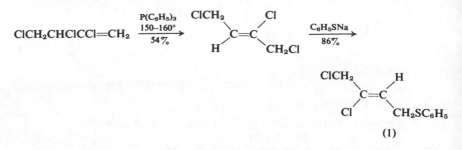

α,β-Unsaturated carbonyl compounds.[1] The reagent (1) is a latent equivalent of 1-chloro-3-butene-2-one, $ClCH_2COCH=CH_2$. For example, it has been used in an interesting synthesis of (5), an obvious steroid precursor. The conversion of (3) to (4) involves a [2,3]sigmatropic rearrangement to an α,β-unsaturated ketone.

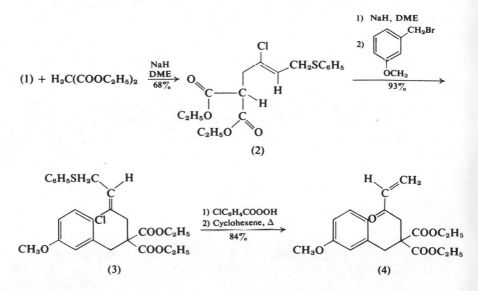

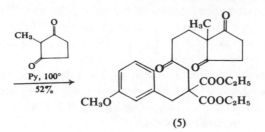

(5)

[1] P. T. Lansbury and R. W. Britt, *Am. Soc.*, **98**, 4577 (1976).

Dichlorotris(triphenylphosphine)ruthenium(II), 4, 564; 5, 740–741; 6, 654–655.

Reduction of cyclic anhydrides. Hydrogenation of unsymmetrical cyclic anhydrides with this catalyst selectively reduces the less hindered carbonyl group (equation I).[1] Note that reduction with sodium borohydride results in

(I)

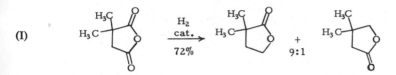

attack mainly at the carbonyl group adjacent to the more highly hindered carbon atom (3, 263–264).

[1] P. Morand and M. Kayser, *J.C.S. Chem. Comm.*, 314 (1976).

Dicobalt octacarbonyl (Octacarbonyldicobalt), 1, 224–225; 3, 89; 4, 139; 5, 204–205; 6, 172.

Conjugated enynes. The direction of cleavage of cyclopropane rings adjacent to an alcohol function is markedly influenced by complexation with dicobalt

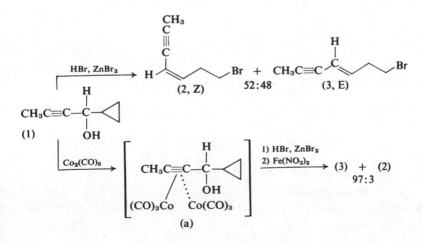

(a)

octacarbonyl. Thus treatment of (1) with HBr and ZnBr$_2$ at -20 to $0°$ gives about equal mixtures of the (Z)- and (E)-enynes, (2) and (3). However, if (1) is first complexed with dicobalt octacarbonyl to form (a), the (E)-isomer (3) is almost the exclusive product.[1]

Indole synthesis. A novel synthesis of 2-styrylindoles from 2-arylazirines is formulated in equation (I).[2] The same reaction has also been effected with bis(chlorodicarbonylrhodium), [Rh(CO)$_2$Cl]$_2$, **5**, 572–574, and with chloro-carbonylbis(triphenylphosphine)rhodium.[3]

(I)

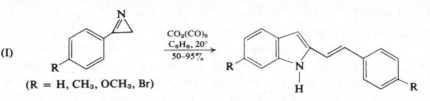

(R = H, CH$_3$, OCH$_3$, Br)

[1] C. Descoins and D. Samain, *Tetrahedron Letters*, 745 (1976).
[2] H. Alper and J. E. Prickett, *Tetrahedron Letters*, 2589 (1976).
[3] *Idem, J.C.S. Chem. Comm.*, 483 (1976).

2,3-*Dicyanobutadiene*. On thermolysis, 1,2-dicyanocyclobutene is converted into 2,3-dicyanobutadiene, which slowly polymerizes. The butadiene, however, can be trapped successfully when generated in the presence of olefins, particularly electron-rich olefins. An example is the cycloaddition with indene (equation I).[1]

(I)

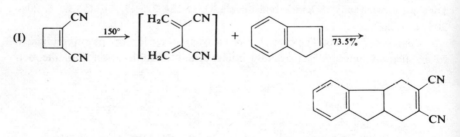

[1] D. Belluš, H. Sauter, and C. D. Weis, *Org. Syn.*, submitted (1976).

Dicyclohexylcarbodiimide (DCC), 1, 231–236; **2**, 126; **3**, 91; **4**, 141; **5**, 206–207; **6**, 174.

Thiol esters. These esters can be prepared by the reaction of acids and thiols mediated by DCC. The yields are low with aliphatic thiols (20–55%), but reasonable with thiophenols (35–75%).[1]

Sulfinamides. Furukawa and Okawara[2] have published two methods for preparation of sulfinamides (equations I and II). Yields in both methods are about the same.

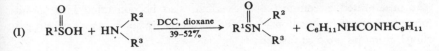

(I) $R^1SOH + HN\begin{smallmatrix}R^2\\R^3\end{smallmatrix}$ $\xrightarrow[\text{39-52\%}]{\text{DCC, dioxane}}$ $R^1SN\begin{smallmatrix}R^2\\R^3\end{smallmatrix}$ $+ C_6H_{11}NHCONHC_6H_{11}$

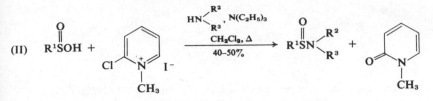

(II) $R^1SOH +$ [structure] $\xrightarrow[\substack{CH_2Cl_2, \Delta\\40-50\%}]{\substack{HN\begin{smallmatrix}R^2\\R^3\end{smallmatrix}, N(C_2H_5)_3}}$ $R^1SN\begin{smallmatrix}R^2\\R^3\end{smallmatrix}$ $+$ [structure]

[1] J. R. Grunwell and D. L. Forest, *Syn. Comm.*, **6**, 453 (1976).
[2] M. Furukawa and T. Okawara, *Synthesis*, 339 (1976).

Dicyclohexylmethylthioborane, $(C_6H_{11})_2BSCH_3(1)$. Mol. wt. 224.21, stable.
    Preparation[1]:

$$(C_6H_{11})_3B + CH_3SH \xrightarrow{h\nu} (C_6H_{11})_2BSCH_3$$

Conjugated diynes, $R^1C\equiv C-C\equiv CR^2$. Pelter *et al.*[2] have extended the synthesis of symmetrical conjugated diynes (*6*, 294) to a synthesis of unsymmetrical conjugated diynes (equation I) in which this reagent is used as the carrier borane.

$$(1) + LiC\equiv CR^1 \xrightarrow{THF} (C_6H_{11})_2B^-(C\equiv CR^1)(SCH_3)Li^+ \underset{-LiSCH_3}{\rightleftharpoons}$$

$$(C_6H_{11})_2BC\equiv CR^1 \xrightarrow{LiC\equiv CR^2} \overset{+}{Li}(C_6H_{11})_2B^-(C\equiv CR^1)(C\equiv CR^2) \xrightarrow[\substack{50-60\%\\\text{overall}}]{I_2}$$

$$R^1C\equiv C-C\equiv CR^2$$

[1] A. Pelter, K. Rowe, D. N. Sharrocks, and K. Smith, *J.C.S. Chem. Comm.*, 531 (1975).
[2] A. Pelter, R. Hughes, K. Smith, and M. Tabata, *Tetrahedron Letters*, 4385 (1976).

Di(η^5-cyclopentadienyl)(chloro)hydridozirconium (IV), *6*, 175–179.
    Hydrozirconation. This process has been reviewed (77 references).[1]
    γ,δ-Unsaturated aldehydes.[2] This zirconium hydride reacts with 1,3-dienes via 1,2-addition to the less hindered double bond to give homoallylic complexes. These complexes insert carbon monoxide readily to form unsaturated aldehydes.
    Examples:

$CH_3\diagup\diagdown CH_2 + (1) \xrightarrow{ca.\ 85\%} CH_3\diagup\diagup\diagdown (Zr) \xrightarrow[81\%]{\substack{1)\ CO\\2)\ H_3O^+}} CH_3\diagup\diagup\diagdown CHO$

$H_3C\diagup\diagdown CH_2 \xrightarrow{(1)} \xrightarrow{CO}_{98\%} CH_3\diagup\diagdown CHO$

The unsaturated zirconium complexes can also be converted into γ,δ-unsaturated bromides by reaction with NBS; however, in some cases cyclopropylcarbinyl bromides are also formed:

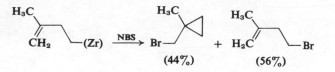

[1] J. Schwartz and J. A. Labinger, *Angew. Chem., Int. Ed.*, **15**, 333 (1976).
[2] C. A. Bertelo and J. Schwartz, *Am. Soc.*, **98**, 262 (1976).

2,2-Diethoxyvinylidenetriphenylphosphorane, 3, 93–94.

Synthesis of γ,δ-unsaturated 1,3-dicarbonyl compounds. The definitive paper has been published.[1]

[1] H.-J. Bestmann and R. W. Saalfrank, *Ber.*, **109**, 403 (1976).

Diethyl acetylmethylmalonate, $CH_3COCH_2CH(COOC_2H_5)_2$. Mol. wt. 216.24, b.p. 102–104°/0.9 mm.

Preparation.[1] This keto ester is prepared by condensation of diethyl sodiomalonate with α-chloroacetone in DMF (55% yield).

Michael additions.[2] ω-Acetoxymalonic esters undergo Michael additions to cyclic α,β-unsaturated aldehydes[3] (equation I) and acrolein (equation II).

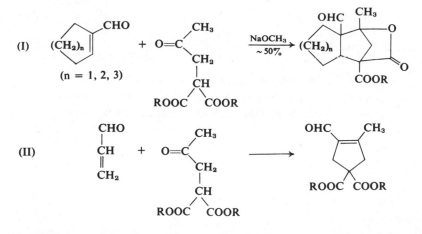

[1] C. Mercier, A. R. Addas, and P. Deslongchamps, *Canad. J. Chem.*, **50**, 1882 (1972).
[2] H. Marschall, F. Vogel, and P. Weyerstahl, *Tetrahedron Letters*, 175 (1976).
[3] Preparation: H. Taguchi, S. Tanaka, H. Yamamoto, and H. Nozaki, *ibid.*, 2465 (1973).

Diethylcarbo-*t*-butoxymethylalane, $(C_2H_5)_2AlCH_2COOC(CH_3)_3$. Mol. wt. 200.26. This aluminum derivative of *t*-butyl acetate is prepared as formulated.

$$LiCH_2COOC(CH_3)_3 + (C_2H_5)_2AlCl \xrightarrow[\substack{-40° \\ -LiCl}]{Toluene} (C_2H_5)_2AlCH_2COOC(CH_3)_3$$

$$(1)$$

Reaction with epoxides. *t*-Butyl lithioacetate (**5**, 371) reacts with cyclohexene oxide (**2**) to form (**3**) in at best 8% yield. On the other hand, the aluminum derivative (**1**) of *t*-butyl acetate reacts with (**2**) to form (**3**) in 68% yield (compare Fried's work with dialkylalkynylalanes, **4**, 144–145). However, this epoxide

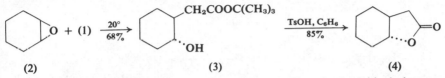

(2)   (3)   (4)

cleavage reaction does not appear to be general. Thus 2α,3α-oxidocholestane does not react with (**1**).

The aluminum derivative of ethoxyacetylene, $(C_2H_5)_2AlC\equiv C—OC_2H_5$ (**5**), is apparently more reactive, since this reagent reacts satisfactorily with 2α,3α-oxidocholestane (63% yield).

Example:

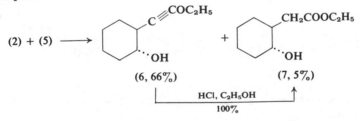

(6, 66%)   (7, 5%)

HCl, C_2H_5OH

100%

[1] S. Danishefsky, T. Kitahara, M. Tsai, and J. Dynak, *J. Org.*, 1669 **41**, (1976).

Diethyl dibromophosphoramidate, $(C_2H_5O)_2\overset{\overset{O}{\|}}{P}—NBr_2$ (**1**), Mol. wt. 310.94, n_D^{20} 1.5244. Stable in the dark at 0–5°.

Preparation[1]:

$$(C_2H_5O)_2\overset{\overset{O}{\|}}{P}—NH_2 + 2\,Br_2 \xrightarrow[92\%]{K_2CO_3,\ H_2O,\ 0^0} \quad (1)$$

β-*Bromo amines.*[2] In the presence of BF_3 etherate, the reagent (**1**) adds to styrene and to cyclohexene to form addition products that are reduced ($NaHSO_3$) to N-(β-bromoalkyl)phosphoramides. These products can be transformed into aziridines or into β-bromo amines as shown in the formulations. Both Markovnikov addition reactions are regio- and stereospecific.

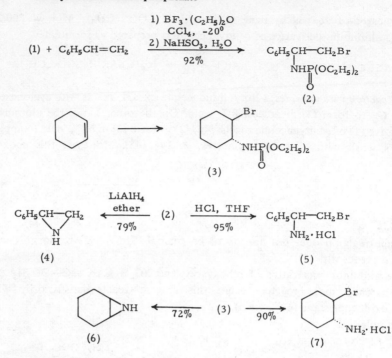

[1] A. Zwierzak and S. Zawadzki, *Synthesis*, 323 (1971); *idem*, *Tetrahedron*, **29**, 315 (1973).
[2] A. Zwierzak and K. Osowska, *Angew. Chem., Int. Ed.*, **15**, 302 (1976).

Diethyl dibromomethanephosphonate, $(C_2H_5O)_2\overset{\overset{O}{\|}}{P}\!-\!CHBr_2$. Mol. wt. 309.93.
  Preparation:

$$(C_2H_5O)_2\overset{\overset{O}{\|}}{P}CH_2Cl \xrightarrow[-HCBr_3]{\substack{1)\ n\text{-}C_4H_9Li,\ THF \\ 2)\ CBr_4,\ -70°}} \left[(C_2H_5O)_2\overset{\overset{O}{\|}}{P}\!-\!\overset{\overset{Li^+}{}}{\underset{Cl}{C}}{\overset{}{\diagdown}}Br \right] \xrightarrow{LiBr,\ H_2O}$$

$$(C_2H_5O)_2\overset{\overset{O}{\|}}{P}\!-\!CHBr_2 + (C_2H_5O)_2\overset{\overset{O}{\|}}{P}\!-\!CH\overset{\diagup Br}{\diagdown Cl}$$

$$(1,\ 90\%) \qquad\qquad\qquad (2,\ 10\%)$$

1,1-Dibromoalkenes.[1] Treatment of the mixture of (1) and (2) obtained as formulated above with LDA in THF at $-70°$ leads to an anion that reacts with aldehydes and ketones to form 1,1-dibromoalkenes in 40–70% yield. For the preparation of 1,1-dichloralkenes by a related method *see* **6**, 189.

[1] P. Savignac and P. Coutrot, *Synthesis*, 197 (1976).

Diethyl lithiomorpholinomethylphosphonate, (1). Mol. wt. 243.17.
    Preparation[1]:

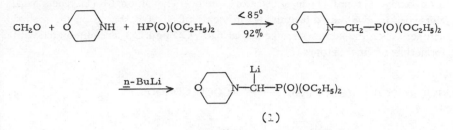

$$CH_2O \ + \ \text{(morpholine)} NH \ + \ HP(O)(OC_2H_5)_2 \ \xrightarrow[92\%]{<85^0} \ \text{(morpholine)} N-CH_2-P(O)(OC_2H_5)_2$$

$$\xrightarrow{n-BuLi} \ \text{(morpholine)} N-\overset{\overset{\text{Li}}{|}}{C}H-P(O)(OC_2H_5)_2$$

(1)

Spiroannelation.[2] This phosphonate reacts with ketones to form enamines (2). These products react with methyl vinyl ketone (followed by hydrolysis) to give δ-keto aldehydes (3), which are converted spontaneously into 4,4-disubstituted 2-cyclohexene-1-ones (4).

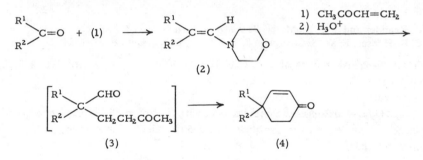

Examples:

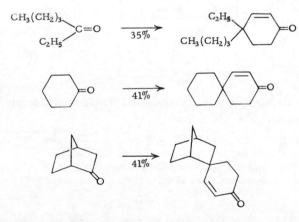

[1] E. K. Fields, *Am. Soc.*, **74**, 1528 (1952).
[2] S. F. Martin, *J. Org.*, **41**, 3337 (1976).

Diethyl oxalate, 1, 250–251; **2,** 132.

Vinyl ketones.[1] Enolates of methyl ketones react with diethyl oxalate to form oxalyl ketones (1) in satisfactory yield.[2] The anion of (1) reacts with aldehydes to give diketo lactones (2) in yields of 60–90%, based on the methyl ketone. When the lactones are heated at 620°, CO and CO_2 are eliminated with formation of vinyl ketones (3).

(I) $RCOCH_3 + (COOC_2H_5)_2$ $\xrightarrow{\text{NaH}}$ $RCOCH_2COCOOC_2H_5$ $\xrightarrow[\text{60–90\%}]{\overset{\text{Base}}{\text{R}^1\text{CHO}}}$

(1)

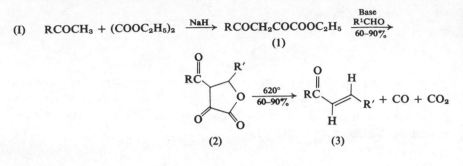

(2)   (3)

[1] G. Ksander and J. E. McMurry, *Tetrahedron Letters*, 4691 (1976).
[2] C. R. Hauser, F. W. Swamer, and J. T. Adams, *Org. React.*, **8**, 59 (1954).

Diethyl N-phenylamidophosphate, $C_6H_5NHP(O)(OC_2H_5)_2$ (1). Mol. wt. 229.22, m.p. 96.5°.

Preparation.[1]

Epoxides → aziridines.[2] Epoxides react with the anion of this ester in refluxing toluene or xylene to form aziridines in 30–60% yield. The reaction is slow in DME (b.p. 83°).

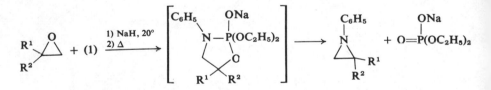

A somewhat similar route has been published by German chemists,[3] who used iminophosphoranes.

[1] H. McCombie, B. C. Saunders, and G. J. Stacey, *J. Chem. Soc.*, 380 (1945); F. R. Atherton, H. T. Openshaw, and A. R. Todd, *ibid.*, 660 (1945).
[2] I. Shahak, Y. Ittah, and J. Blum, *Tetrahedron Letters*, 4003 (1976).
[3] R. Appel and M. Halstenberg, *Ber.*, **109**, 814 (1976).

2-(Diethylphosphono)propionitrile, $(C_2H_5O)_2\overset{\displaystyle O}{\underset{\displaystyle CN}{P\overset{\displaystyle CH_3}{CH}}}$ (1). Mol. wt. 191.17, b.p. 103–107°/1.0 mm.

Wittig-Horner reaction. This phosphonate has been used in a synthesis of progesterone (4) from dehydroepiandrosterone (2). One step in the conversion of (3) into (4) was an Oppenauer oxidation. The usual conditions (aluminum

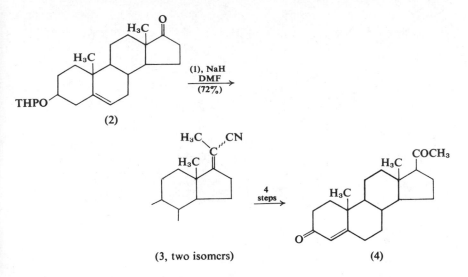

(2)

(3, two isomers)   (4)

isopropoxide and cyclohexanone) were modified by use of 4-methyl-1-piperidone, which is easily removed from the product by acid treatment. This modification was shown to be generally useful in the steroid field.[1]

[1] M. L. Raggio and D. S. Watt, *J. Org.*, **41**, 1873 (1976).

Diethyl phosphorocyanidate, 5, 217; **6,** 192–193.

Esters. Carboxylic acids react with this reagent and triethylamine in the presence of an alcohol at room temperature or below to form esters, but reported yields are only in the range of 30–55%.

A similar reaction, but with an amine, results in amides. The yields in this reaction are usually high, and the Young test indicates that racemization is slight in the case of peptides. DMF is a suitable solvent.[1]

[1] T. Shioiri, Y. Yokoyama, Y. Kasai, and S. Yamada, *Tetrahedron*, **32**, 2211 (1976).

N,N-Diethyl-1-propynylamine (N,N-Diethylaminopropyne), 2, 133–134; **3,** 98; **5,** 217–219.

Stereoselective cycloadditions (**5,** 219). Ficini *et al.*[1] have published further examples of stereoselective cycloadditions of this reagent to 4-methylcyclohex-2-enone and 1-cyanocyclopentene-1.

The addition reaction depicted in equation (II) was used in a synthesis of (±)-isodihydronepetalactone (1) and two isomers.[2]

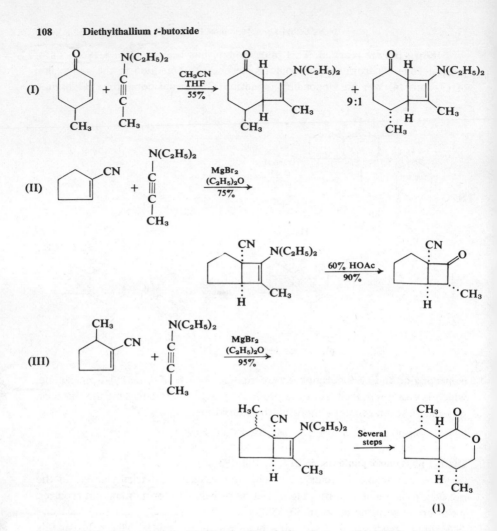

[1] J. Ficini, A. Eman, and A. M. Touzin, *Tetrahedron Letters*, 679 (1976); J. Ficini, J. d'Angelo, A. Eman, and A. M. Touzin, *ibid.*, 683 (1976).
[2] J. Ficini and J. d'Angelo, *ibid.*, 687 (1976).

Diethylthallium *t*-butoxide, $(C_2H_5)_2TlOC(CH_3)_3$ (1).

This thallium compound is prepared in ether or THF solution from potassium *t*-butoxide and diethylthallium bromide.

α-*Halo*-β-*lactams*.[1] Fused β-lactams can be formed by the reaction of (1) with dihaloacetamides in bromobenzene (equation I). Two cepham derivatives

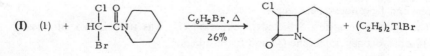

(II)   (1) + 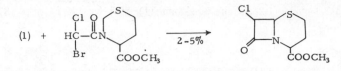 $\xrightarrow[2-5\%]{}$

were prepared in this way, albeit in low yields and as a mixture of *trans*- and *cis*-isomers.

[1] B. Åkermark, J. Chodowska-Palicka, and I. Lagerlund, *J. Organometal. Chem.*, **113** C4 (1976).

Dihydridotetrakis(triphenylphosphine)ruthenium(II), $RuH_2[P(C_6H_5)_3]_4$. Mol. wt. 1152.26.

Preparation.[1]

Transfer hydrogenation.[2] This ruthenium(II) complex is an effective catalyst for transfer of hydrogen from cyclic ethers, benzyl alcohol, and cyclohexanol to aldehydes and ketones. $RuH_2(CO)[P(C_6H_5)_3]_3$ is somewhat less active.

[1] J. J. Levison and S. D. Robinson, *J. Chem. Soc. (A)*, 2947 (1970).
[2] H. Imai, T. Nishiguchi, and K. Fukuzumi, *J. Org.*, **41**, 665 (1976).

Dihydropyrane, 1, 256–257; **3,** 99; **5,** 220.

Caution[1]: Violent explosions have been observed in reactions of tetrahydropyranyl ethers with hydrogen peroxide and with peracids. The explosive contaminant was not detectable by known tests; possibly it is derived from 2-hydroperoxytetrahydropyrane.

Reactions of tetrahydropyranyl ethers. Sonnet has reported the following high-yield conversions of hexadecane-1-ol THP ether (1).[2]

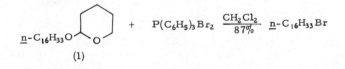

(1)

(1) $\xrightarrow[84\%]{\begin{array}{l}1)\ P(C_6H_5)_3Br_2\\2)\ NaCN\end{array}}$ $\underline{n}\text{-}C_{16}H_{33}CN$

(1) $\xrightarrow[89\%]{CF_3COOH}$ $\underline{n}\text{-}C_{16}H_{33}O_2CCF_3$

[1] A. I. Meyers, S. Schwartzman, G. L. Olson, and H.-C. Cheung, *Tetrahedron Letters*, 2417 (1976).
[2] P. E. Sonnet, *Syn. Comm.*, **6**, 21 (1976).

3,4-Dihydro-2H-pyrido[1,2-*a*]pyrimidine-2-one (1). Mol. wt. 148.17, m.p. 187–188°.
   Preparation[1]:

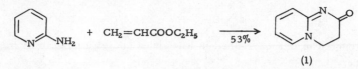

Acid scavenger. Japanese chemists[2] have used this betaine as an acid acceptor in the synthesis of esters by reaction of acids and alcohols with 2-chloro-1-methylpyridinium iodide. This mild, neutral method of esterification is useful for sensitive substrates.

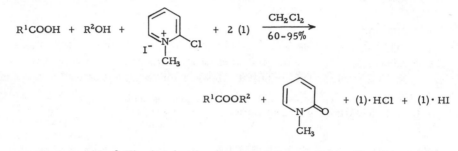

Carboxamides.[3] The betaine is also useful for synthesis of carboxamides, but in this case satisfactory yields are obtained only if tri-*n*-butylamine is also added.

R¹COOH + HNR²R³ +

[1] R. Adams and I. J. Pachter, *Am. Soc.*, **74**, 5491 (1952).
[2] T. Mukaiyama, H. Toda, and S. Kobayashi, *Chem. Letters*, 13 (1976).
[3] T. Mukaiyama, Y. Aikawa, and S. Kobayashi, *ibid.*, 57 (1976).

Diiron nonacarbonyl (Nonacarbonyldiiron), **1**, 259–260; **2**, 139–140; **3**, 101; **4**, 157–158; **5**, 221–224; **6**, 195–198.

   Iron tricarbonyl adduct of ergosterol.[1] Preparation of the iron tricarbonyl complex of ergosterol benzoate (1) with iron pentacarbonyl gives material that is contaminated with 3 5% of the starting diene. It has been reported[2] that arylmethyleneacetone–iron tricarbonyl complexes are efficient for transfer of iron tricarbonyl. And indeed the pure iron tricarbonyl complex of ergosterol

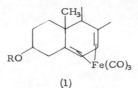

(1)

benzoate can be prepared in 60% yield with *p*-methoxybenzylideneacetone tricarbonyliron. However, a higher yield (80%) is obtained by treatment of the diene with diiron nonacarbonyl and *p*-methoxybenzylideneacetone. The diene unit can be regenerated by oxidative cleavage of the complex with ferric chloride hexahydrate.[3]

The complex was used for the synthesis of 22-dihydroergosterol, a fungal steroid, and for the synthesis of the previously unknown epiergosterol (oxidation to the 3-ketone followed by reduction with lithium tri-*t*-butoxy aluminum hydride[4]).

[1] D. H. R. Barton, A. A. L. Gunatilaka, T. Nakanishi, H. Patin, D. A. Widdowson, and B. R. Worth, *J.C.S. Perkin I*, 821 (1976).
[2] J. A. S. Howell, B. F. G. Johnson, P. L. Josty, and J. Lewis, *J. Organometal. Chem.*, **39**, 329 (1972).
[3] G. F. Emerson, J. E. Mahler, R. Kochhar, and R. Pettit, *J. Org.*, **29**, 3620 (1964).
[4] D. H. R. Barton and H. Patin, *J.C.S. Perkin I*, 829 (1976).

Diisobutylaluminum hydride, 1, 260–262; **2,** 140–142; **3,** 101–102; **4,** 158–160; **5,** 224–225; **6,** 198–202.

β-*Hydroxyalkenes;* β,γ-*unsaturated carbonyl compounds.*[1] Lithium *trans*-alkenyltrialkylaluminates (1), prepared as shown, react with epoxides as formulated in equation (I). Yields are considerably lower (30–50%) when alkenylalanes are used.

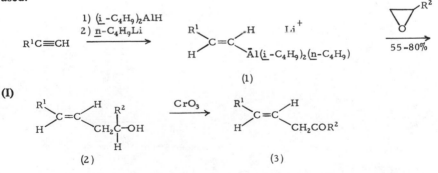

(1)

(I)

(2)   (3)

trans-*Alkenes.* The reaction of lithium *trans*-alkenyltrialkylaluminates (1) with certain alkyl halides produces *trans*-alkenes (2) stereospecifically (equation I). The reaction proceeds well with mesylates and the more reactive halides such as methyl iodide, allyl bromide, benzyl iodide, chloromethyl methyl ether (yields 60–80%). The reaction with tertiary alkyl halides is rapid, but more complex.[2]

(I)

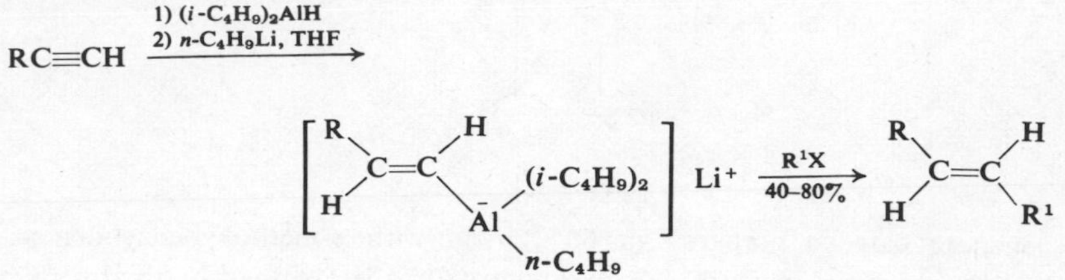

(1) (2)

trans-*Allyl ethyl ethers.* These ethers can be prepared from alkynes by *cis*-monohydroalumination followed by reaction with chloromethyl ethyl ether (**6**, 304). It is not necessary to convert the vinylalanes into ate complexes to obtain satisfactory yields.[3]

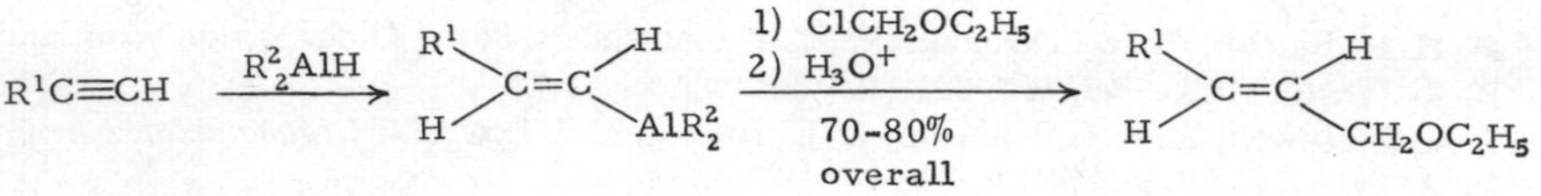

The preparation of *cis*-allyl ethyl ethers from acetylenic acetals has already been mentioned (**5**, 40–41).

Intramolecular cyclizations of polyenes. When 1,5-hexadiene is heated in mineral oil at 115° (N_2) with a catalytic amount of diisobutylaluminum hydride, methylenecyclopentane is formed in 81% yield (equation I).[4]

(I)

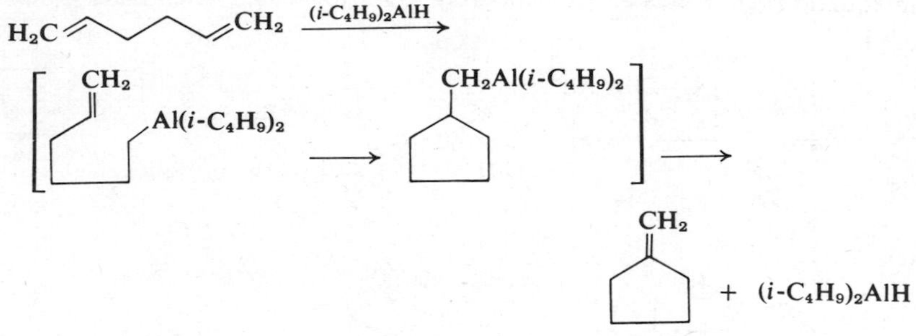

(81%)

Other examples:

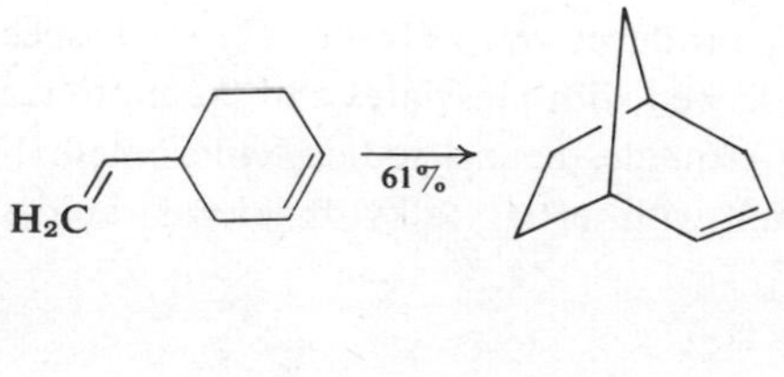

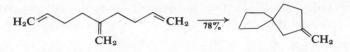

[1] E. Negishi, S. Baba, and A. O. King, *J.C.S. Chem. Comm.*, 17 (1976).
[2] S. Baba, D. E. Van Horn, and E. Negishi, *Tetrahedron Letters*, 1927 (1976).
[3] G. Zweifel and R. A. Lynd, *Synthesis*, 816 (1976).
[4] P. W. Chum and S. E. Wilson, *Tetrahedron Letters*, 1257 (1976).

(−)-2,3;4,6-Di-O-isopropylidene-2-keto-L-gulonic acid hydrate, (−)-DAG, (1).
Available from Commercial Development Dept., Hoffman-LaRoche, Inc.

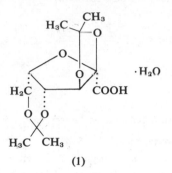

(1)

Resolution of amines. The preparation of this acid and its use for resolution of a variety of amines were first disclosed in a patent.[1] One advantage of this reagent is that it is water-insoluble and thus can be recovered in 91–94% yield. An example of its use is the resolution of α-(1-naphthyl)ethylamine to obtain the S-(−) isomer in 85–90% yield.[2]

[1] C. W. Den Hollander, W. Leimgruber, and E. Mohacsi, U.S. Patent 3,682,925 (1972).
[2] E. Mohacsi and W. Leimgruber, *Org. Syn.*, **55**, 80 (1976).

Dilithioacetate, $LiCH_2COOLi$. Mol. wt. 71.92.
   Preparation.[1]
   Stereoselective epoxide cleavage.[2] The key step in a relatively short synthesis

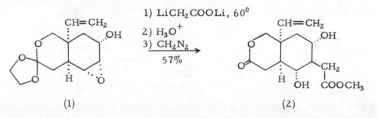

(1)   (2)

of (2), an intermediate in a total synthesis of the sesquiterpenes vernolepin and vernomenin, is cleavage of the epoxide (1) with dilithioacetate (large excess).

The factors involved in this selective (and desired) cleavage are not clear; however, the tetrahydropyranyl ether of (1) is unreactive under the same conditions.

[1] P. L. Creger, *J. Org.*, **37**, 1907 (1972).
[2] S. Danishefsky, T. Kitahara, P. F. Schuda, and S. J. Etheredge, *Am. Soc.*, **98**, 3028 (1976).

Dilithium tetrachlorocuprate, 4, 163–164; 5, 226.

Alkanoic acids.[1] Fatty acids can be prepared by reaction of Grignard reagents and the chloromagnesium salts of ω-bromo acids (used because of solubility in THF) with Li_2CuCl_4 as catalyst (equation I). Usually some amounts of the diacid derived from the bromo acid are also formed.

$$(I) \quad RMgBr + Br(CH_2)_nCOOMgCl \xrightarrow[\text{THF, }-20^0]{Li_2CuCl_4} R(CH_2)_nCOOMgCl + MgBr_2$$

$$75-95\% \downarrow H_3O^+$$

$$R(CH_2)_nCOOH$$

[1] T. A. Baer and R. L. Carney, *Tetrahedron Letters*, 4697 (1976).

Dilithium tetrachloropalladate, Li_2PdCl_4, 5, 413.

π-Allylpalladium compounds. Vinylmercuric chlorides and alkenes form π-allylpalladium compounds in the presence of this salt (equation I).[1]

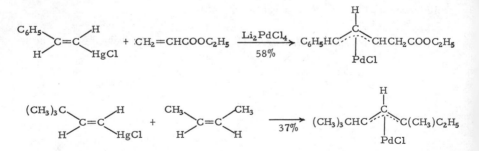

Examples:

Pterocarpins. Japanese chemists[2] have developed a new route to pterocarpins[3] that involves the reaction of 2*H*-chromenes with *o*-chloromercuriphenols in the presence of this palladium catalyst (equation I).

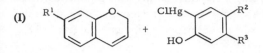

(I)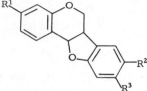

[1] R. C. Larock and M. A. Mitchell, *Am. Soc.*, **98**, 6718 (1976).

[2] H. Horino and N. Inoue, *J.C.S. Chem. Comm.*, 500 (1976).

[3] F. M. Dean, *The Total Synthesis of Natural Products*, Vol. 1, J. ApSimon, ed., Wiley, New York, 1973, pp. 513–525.

Dilithium trimethyl cuprate, $(CH_3)_3CuLi_2$, 6, 386.

Reaction with cyclohexanones. Still and Macdonald[1] have reported further examples of the utility of R_3CuLi_2 reagents for selective synthesis of axial cyclohexanols.

Examples:

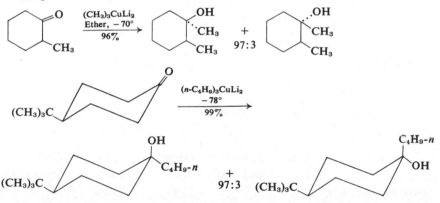

In the original investigations 3 equiv. of R_3CuLi_2 were required, but it is also possible to use 0.25 equiv. of R_2CuLi and 1.25 equiv. of RLi.

R_3CuLi_2 undergoes conjugate addition to 2-cyclohexenones to form β-alkylcyclohexanones (85–90% yield). In fact $(CH_3)_3CuLi_2$ adds preferentially to α,β-enones.

[1] W. C. Still and T. L. Macdonald, *Tetrahedron Letters*, 2659 (1976).

Dimethoxymethane (Methylal), 1, 671–672; 6, 374.

Methoxymethyl ethers (*cf.* 6, 374). Wyeth chemists[1] have published another method using methylal for preparation of these ethers that avoids use of the carcinogenic chloromethyl methyl ether. The method involves *p*-toluenesulfonic acid catalyzed acetal exchange. The methanol formed is removed with a molecular sieve. Methylene chloride, which forms an azeotrope with methanol (b.p. 37.8°), is used as cosolvent.

$$\text{ROH} + \text{H}_2\text{C(OCH}_3)_2 \xrightarrow[\substack{-\text{CH}_3\text{OH}}]{\substack{\text{TsOH, CH}_2\text{Cl}_2 \\ \text{Molecular sieve}}} \text{ROCH}_2\text{OCH}_3$$
$$(60\text{--}80\%)$$

Protection of carboxyl groups. Methoxymethyl esters of carboxylic acids can also be prepared from methylal as formulated for the preparation of methoxymethyl benzoate. The method is applicable to phenols and thiols[2] as well.

$$\text{C}_6\text{H}_5\text{COOH} + \text{BrZnCH}_2\text{COOC}_2\text{H}_5 \longrightarrow [\text{C}_6\text{H}_5\text{COOZnBr}] \xrightarrow[\text{CH}_3\text{COCl}]{\text{H}_2\text{C(OCH}_3)_2}$$

$$\text{C}_6\text{H}_5\text{COOCH}_2\text{OCH}_3 + \text{ZnBrCl} + \text{CH}_3\text{COOCH}_3$$
$$(72\%)$$

[1] J. P. Yardley and H. Fletcher, III, *Synthesis*, 244 (1976).
[2] F. Dardoize, M. Gaudemar, and N. Goasdoue, *Synthesis*, 567 (1977).

$$\text{CH}_3\text{O} \quad \text{OCH}_3$$

1,2-Dimethoxyvinyllithium, $\text{HC}{=}\text{CLi}$ (1). Mol. wt. 94.04.

Preparation:

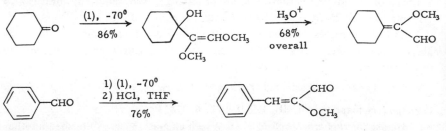

α-*Methoxy-α,β-unsaturated aldehydes.* The reagent reacts with aldehydes or ketones at $-70°$ to form allylic alcohols which usually rearrange directly to the unsaturated aldehydes.[1]

Examples:

[1] C. N. Skold, *Syn. Comm.*, **6**, 119 (1976).

Dimethyl acetylenedicarboxylate (DMAD), **1**, 272–273; **2**, 145–146; **3**, 103–104; **4**, 168–172; **5**, 227–230; **6**, 206.

2,3-Dicarbomethoxy-4-hydroxy-1,4-benzoxazines. These heterocycles (2) can be prepared[1] by reaction of dimethyl acetylenedicarboxylate with bis copper(II) complexes (1) of *o*-nitrosophenols. These complexes can be prepared either by nitrosation of a phenol in the presence of copper(II) sulfate[2] or by reaction of an *o*-nitrosophenol with copper(II) acetate in aqueous acetic acid.[3] No cycloaddition

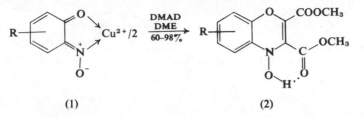

(1)   (2)

is observed with *o*-nitrosophenols themselves or with cobalt(III) complexes of *o*-nitrosophenols.

Benzoannelation. Paquette *et al.*[4] have devised a method for benzoannelation of a ketone with an adjacent CH_2 group. Two examples of this transformation are formulated.

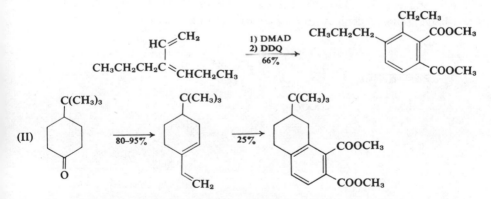

(I) $(CH_3CH_2CH_2)_2C{=}O + CH_2{=}CHMgBr \xrightarrow[93\%]{THF} (CH_3CH_2CH_2)_2C\overset{OH}{\underset{CH=CH_2}{<}} \xrightarrow[74\%]{I_2, 95°}$

Reaction with macrocyclic annulenes. The aromaticity of these systems has been determined mainly on the basis of magnetic properties. Sondheimer and co-workers[5] have noted that such assignments can be supported by the reactivity in Diels-Alder reactions with this dienophile. Thus (1) reacts readily even at 20°, whereas (3) does not react even after 7 days at 80°. Presumably the annulene

(3) does not react because the adduct (and probably the transition state) is a dehydro[4n]annulene and is destabilized (antiaromatic), whereas the adduct from (1) is a dehydro[4n + 2]annulene.

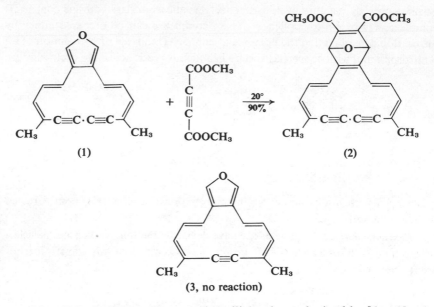

[6 + 2]Cycloaddition. The pentalene (2) has been obtained by [6 + 2]cyclo-addition of dimethyl acetylenedicarboxylate with 1,3-di-*t*-butyl-6-dimethyl-aminofulvene (1) in benzene at room temperature.[6] The fulvene lacking the two *t*-butyl groups undergoes Michael addition with the reagent.

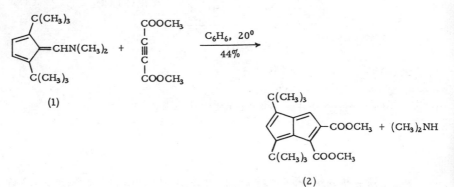

[1] A. McKillop and T. S. B. Sayer, *J. Org.*, **41**, 1079 (1976).
[2] G. Cronheim, *ibid.*, **12**, 1 (1947).
[3] J. Charalambous, M. J. Frazer, and F. B. Taylor, *J. Chem. Soc. (A)*, 2787 (1969).
[4] L. A. Paquette, W. P. Melega, and J. D. Kramer, *Tetrahedron Letters*, 4033 (1976).

[5] R. H. Wightman, T. M. Cresp, and F. Sondheimer, *Am. Soc.*, **98**, 6052 (1976).
[6] K. Hafner and M. Suda, *Angew. Chem., Int. Ed.*, **15**, 314 (1976).

Dimethylamine, $(CH_3)_2NH$. Mol. wt. 45.09, b.p. 7°.

Protection of the α,β-unsaturated methylene group of γ-lactones. This unit can be protected by addition of dimethylamine; regeneration is accomplished by quaternization and treatment with bicarbonate (equation I). The method was

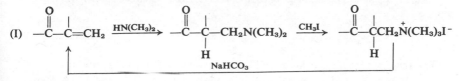

(I)

used in a synthesis of dehydrosaussuria lactone (4) from costunolide (1) by thermolysis of the amine adduct.[1]

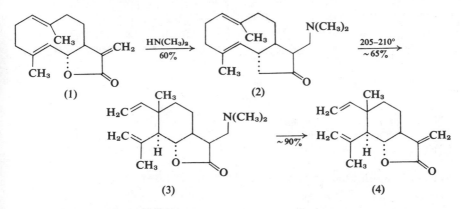

(1)   (2)   (3)   (4)

[1] T. C. Jain, C. M. Banks, and J. E. McCloskey, *Tetrahedron*, **32**, 765 (1976).

N,N-Dimethylbenzeneselenenamide, $(CH_3)_2NSeC_6H_5$ (1). Mol. wt. 200.14, b.p. 39–40°/0.1 mm., easily hydrolyzed.

This substance is prepared in 62% yield by the reaction of dimethylamine with C_6H_5SeCl in hexane at 0°.[1]

Michael additions.[1] The reagent undergoes conjugate additions to the more reactive Michael acceptors. N,N-Dimethylbenzenesulfinamide does not add to enones.

Examples:

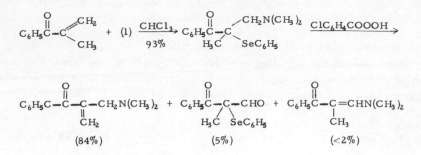

$$C_6H_5C-C-C-CH_2N(CH_3)_2 + C_6H_5C-C-CHO + C_6H_5C-C=CHN(CH_3)_2$$

(84%)   (5%)   (<2%)

The reagent also adds to dimethyl acetylenedicarboxylate to form the isomeric esters (2) and (3) in about equal yield.[2] A kinetic study shows that (2) is formed exclusively in the early stages. Apparently (3) is formed from (2) by thermal rotation around the double bond.

$CH_3OOCC{\equiv}CCOOCH_3 + (1) \longrightarrow$

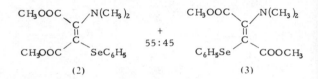

55:45

(2)   (3)

[1] H. J. Reich and J. M. Renga, *J. Org.*, **40**, 3313 (1975).
[2] H. J. Reich, J. M. Renga, and J. E. Trend, *Tetrahedron Letters*, 2217 (1976).

Dimethylcopperlithium (Lithium dimethylcuprate), **2**, 151–153; **3**, 106–113; **4**, 177–183; **5**, 234–244; **6**, 209–215.

   Structure. The reagent was prepared in ether solution free from contaminants as shown in equation (I). Vapor-pressure measurements, NMR, and X-ray

(I) $CuI + LiCH_3 \xrightarrow[\substack{Ether \\ -78^0}]{} (CuCH_3)_x + LiI$

$$\text{Ether} \downarrow \text{LiCH}_3$$
$$-78^0$$

$$(CH_3)_2CuLi$$

scattering indicate that the reagent is a dimer in ether solution. Indeed the dimeric structure may be the reason for the unusual reactivity.[1]

   Reaction with an allylic acetate. The reaction of *cis*- and *trans*-5-methyl-2-cyclohexenyl acetate, (1) and (2), with this cuprate is highly stereoselective. Experiments with deuterium-labeled (1) and (2) show involvement of a symmetrical intermediate in which the two allylic positions are equivalent.[2]

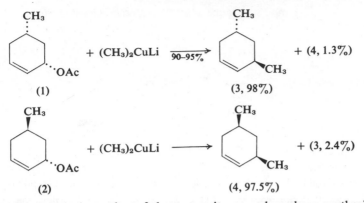

(3, 98%)

+ (4, 1.3%)

(4, 97.5%)

+ (3, 2.4%)

Spirovetivanes. A number of these sesquiterpenes have been synthesized by Büchi *et al.*[3] by the reaction of dimethylcopperlithium with the fulvene (1) in ether at −20°. The product is the carbinol (2). Selective reaction of one double

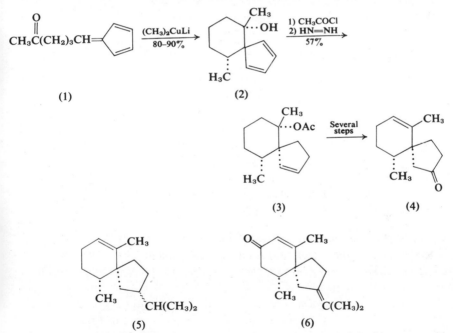

bond of (2) with a number of reagents failed but was accomplished in reasonable yield with diimide (**1**, 257–258; **2**, 139; **3**, 99–101; **4**, 154–155; **5**, 220). The product (3) was converted in several steps into the C_{12}-ketone (4). This product was converted by known reactions into hinesol (5) and β-vetivone (6).

Allenes (**2**, 152; **3**, 106). The reaction of dialkylcopperlithium reagents with acetylenic acetates (1) under standard conditions gives both alkylated and non-alkylated allenes (2) and (3). Crabbé *et al.*[4] have now defined conditions leading to either (2) or (3). If the reaction is conducted at about $-10°$ in ether, (2) is the

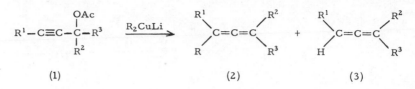

major product: $(2)/(3) = 95:5$. If the reaction is conducted at $-75°$ (2 equiv. R_2CuLi) and the reaction mixture is then treated with $LiAlH_4$ in THF, the allenes (3) are the major products. Yields are low if the reduction step is omitted.

Cleavage of cyclopropanes. Trost *et al.*[5] have disclosed a new method for synthesis of the acyclic C_7 side chain characteristic of steroids and some other natural products. The key step involves cleavage of a cyclopropane ring with dimethylcopperlithium, (1) → (2). The paper cites some relevant references to cleavage of cyclopropane rings with organocuprates.

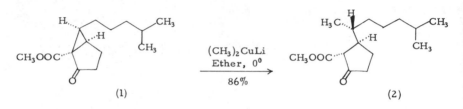

[1] R. G. Pearson and C. D. Gregory, *Am. Soc.*, **98**, 4098 (1976).

[2] H. L. Goering and V. D. Singleton, Jr., *Am. Soc.*, **98**, 7854 (1976).

[3] G. Büchi, D. Berthet, R. Decorzant, A. Grieder, and A. Hauser, *J. Org.*, **41**, 3208 (1976).

[4] P. Crabbé, E. Barreiro, J.-M. Dollat, and J.-J. Luche, *J.C.S. Chem. Comm.*, 183 (1976).

[5] B. M. Trost, D. F. Taber, and J. B. Alper, *Tetrahedron Letters*, 3857 (1976).

6,6-Dimethyl-5,7-dioxaspiro[2.5]octane-4,8-dione, (1), 6, 216–217.

Details for the preparation of this interesting compound have been submitted to *Organic Synthesis.*[1] Yields are somewhat lower than those reported previously for smaller-scale preparations.

α-Methylene-γ-butyrolactone.[2] Treatment of (1) with 2:1 acetone–water under reflux gives the two products (2) and (3). The latter product was converted into α-methylene-γ-butyrolactone (4) by a known route (3, 190).

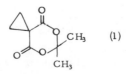

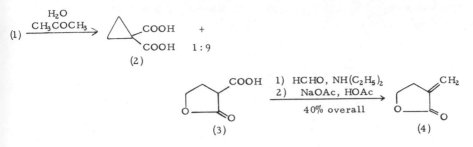

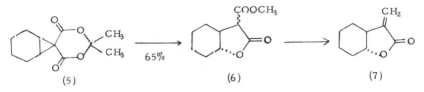

This solvolysis was extended to (5) and was shown to occur with inversion.

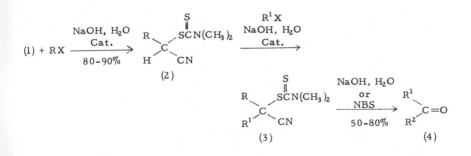

[1] R. K. Singh and S. Danishefsky, *Org. Syn.* submitted (1976).
[2] R. K. Singh and S. Danishefsky, *J. Org.*, **41**, 1668 (1976).

N,N-Dimethyldithiocarbamoylacetonitrile, $NCCH_2SCN(CH_3)_2$ (1).   Mol.   wt.
160.26, m.p. 73–74°.

The α-cyanodithioester is obtained by the reaction of chloroacetonitrile and
sodium N,N-dimethyldithiocarbamate in CH_3OH at room temperature.

Ketone synthesis. The reagent (1) can be mono- or dialkylated in aqueous
NaOH under phase-transfer catalysis [$(n\text{-}C_4H_9)_4N^+I^-$] in high yields. The latter
products (3) can be hydrolyzed to ketones (4) by either aqueous NaOH or NBS.
The method is particularly useful for synthesis of unsymmetrical ketones.[1]

[1] Y. Masuyama, Y. Ueno, and M. Okawara, *Tetrahedron Letters*, 2967 (1976).

N,N-Dimethyl-O-ethylphenylpropiolamidium tetrafluoroborate (1). Mol. wt. 289.08, colorless air-stable needles.

Preparation:

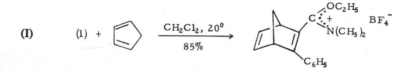

(1)

Cycloaddition. Propiolamidium salts such as (1) undergo facile Diels-Alder reaction with cyclopentadiene (equation I). The corresponding propiolamide is unreactive under the same conditions.[1]

(I)   (1) + [diagram] $\xrightarrow[85\%]{CH_2Cl_2, \, 20^0}$ [diagram]

[1] J. S. Baum and H. G. Viehe, *J. Org.*, **41**, 183 (1976).

Dimethylformamide, **1,** 278–281; **2,** 153–154; **3,** 115; **4,** 184; **5,** 247–249.

α,β-*Unsaturated aldehydes.* Dutch chemists[1] have modified the Shapiro-Heath olefin synthesis (**2,** 418–419; **4,** 511; **5,** 678–679; **6,** 598–599) by addition of DMF to trap the alkenyllithium intermediate (equation I). α,β-Unsaturated aldehydes are obtained in 50–65% yield. In some cases this method is superior

(I)   [diagram] $\xrightarrow[-Ts, \, -N_2]{\substack{n-BuLi, \, -78^0 \\ TMEDA}}$ [diagram] $\xrightarrow[50-65\%]{\substack{1) \, DMF \\ 2) \, H_2O}}$ [diagram]

to Vilsmeier formylation. The reaction proceeds in only low yield (10%) with the hindered tosylhydrazone of camphor.

Demethylation of quaternary pyridinium salts.[2] The rate of demethylation of these salts by DMF (**4,** 184) is accelerated by addition of triphenylphosphine.[3] The combination of the phosphine and DMF was used in a study of electronic and steric factors in demethylation of quaternary pyridinium salts.

[1] P. C. Traas, H. Boelens, and H. J. Takken, *Tetrahedron Letters*, 2287 (1976).
[2] U. Berg, R. Gallo, and J. Metzger, *J. Org.*, **41**, 2621 (1976).
[3] T.-L. Ho, *Syn. Comm.*, **3**, 99 (1973).

Dimethylformamide diethyl acetal, 1, 281–282; **2,** 154; **3,** 115–116; **4,** 184; **5,** 253–254.

Reaction with t-*ethynyl alcohols.* DMF acetals react with *t*-ethynyl alcohols (120–145°) to form dienamines and/or enamine orthoformates.[1]

Examples:

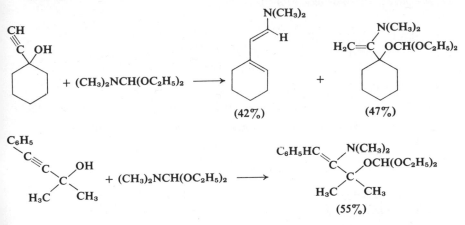

(42%)   (47%)

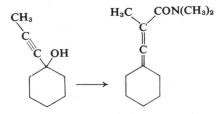

(55%)

A reaction analogous to that of allylic alcohols with DMF acetals (**5,** 253–254) has also been observed (equation I).

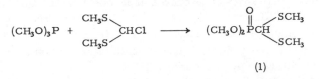

[1] K. A. Parker, R. W. Kosley, Jr., S. L. Buchwald, and J. J. Petraitis, *Am. Soc.,* **98,** 7104 (1976).

O,O-Dimethyl formylphosphonate S,S-dimethyl thioacetal, (1). Mol. wt. 216.26.
   Preparation:

$$(CH_3O)_3P \ + \ \underset{CH_3S}{\overset{CH_3S}{\diagdown}}CHCl \ \longrightarrow \ (CH_3O)_2 \overset{\overset{O}{\|}}{P}C\underset{\diagdown SCH_3}{\overset{\diagup SCH_3}{H}}$$

(1)

Ketene thioacetals. Ketene thioacetals can be prepared in 80–95% yield by reaction of the anion of (1) with carbonyl compounds (Wittig-Horner reaction,

(I)   (1) $\dfrac{\text{1) } \underline{n}\text{-BuLi, THF, } -78^0}{\text{2) } R^1COR^2 \qquad\qquad\qquad}$

$\qquad\qquad\qquad$ 80-95%

$$R^1 \!\!\!\!\diagdown \atop R^2 \!\!\!\!\diagup \!\!\! C\!\!=\!\!C \!\! {\diagup SCH_3 \atop \diagdown SCH_3}$$

equation I). The reaction can also be carried out in a two-phase system under phase-transfer catalysis (equation II).

(II)   (1) + C_6H_5CHO $\dfrac{\text{NaOH, } H_2O, \ CH_2Cl_2}{\underset{88\%}{(C_2H_5)_3\overset{+}{N} \ CH_2C_6H_5Cl^-}}$ $C_6H_5CH\!\!=\!\!C(SCH_3)_2$

Note that the Wittig synthesis of ketene thioacetals can only be used with aldehydes. Of course, the reaction can be used for preparation of other thioacetals.[1]

[1] M. Mikołajczyk, S. Grzejszczak, A. Zatorski, and B. Młotkowska, *Tetrahedron Letters*, 2731 (1976).

5,5-Dimethylhydantoin, (1). Mol. wt. 128.13, m.p. 176–178°.
Supplier: Aldrich.

N-Methylarylamines. These compounds can be prepared from anilines as shown in equation (I).[1]

(I) $ArNH_2$ + HCHO + (1) $\xrightarrow{\ C_2H_5OH, \ \Delta\ }$

$ArNH\!\!-\!\!CH_2\!\!-\!\!N \qquad \xrightarrow[\underset{35-80\%}{DMSO}]{NaBH_4} ArNH\!\!-\!\!CH_3$

[1] K. Horiki, *Heterocycles*, **5**, 203 (1976).

N,N-Dimethylhydrazine, 1, 289–290; **2,** 154–155, **3,** 117; **5,** 254; **6,** 223.

Oxidative cleavage. N,N-Dimethylhydrazones have been used only to a limited extent as derivatives of carbonyl compounds because oxidative cleavage is sometimes unsatisfactory. This reaction can be accomplished in high yield with aqueous sodium periodate in methanol or THF at pH 7 at 20–25°.[1] Hydrolysis can also be effected in high yield under even milder conditions with copper(II) acetate or chloride (2 equiv.) in aqueous THF.[2]

Alkylation of N,N-dimethylhydrazones.[1] The DMH derivatives of enolizable aldehydes and ketones can be metallated by LDA in THF at 0° or by *n*-butyllithium in THF at −78°. These lithio derivatives react readily with electrophiles (alkyl halides, oxiranes, carbonyl compounds).[1,3]

Examples:

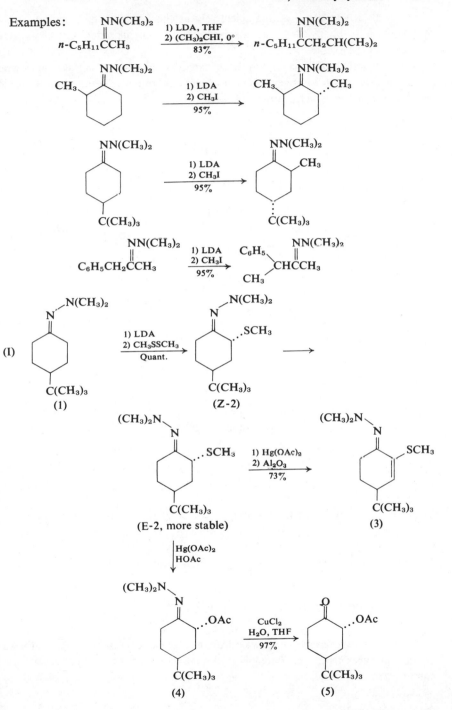

Generally, metallation occurs selectively at the less alkylated carbon. Axial methylation is highly favored; no isomeric or polyalkylated products were observed in the examples cited.

This alkylation of ketones has been extended by Corey and Knapp,[4] who used the DMH of 4-*t*-butylcyclohexanone (1) as the model compound. Some of the interesting new reactions reported are formulated in equations (I) and (II).

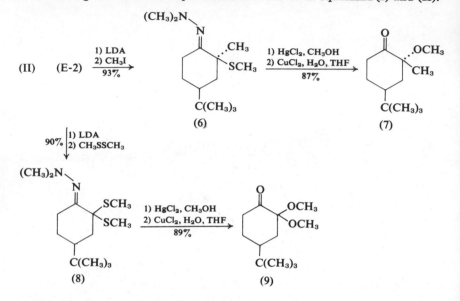

(II)   (E-2) $\xrightarrow[93\%]{\begin{array}{l}1)\ LDA\\2)\ CH_3I\end{array}}$ (6) $\xrightarrow[87\%]{\begin{array}{l}1)\ HgCl_2,\ CH_3OH\\2)\ CuCl_2,\ H_2O,\ THF\end{array}}$ (7)

90% $\begin{array}{l}1)\ LDA\\2)\ CH_3SSCH_3\end{array}$

(8) $\xrightarrow[89\%]{\begin{array}{l}1)\ HgCl_2,\ CH_3OH\\2)\ CuCl_2,\ H_2O,\ THF\end{array}}$ (9)

α,β-*Unsaturated aldehydes*.[5] A new method for synthesis of these aldehydes from a saturated aldehyde is formulated in equation (III). Use of a ketone,

(III) $R'CH_2CH=NN(CH_3)_2$ $\xrightarrow[]{\begin{array}{l}1)\ LDA\\2)\ (CH_3)_3SiCl\end{array}}$ $(CH_3)_3SiCHCH=NN(CH_3)_2$
                                                                          $\underset{R'}{|}$
$\xrightarrow[85-95\%]{\begin{array}{l}1)\ LDA,\ -20°\\2)\ RCHO,\ -78°\end{array}}$

$\underset{R'}{\overset{}{R}}CH=CCH=NN(CH_3)_2$ $\longrightarrow$ $RCH=\underset{R'}{\overset{}{C}}CHO$

$RCOR''$, results in $RR''C=\underset{R'}{\overset{}{C}}CHO$ as the final product. This sequence has also been carried out with the *t*-butylimines, $(CH_3)_3SiCHCH=NC(CH_3)_3$.
                                                                         $\underset{R'}{|}$

Various polyfunctional compounds.[6] Corey et al. have reported useful syntheses of various types from metallated N,N-dimethylhydrazones. Typical results are summarized in the equations.

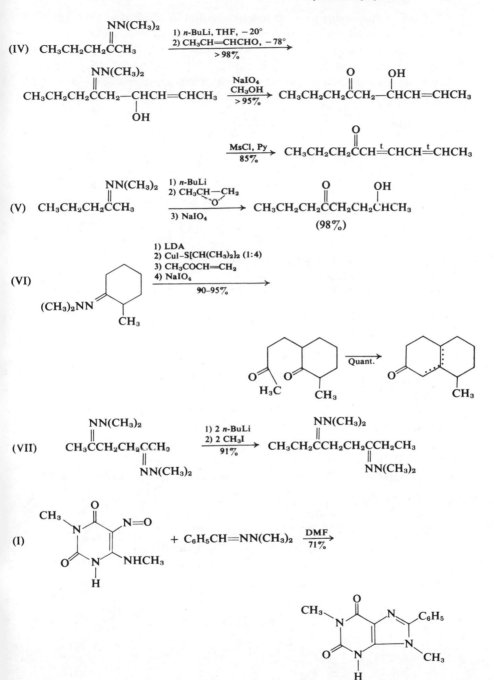

For use of dimethylhydrazones in a conversion of aldehydes into nitriles *see* ***Lithium diethylamide–Hexamethylphosphoric triamide*** (this volume).

Xanthines; purines. The reaction of a nitrosouracil with an aldehyde and N,N-dimethylhydrazine results in formation of a xanthine (equation I, p.129). Purines are obtained by a similar reaction with a nitrosopyrimidine (equation II).[7]

(II)

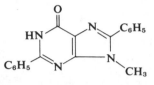

$$+ C_6H_5CH=NN(CH_3)_2 \xrightarrow[49\%]{DMF}$$

[1] E. J. Corey and D. Enders, *Tetrahedron Letters*, 3 (1976).
[2] E. J. Corey and S. Knapp, *Tetrahedron Letters*, 3667 (1976).
[3] A similar reaction has been mentioned by G. Stock and J. Benaim, *Am. Soc.*, **93**, 5938 (1971).
[4] E. J. Corey and S. Knapp, *ibid.*, 4687 (1976).
[5] E. J. Corey, D. Enders, and M. G. Bock, *ibid.*, 7 (1976).
[6] E. J. Corey and D. Enders, *ibid.*, 11 (1976).
[7] F. Yoneda and T. Nagamatsu, *J.C.S. Perkin I*, 1547 (1976).

Dimethyl(methylene)ammonium salts, $H_2C=N^+(CH_3)_2A^-$ (1), **3,** 114–115; **4,** 186–187.

Lactone methylenation.[1] Dimethylaminomethylation of lactone enolates can be carried out with (1, $A^- = I^-$). A hydroxyl group need not be protected; thus (2) is converted into (3) without the usual necessity of protection of the OH group (equation I). In the same way, bisnorvernolepin was converted into *dl*-vernolepin in 31% overall yield. The yield is decreased to 18% if the hydroxyl group is

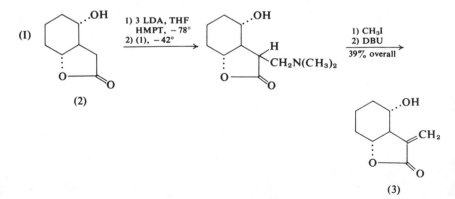

protected as the THP ether. The transformation of (4) into *dl*-vernomenin (5) was also reported.

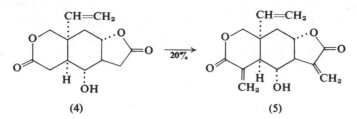

(4)   (5)

Reaction with silyl enol ethers.[1] The Mannich reagent (1) reacts with 1-trimethylsilyloxycyclohexene (2) to form a salt (3), which is converted into the Mannich base (5) in high overall yield.

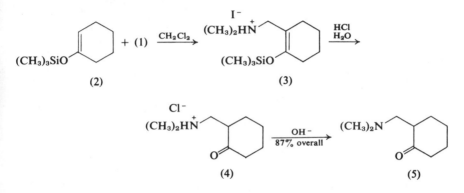

(2)   (3)

(4)   (5)

Cycloaddition of (1) with the diene (6) at 20° affords (7), converted into the useful Mannich base (8) in 95% overall yield.[1]

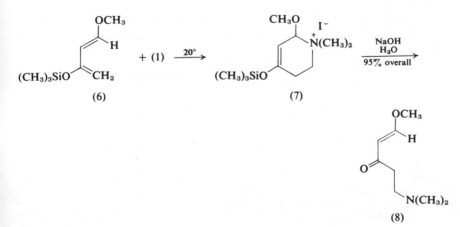

(6)   (7)

(8)

Related Mannich reactions have been reported by Holy and Wang.[2] These chemists generated the silyl enol ethers under either thermodynamic or kinetic control, but cleaved the ether with methyllithium to the same lithium enolate and then added the Mannich salt. Product distributions demonstrated that the addition reaction is regiospecific. They also found that the reaction can be conducted by the trapping technique of conjugate addition of dimethylcopper-lithium to cyclohexenone followed by addition of the immonium salt (equation I.)

(I)

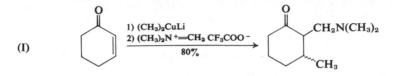

They also used this general method for synthesis of α-methylene-γ-butyro-lactone (equation II).

(II)

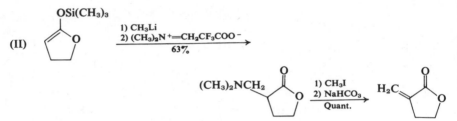

Poulter et al.[3] in the case of cyclohexanone generated the enolate with potassium hydride; addition of the Mannich reagent gave an 88% yield of 2-dimethyl-aminomethylcyclohexanone. In the case of esters and lactones, they used LDA for generation of the enolate, to which the Mannich salt was then added. The yield of α-methylene-γ-butyrolactone by this procedure was about the same as that obtained with the silyl enol ether.

t-Amines. The trifluoroacetate salt (b.p. 130–140°, 0.05 mm.) reacts with Grignard reagents or alkyllithiums to form alkyldimethylamines.[4]

Examples:

$$C_6H_5MgBr \xrightarrow[85\%]{CH_2=N^+(CH_3)_2CF_3COO^-} C_6H_5CH_2N(CH_3)_2$$

$$n\text{-}C_4H_9Li \xrightarrow{72\%} CH_3(CH_2)_4N(CH_3)_2$$

$$(p)\text{-}CH_3C_6H_4MgBr \xrightarrow{72\%} (p)\text{-}CH_3C_6H_4CH_2N(CH_3)_2$$

[1] S. Danishefsky, T. Kitahara, R. McKee, and P. F. Schuda, *Am. Soc.*, **98**, 6715 (1976).
[2] N. L. Holy and Y. F. Wang, *ibid.*, **99**, 944 (1977).
[3] J. L. Roberts, P. S. Borromeo, and C. D. Poulter, *Tetrahedron Letters*, 1621 (1977).
[4] N. L. Holy, *Syn. Comm.*, **6**, 539 (1976).

Dimethyloxosulfonium methylide, 1, 315–318; **2,** 171–173; **3,** 125–127; **4,** 197–199; **5,** 254–257.

1-Benzoxepins. A new route to these oxygen heterocycles is formulated in equation (I).[1]

(I)

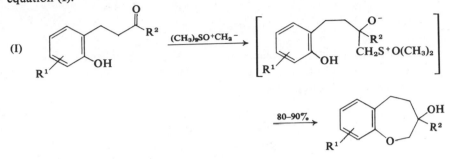

[1] P. Bravo, C. Ticozzi, and D. Maggi, *J.C.S. Chem. Comm.,* 789 (1976).

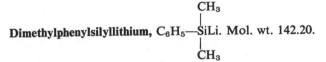

Dimethylphenylsilyllithium, C_6H_5—$\overset{\overset{\displaystyle CH_3}{|}}{\underset{\underset{\displaystyle CH_3}{|}}{Si}}$Li. Mol. wt. 142.20.

Preparation.[1]

Deoxygenation of epoxides.[2] Epoxides are converted into olefins by reaction with this reagent in THF at 20°. The reaction is stereospecific. Thus the oxide of *trans*-stilbene is converted into *cis*-stilbene (75% yield, >97% stereospecific), and the oxide of *cis*-stilbene is converted into *trans*-stilbene (83% yield, >99% stereospecific).

[1] H. Gilman and G. D. Lichtenwalter, *Am. Soc.,* **80,** 608 (1958).
[2] M. T. Reetz and M. Plachky, *Synthesis,* 199 (1976).

Dimethyl sulfoxide, 1, 296–310; **2,** 157–158; **3,** 119–123; **4,** 192–194; **5,** 263–266; **6,** 225–229.

Solvent Effects

Solvent-dependent Michael reaction. A key step in the first total synthesis of gliotoxin (5) is the Michael reaction of (1) with (2) with Triton B as base. Two alcohols are obtained, but the ratio of (3) to (4) is markedly dependent on the

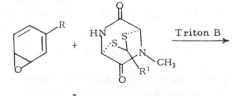

$\left[1, \text{R} = \text{COOC}(CH_3)_3\right]$ (2, R^1 = $C_6H_4OCH_3$-$\underline{p}$)

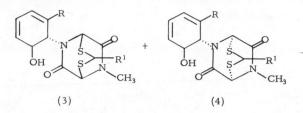

(3)   (4)

solvent. The alcohol (3) predominates when CH_2Cl_2 is the solvent; a 3:1 ratio of (4) to (3) is obtained in DMSO. The isomeric alcohol (4) was converted by more or less standard reactions into *dl*-(5).[1]

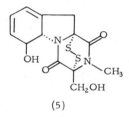

(5)

Finkelstein reaction.[2] Iodomethyl sulfoxides can be prepared in 80–85% yield by reaction of chloromethyl sulfoxides with KI in DMSO at 100–110°.[3]

Cleavage of acenaphthenequinone. This reaction was originally carried out

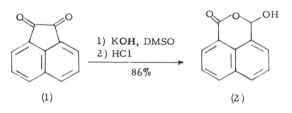

(1)   (2)

with 30% aqueous KOH at 150°. The reaction can be carried out at room temperature if DMSO is used as cosolvent. DMF is less satisfactory.[4]

Reactions

Oxidation of benzylammonium bromides. A few years ago Traynelis and Ode[5] noted that benzylammonium chlorides are oxidized to carbonyl compounds in low to fair yields when heated in DMSO. The reaction proceeds more readily and in higher yield (50–85%) with benzylammonium bromides.[6] The reaction

$$ArCH_2\overset{+}{N}H_3\,\overset{-}{Br} \xrightarrow{\underset{\Delta}{DMSO}} ArCHO$$

may involve a complex of bromine with DMSO (compare *DMSO–Cl$_2$*, **4**, 200). Benzophenone was made in this way in 88% yield.

Oxidation of indoles. Tryptophan and related 3-substituted indoles are oxidized to oxindoles in high yield by DMSO and conc. HCl (equation I). This reaction can be used for modification of tryptophan-containing peptides and proteins.[7]

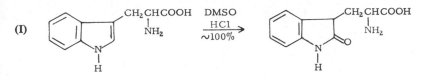

(I)

[1] T. Fukuyama and Y. Kishi, *Am. Soc.*, **98**, 6723 (1976).
[a] H. Finkelstein, *Ber.*, **43**, 1528 (1910).
[3] M. Hojo, R. Masuda, T. Saeki, and S. Uyeda, *Synthesis*, 697 (1976).
[4] H. Bader and Y. H. Chiang, *ibid.*, 249 (1976).
[5] V. J. Traynelis and R. H. Ode, *J. Org.*, **35**, 2207 (1970).
[6] P. A. Zoretic, P. Soja, M. Jodoin, and R. Levine, *Org. Prep. Proc. Int.*, **8**, 33 (1976).
[7] W. E. Savige and A. Fontana, *J.C.S. Chem. Comm.*, 599 (1976).

Dimethyl sulfoxide–Acetic anhydride, 1, 305; **2,** 163–165; **3,** 121–122; **4,** 199.

Methylthiomethyl ethers. Methylthiomethyl ethers have been obtained as by-products in the oxidation of alcohols with DMSO–Ac$_2$O (**2,** 163). If HOAc is deliberately added, these ethers can become major products, even in the oxidation of *sec*- and *tert*-alcohols, in addition to primary alcohols.[1] This method is superior to that of Corey and Bock, which is only suitable in the case of primary alcohols (**6,** 109–110).

$$-\overset{|}{\underset{|}{C}}-OH \xrightarrow[80-90\%]{\substack{DMSO \\ Ac_2O}} -\overset{|}{\underset{|}{C}}-OCH_2SCH_3$$

The ethers are reduced by Raney nickel to methyl ethers in high yield. They are cleaved conveniently by methyl iodide in acetone (NaHCO$_3$ can be added for acid-sensitive alcohols); *cf.* **4,** 85. The derivatives are also cleaved by mercuric chloride and cadmium carbonate in aqueous acetonitrile (50°, 3 hours).

For protection of primary alcohols as the methylthiomethyl ether, *see Chloro-methyl methyl sulfide,* this volume.

[1] K. Yamada, K. Kato, H. Nagase, and Y. Hirata, *Tetrahedron Letters*, 65 (1976); P. M. Pojer and S. J. Angyal, *ibid.*, 3067 (1976).

Dimethyl sulfoxide–*t*-Butyl hypochlorite.

N-Arylsulfoximines. Dimethyl sulfoxide and *t*-butyl hypochlorite form a complex at −60° formulated as (1). This complex converts aryl amines into azasulfoxonium chlorides (2), which yield N-arylsulfoximines (3) on treatment with base. The method fails with aryl amines containing strong electron-with-drawing or electron-donating groups.[1]

$$(CH_3)_2\overset{O}{\underset{Cl^-}{\overset{\|+}{S}}}-O-C(CH_3)_3 \ + \ ArNH_2 \ \longrightarrow \ (CH_3)_2\overset{O}{\underset{Cl^-}{\overset{\|+}{S}}}-NHAr \ \xrightarrow[30-70\%]{N(C_2H_5)_3} \ ArN=\overset{O}{\overset{\|}{S}}(CH_3)_2$$

(1)   (2)   (3)

[1] R. W. Heintzelman, R. B. Bailey, and D. Swern, *J. Org.*, **41**, 2207 (1976).

Dimethyl sulfoxide–Trifluoroacetic anhydride, 5, 266–267.

Oxidation of alcohols. The complete paper is available.[1] Under optimum conditions, aldehydes are obtained in 60% yield and ketones in 80–85% yield. Benzylic and certain allylic alcohols are oxidized in the highest yields (80–100%). The reagent may be useful for selective oxidations.

This new reagent is particularly effective for oxidation of hindered alcohols.[2] Examples:

$$(CH_3)_2CHCHOHCH(CH_3)_2 \ \xrightarrow{86\%} \ (CH_3)_2CH\overset{O}{\overset{\|}{C}}CH(CH_3)_2$$

[1] K. Omura, A. K. Sharma, and D. Swern, *J. Org.*, **41**, 957 (1976).
[2] S. L. Huang, K. Omura, and D. Swern, *ibid.*, **41**, 3329 (1976).

Diperoxo-oxohexamethylphosphoramidomolybdenum(VI), 4, 203–204; **5,** 269–270.

Cyanohydrins. The anions of certain nitriles are converted into cyanohydrins by the reagent.[1]

$$CH_3(CH_2)_{15}CH_2CN \ \xrightarrow[55-60\%]{\substack{1)\ LDA \\ 2)\ MoO_5\cdot Py\cdot HMPT}} \ CH_3(CH_2)_{15}\overset{OH}{\overset{|}{C}}HCN$$

$$(C_6H_5)_2CHCN \ \longrightarrow \ \underbrace{(C_6H_5)_2\overset{OH}{\overset{|}{C}}CN \ + \ (C_6H_5)C=O}_{(71\%)} \ + \ (C_6H_5)_2\overset{NC}{\overset{|}{C}}-\overset{CN}{\overset{|}{C}}(C_6H_5)_2$$

                                                      (11%)

[1] E. Vedejs and J. E. Telschow, *J. Org.*, **41**, 740 (1976).

Diphenyl diselenide, 5, 272–276; **6,** 235.

α-*Methylene lactones* (**5,** 275). Selenenylation was used for introduction of the α-methylene unit in a total synthesis of (±)-tuberiferine (4), a terpene isolated from *Sonchus tuberifer Svent.* Selenenylation was also used to introduce the 1,2-double bond.[1]

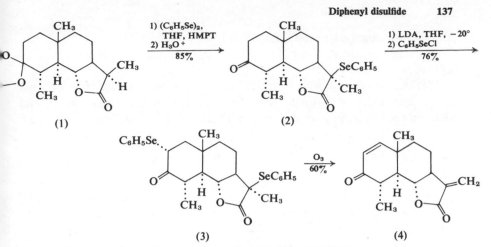

(1)

(2)

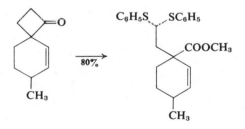

(3)

(4)

[1] P. A. Grieco and M. Nishizawa, *J.C.S. Chem. Comm.*, 582 (1976).

Diphenyl disulfide, 5, 276–277; **6,** 235–238.

Secosulfenylation. Treatment of a cyclobutanone with diphenyl disulfide (excess) in methanol containing sodium methoxide (excess) results in slow bissulfenylation and ring cleavage (equation I).[1] This secosulfenylation is an

(I)

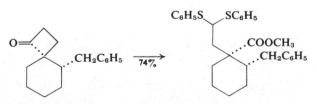

important addition to Trost's geminal alkylation method (**4,** 212–213; **5,** 71–73, 372–373).

Examples:

[1] B. M. Trost and J. H. Rigby, *J. Org.*, **41,** 3217 (1976).

Diphenylphosphinyl chloride, $(C_6H_5)_2P(O)Cl$. Mol. wt. 236.63, b.p. 199–201°/8 mm.

This acid chloride is prepared by bubbling O_2 through a solution of $(C_6H_5)_2PCl$ in benzene (reflux).[1]

Protection of amino groups.[2] In the presence of a base (N-methylmorpholine) this reagent reacts with amino acid esters and is then hydrolyzed under mild alkaline conditions to give diphenylphosphinamide (Dpp) derivatives of amino

$$(I) \quad (C_6H_5)_2P(O)Cl \; + \; \underset{NH_2}{RCHCOOC_2H_5} \quad \xrightarrow[\text{2) } OH^-]{\text{1) Base}} \quad \underset{R}{(C_6H_5)_2PONHCHCOOH}$$

acids (equation I). The derivatives are slightly more acid-labile than BOC-derivatives, being cleaved by HOAc–HCOOH, TFA, or p-TsOH.

Peptide synthesis.[3] Diphenylphosphinic mixed anhydrides, CBZ—NH—CHRCOOPO$(C_6H_5)_2$, have been used to activate amino acids for formation of peptide bonds.

[1] D. A. Tyssee, L. P. Bausher, and P. Haake, *Am. Soc.*, **95**, 8066 (1973).
[2] G. W. Kenner, G. A. Moore, and R. Ramage, *Tetrahedron Letters*, 3623 (1976).
[3] A. G. Jackson, G. W. Kenner, G. A. Moore, R. Ramage, and W. D. Thorpe, *ibid.*, 3627 (1976).

Diphenylphosphorazidate (DPPA), **4**, 210–211; **5**, 280; **6**, 193.

Ring contractions. This azide can function in 1,3-dipolar cycloaddition to enamines of cyclic ketones (1). The resulting Δ^2-triazolines (a) have not been isolated since they undergo loss of nitrogen with ring contraction to form the ring-contracted products (2). These can be converted into cycloalkanoic acids (3).[1]

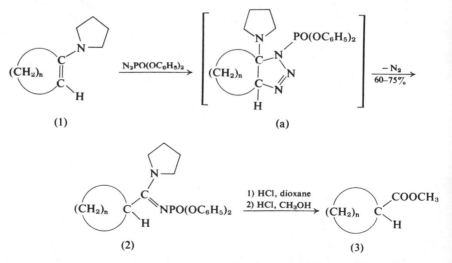

[1] S. Yamada, Y. Hamada, K. Ninomiya, and T. Shiori, *Tetrahedron Letters*, 4749 (1976).

Diphenyl-4-pyridylmethanol, (1). Mol. wt. 261.32, m.p. 235°.
Supplier: Aldrich.

Preparation.[1]

Protection of cysteine and histidine.[2] The S-diphenyl-4-pyridyl derivative of L-cysteine is more stable to acid than the S-trityl derivative. The former derivative is stable to TFA or 45% HBr in HOAc at 20°. The group is useful for protection of cysteine residues during peptide syntheses. It is removed by $Hg(OAc)_2$ in HOAc–H_2O, by I_2 in 80% HOAc, by Zn–HOAc, or by electrolytic reduction. The same group can be used for protection of the imidazole nitrogen of histidine. In this case hydrogenolysis, Zn–HOAc, or electrolytic reduction can be used for deprotection. Derivatives of cysteine are prepared with diphenyl-4-pyridyl-methanol, derivatives of histidine with diphenyl-4-pyridylmethyl chloride.[3]

[1] A. E. Tschitschibabin and S. W. Benewolenskaja, *Ber.*, **61**, 547 (1928).
[2] S. Coyle and G. T. Young, *J.C.S. Chem. Comm.*, 980 (1976).
[3] V. J. Traynelis and J. N. Rieck, *J. Org.*, **38**, 4334 (1973).

Diphenylseleninic anhydride, 6, 240–241.

o-Quinones. Phenols, even those unblocked at the *p*-position, are oxidized to *o*-quinones by diphenylseleninic anhydride in THF at 50° (15 minutes). In the case of α-naphthol, the *o*-quinone is obtained in 62% yield and the *p*-quinone in 10% yield. In other cases only the *o*-quinone was observed. Yields are generally 60–70%.[1]

Oxidation of amines to ketones. Amines are oxidized in several cases to ketones by treatment with this reagent. Unrecognizable products are obtained from amines that can form enamines.[2]

Examples:

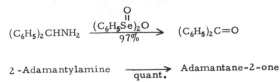

2-Adamantylamine $\xrightarrow{\text{quant.}}$ Adamantane-2-one

[1] D. H. R. Barton, A. G. Brewster, S. V. Ley, and M. N. Rosenfeld, *J.C.S. Chem. Comm.*, 985 (1976).
[2] M. R. Czarny, *ibid.*, 81 (1976).

Diphenylsulfonium cyclopropylide, 4, 211–214; **5,** 281; **6,** 242–243.

Reaction with steroidal 4,6-diene-3-ones. The reaction of these dienones (1) with this ylide (2) leads to a mixture of two products, (3) and (4), both of which incorporate two molecules of (2). The expected product (a) is suggested as an intermediate.[1]

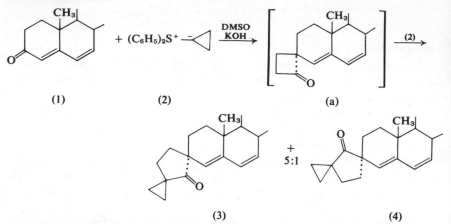

(1)   (2)   (a)

(3)   (4)

[1] M. J. Green, H.-J. Shue, A. T. McPhail, and R. W. Miller, *Tetrahedron Letters,* 2677 (1976).

4,6-Diphenylthieno[3,4-*d*]dioxol-2-one 5,5-dioxide,   (1).
Mol. wt. 326.31, m.p. >250° dec.

This reagent is prepared by reaction of 4-hydroxy-3-oxo-2,5-diphenyl-2,3-dihydrothiophene 1,1-dioxide (4)[1] with phosgene (pyridine, 73% yield).

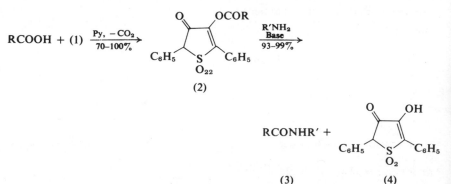

(2)

(3)   (4)

Peptide synthesis.[2] This cyclic carbonate reacts with acids (pyridine) to form stable activated esters (2), which afford amides (3) within minutes on reaction with amines (pyridine or triethylamine). The ester (2) need not be isolated. The method has been used only for synthesis of four dipeptides; no racemization was noted in at least one case.

[1] M. Chaykovsky, M. H. Lin, and A. Rosowsky, *J. Org.*, **37**, 2018 (1972).
[2] O. Hollitzer, A. Seewald, and W. Steglich, *Angew. Chem., Int. Ed.*, **15**, 444 (1976).

Di-(E)-propenylcopperlithium (Lithium di-(E)-propenyl cuprate),

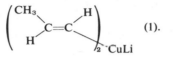

(1).

(E)-Alkenes. The preparation and use of this cuprate is exemplified by a synthesis of (E)-2-undecene (equation I). An alternative preparation is the reaction of (E)-propenyllithium directly with 1-iodooctane in about the same yield.[1]

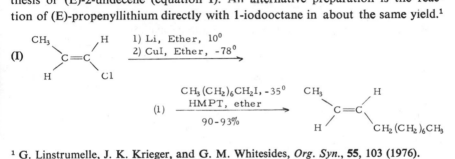

[1] G. Linstrumelle, J. K. Krieger, and G. M. Whitesides, *Org. Syn.*, **55**, 103 (1976).

2,2′-Dipyridyl disulfide–Triphenylphosphine, 5, 285–286; 6, 246–247.

Lactonization. Corey *et al.* have reported further studies on this method for lactonization directed mainly toward elucidation of the mechanism. One useful observation is that lactones containing from 12 to 21 members are formed at about the same rate, but at a rate much lower than that for formation of 5- or 6-membered lactones.[1]

A continuing search for superior reagents for this cyclization has revealed that various bis-2-imidazolyl disulfides offer significant improvements in both yields and rates of reaction. The disulfide (1) is particularly promising.[2]

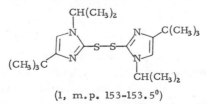

(1, m.p. 153–153.5°)

The "double activation" method has been used for synthesis of the natural macrolide (4) from *Cephalosporium recifei.*[3]

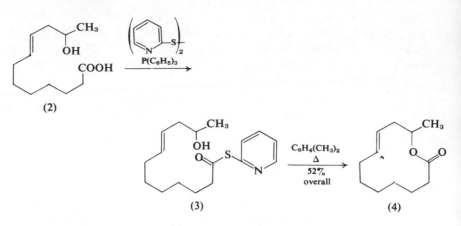

(2)   (3)   (4)

Ricinelaidic acid lactone. Details are available for isomerization of ricinoleic acid (1) to ricinelaidic acid (2) and for lactonization of the latter acid promoted by silver perchlorate or silver trifluoromethanesulfonate.[4]

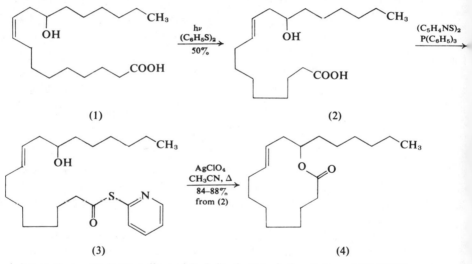

(1)   (2)

(3)   (4)

[1] E. J. Corey, D. J. Brunelle, and P. J. Stork, *Tetrahedron Letters*, 3405 (1976).
[2] E. J. Corey and D. J. Brunelle, *ibid.*, 3409 (1976).
[3] E. J. Corey, P. Ulrich, and J. M. Fitzpatrick, *Am. Soc.*, **98**, 222 (1976).
[4] A. Thalmann, K. Oertle, and H. Gerlach, *Org. Syn.*, submitted (1976).

1,3-Dithiane, 2, 187; **3,** 135–136; **4,** 216–218; **6,** 248.

3β,20,22-Δ⁵-Cholestenetriols. The 1,3-dithiane anion system has been used in two syntheses of these triols, which are precursors to the four possible 20,22-epoxycholesterols of interest for the study of metabolism of cholesterol to pregnenolone. The syntheses are outlined in equation (I) and (II).[1]

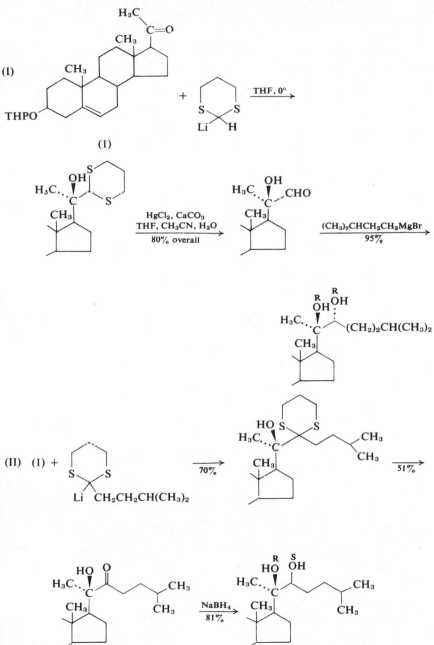

(I)

(1)

(II)   (1) +

[1] M. Koreeda, N. Koizumi, and B. A. Teicher, *Tetrahedron Letters*, 4565 (1976).

Dodecacarbonyltriiron, $Fe_3(CO)_{12}$. Mol. wt. 503.66, m.p. 140° dec. Suppliers: Alfa, ROC/RIC.

Cyclotrimerization of acetylenes. Diphenylacetylene trimerizes to hexaphenylbenzene at 180° (sealed tube) in the presence of this catalyst (equation I).[1]

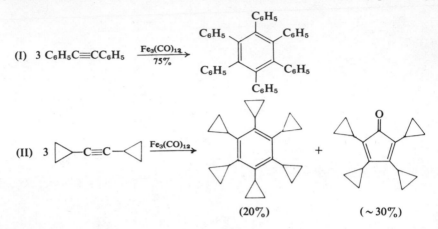

(I) $3 C_6H_5C{\equiv}CC_6H_5$ $\xrightarrow[75\%]{Fe_3(CO)_{12}}$

(II) 3 $\xrightarrow{Fe_3(CO)_{12}}$ +

(20%)   (~30%)

A similar cyclotrimerization of dicyclopropylacetylene has been observed (equation II).[2] Hexacyclopropylbenzene has an unusually high melting point (266–270°), but the NMR spectrum is consistent with that of a simple benzene with freely rotating substituents.

[1] W. Hübel and C. Hoogzand, *Ber.*, **93**, 103 (1960).
[2] V. Usieli, R. Victor, and S. Sarel, *Tetrahedron Letters*, 2705 (1976).

(−)-N-Dodecyl-N-methylephedrinium bromide, 6, 249.

Borohydride reduction of ketones. The reduction of ketones to alcohols by sodium borohydride in benzene–water has been reported to be catalyzed by this salt (**6**, 249). However, no asymmetric induction was noted in the case of 2-octanone, 1-phenyl-1-propanone, or acetophenone.[1] More recently, asymmetric induction has been observed with more hindered ketones. The maximum was observed with *t*-butyl phenyl ketone when an enantiomeric excess of 14% of the (R)-alcohol could be obtained. The enantiomeric excess is only 3.6% in the reduction of isopropyl phenyl ketone. No asymmetric induction was observed when (−)(R)-N,N-dimethyl-N-dodecylamphetaminium bromide was used as catalyst. The structure of the chiral catalyst is evidently important.[2]

[1] S. Colonna and R. Fornasier, *Synthesis*, 531 (1975).
[2] J. Balcells, S. Colonna, and R. Formasier, *ibid.*, 266 (1976).

E

Ethoxyacetylene–Magnesium bromide.
This combination reacts with (1) to form the carbinol (2) in high yield. The product was converted into homosafranic acid (3).[1]

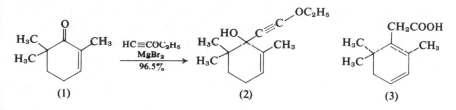

| (1) | (2) | (3) |

[1] C. Schmidt, *Canad. J. Chem.*, **54**, 2310 (1976).

Ethoxycarbonylformonitrile oxide, (1). Mol. wt. 115.09, b.p. 120–124°/2 mm.
  Preparations[1]:

$$N_2CHCOOC_2H_5 + N_2O_3 \longrightarrow O \leftarrow N \equiv CCOOC_2H_5 + HNO_2 + N_2$$
$$(1)$$

$$N_2CHCOOC_2H_5 + NOCl \longrightarrow O \leftarrow N \equiv CCOOC_2H_5 + HCl + N_2$$
$$(1)$$

Cleavage of C=C bonds. Swiss chemists[2] have carried out this reaction by 1,3-dipolar addition of (1) to an alkene to form a Δ^2-isoxazoline (2). The adduct is hydrolyzed and the free acid (3) on thermal decarboxylation is cleaved to a carbonyl compound (4) and a nitrile (5). The sequence is illustrated for the case of *trans*-stilbene. The same sequence with styrene, $C_6H_5CH=CH_2$, did not follow

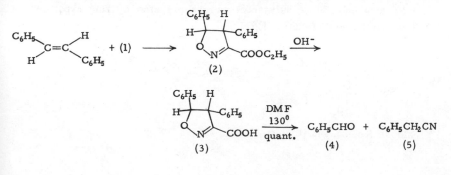

the desired pathway and decarboxylation led to $C_6H_5CHOHCH_2CN$; however, the method was used successfully for some steroids. A disadvantage is that fairly high temperatures are required in the final step.

If triphenylacetonitrile oxide, $(C_6H_5)_3C$—$C\equiv N\rightarrow O$, is used in place of (1), then the resulting 3-triphenylmethyl-Δ^2-isoxazoline can be cleaved to a carbonyl compound and a nitrile by ultraviolet irradiation.[3]

[1] G. S. Skinner, *Am. Soc.*, **46**, 731 (1924).
[2] J. Kalvoda and H. Kaufmann, *J.C.S. Chem. Comm.*, 209 (1976).
[3] *Idem, ibid.*, 210 (1976).

Ethoxycarbonyl isothiocyanate, 4, 223–224.

Aryl thioamides. The reagent reacts with arenes at 0–3° in the presence of about 1.5 equiv. of $AlCl_3$ to form N-ethoxycarbonylthioamides (1); reactions run in an excess of the arene as solvent at 25° or higher result in thioamides (2).[1]

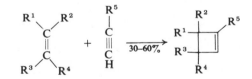

[1] E. P. Papadopoulos, *J. Org.*, **41**, 962 (1976).

Ethylaluminum dichloride, 6, 251–252.

[2 + 2]Cycloadditions. In the presence of this Lewis acid, alkenes and 1-alkynes are converted into cyclobutenes (equation I); a side reaction is trimerization of the alkyne to a trialkylbenzene. With disubstituted alkynes this trimerization is the only observed reaction.[1]

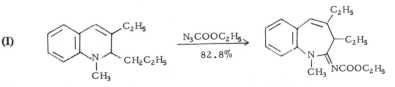

[1] J. H. Lukas, F. Baardman, and A. P. Kouwenhoven, *Angew. Chem., Int. Ed.*, **15**, 369 (1976).

Ethyl azidoformate, 1, 363–364; 2, 191–192; 3, 138; 4, 225.

1-Benzazepines. These substances can be prepared by ring expansion of 1,2-dihydroquinolines (equation I).[1]

(I)

[1] Y. Sato, H. Kojima, and H. Shirai, *J. Org.*, **41**, 195 (1976).

Ethyl *trans*-1,3-butadiene-1-carbamate, 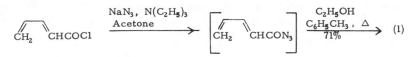 **(1). Mol. wt. 141.17, m.p. 44–45°, stable for several weeks at −20°.**

Preparation[1]:

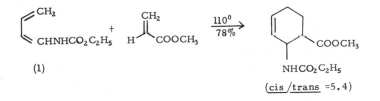

Diels-Alder reactions. N-Acyl-1-amino-1,3-dienes such as **(1)** are useful dienes for Diels-Alder reactions, as formulated in equation (I). They are more active than N-trichloroacetyl-1-amino-1,3-dienes.[2]

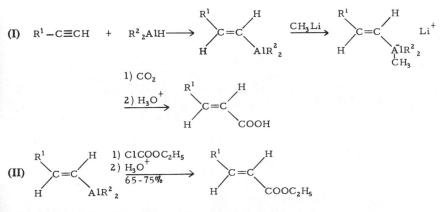

[1] L. E. Overman, G. F. Taylor, and P. J. Jessup, *Tetrahedron Letters*, 3089 (1976).
[2] L. E. Overman and L. A. Olizbe, *Am. Soc.*, **98**, 2352 (1976).

Ethyl chloroformate, 1, 364–367; **2,** 193; **4,** 228; **5,** 294–295.

α,β-*Unsaturated esters.* α,β-Unsaturated acids (*trans*) have been prepared[1] from *trans*-vinylalanes as shown in equation (I). An even simpler route has now been reported: reaction of the vinylalane directly with ethyl or methyl chloroformate, followed by hydrolysis (equation II).[2]

(I) $R^1-C\equiv CH$ + $R^2{}_2AlH \longrightarrow$ ...

(II) ...

[1] G. Zweifel and R. B. Steele, *Am. Soc.*, **89**, 2754 (1967).
[2] G. Zweifel and R. A. Lynd, *Synthesis*, 625 (1976).

Ethyldiisopropylamine, 1, 371; 4, 230.

Bicyclo[4.n.1]enones.[1] These substances (3) can be obtained by cyclodialkylation of pyrrolidine enamines of cycloalkanones (1) with 1,4-dichloro-2-butene (2)[2] in the presence of potassium iodide, DMF, and this hindered amine. Yields of (3) are about 40% when n = 2 and 3, but only about 10% when n = 5. The reaction with the enamine of 4-methylcyclohexanone is improved by addition of HMPT.

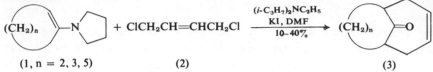

(1, n = 2, 3, 5)   (2)   (3)

[1] W. C. Still, *Synthesis*, 453 (1976).
[2] J. M. Bobbitt, L. H. Amundsen, and R. I. Steiner, *J. Org.*, **25**, 2230 (1960).

3,3-Ethylenedioxybutylmagnesium bromide,   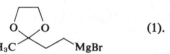   (1).
Mol. wt. 219.38.

This Grignard reagent is prepared by reaction of the corresponding bromide with a threefold excess of Mg and a trace of 1,2-dibromoethane in THF at 25°.

Annelation. This reagent reacts slowly, but in high yield, with cyclohexanone and cyclopentanone to form a tertiary alcohol. The annelation of cyclohexanone to form $\Delta^{1,9}$-octalone is illustrated (equation I). This annelation differs from a Robinson annelation in that the 3-ketobutyl group is introduced at the site of the carbonyl group rather than in the α-position.[1]

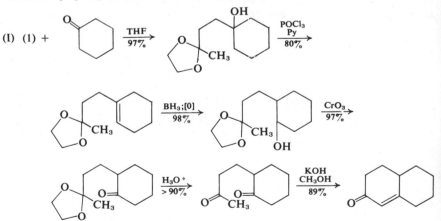

[1] A. A. Ponaras, *Tetrahedron Letters*, 3105 (1976).

O-Ethyl-S-ethoxycarbonylmethyl dithiocarbonate, $C_2H_5OC{-}S{-}CH_2COOC_2H_5$.
Mol. wt. 176.24. $\overset{\|}{S}$

This dithiocarbonate is prepared in 92% yield by the reaction of ethyl bromo-acetate with carbon disulfide, ethanol, and KOH.

α,β-Unsaturated esters. The carbanion (1), prepared with LDA, reacts with aldehydes and ketones to form, after hydrolytic work up, α,β-unsaturated esters (2) in 65–88% yield. The related carbanion (3) also reacts with aldehydes to form α,β-unsaturated esters (4).

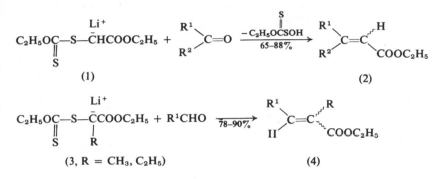

(1)   (2)

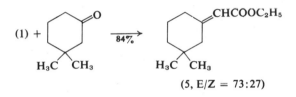

(3, R = CH₃, C₂H₅)   (4)

This reaction was used to synthesize the sex pheromone of the boll weevil (5).[1]

(1) +   84% →

(5, E/Z = 73:27)

[1] K. Tanaka, R. Tanikaga, and A. Kaji, *Chem. Letters*, 917 (1976).

Ethylmagnesium bromide Cuprous iodide (?·1)

cis-Reduction of acetylenes. A copper hydride species prepared from ethyl- or *n*-butylmagnesium bromide in ether reduces disubstituted alkynes to *cis*-alkenes. When THF is used as solvent, addition of the alkylcopper reagent becomes a competing reaction.[1]

[1] J. K. Crandall and F. Collonges, *J. Org.*, **41**, 4089 (1976).

Ethyl α-trifluoromethylsulfonyloxyacetate, $CF_3SO_2OCH_2COOC_2H_5$ (1). Mol. wt. 236.17, m.p. 22–23°, lachrymator.

The triflate is prepared by reaction of ethyl diazoacetate with triflic acid in liquid SO_2 at $-78°$ (73% yield).

Alkene synthesis.[1] Sulfides can be converted into stabilized ylides (3) by reaction with (1) in acetonitrile followed by deprotonation with DBU. The ylides fragment at 20–50° to ethyl α-methylsulfenyl acetate and an alkene (4, equation I).

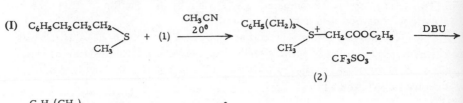

(I) $C_6H_5CH_2CH_2CH_2$

(2)

$C_6H_5(CH_2)_3$

(3)

$\xrightarrow{20-50^0}$ $C_6H_5CH_2CH=CH_2$ + $CH_3SCH_2COOC_2H_5$

(4, 69%)

This reaction has also been applied to α-sulfenyl esters (equation II) and α-sulfenyl lactones (equation III).

(II) $C_5H_{11}CHCOOCH_3$ + (1) $\longrightarrow$
         |
         SCH_3

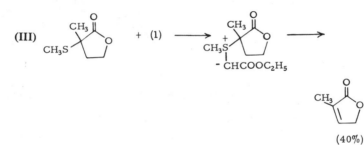

$\xrightarrow[88\%]{25^0}$ $C_4H_9CH=CHCOOCH_3$

(III)

(40%)   +   (47%)

[1] E. Vedejs and D. A. Engler, *Tetrahedron Letters,* 3487 (1976).

Ethyl trimethylsilylacetate, $(CH_3)_3SiCH_2COOC_2H_5$. Mol. wt., 160.29, b.p. 76–77°/40 mm.

This reagent is prepared in 70–75% yield by the reaction of chlorotrimethyl-silane with ethyl bromoacetate in the presence of zinc dust.[1]

Silylation.[2] Alcohols and ketones are silylated in high yield by ethyl tri-methylsilylacetate with catalysis by tetra-*n*-butylammonium fluoride (sensitive to moisture) as base. The side product is the volatile ethyl acetate (equation I).

(I) ROH + $(CH_3)_3SiCH_2COOC_2H_5$ $\xrightarrow[25^0]{(C_4H_9)_4NF}$ $ROSi(CH_3)_3$ + $CH_3COOC_2H_5$

(~90%)

Ketones can be silylated in this way with high regioselectivity comparable to that obtained with chlorotrimethylsilane (equation II).

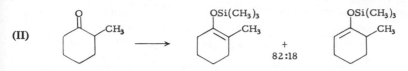

(II)

+
82:18

Phenylacetylene also is silylated by this procedure. In this case, the actual reagent is $(CH_3)_3SiF$ (equation III).

$$(III) \quad C_6H_5C{\equiv}CH + (CH_3)_3SiCH_2COOC_2H_5 \xrightarrow[\substack{88\%}]{\substack{(C_4H_9)_4NF \\ THF, 20^0}}$$

$$C_6H_5C{\equiv}CSi(CH_3)_3 + CH_3COOC_2H_5$$

Aldehydes also react smoothly to afford adducts in 75–85% yield (equation IV). In fact aldehydes are more reactive than ketones.

$$(IV) \quad RCHO \longrightarrow R\overset{OSi(CH_3)_3}{\underset{}{C}}HCH_2COOC_2H_5 \xrightarrow{CH_3OH} R\overset{OH}{\underset{}{C}}HCH_2COOC_2H_5$$

Epoxides, esters, and nitriles do not react with the reagent.

Reformatsky-type reactions. The reaction of ethyl trimethylsilylacetate and a carbonyl compound in the presence of $Bu_4N^+F^-$ in THF results in formation of β-siloxy esters, which can be hydrolyzed to β-hydroxy esters (equation I).

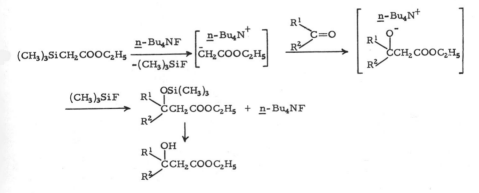

Examples:

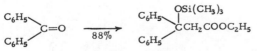

$$C_6H_5CH{=}CHCHO \xrightarrow{81\%} C_6H_5CH{=}CH\overset{OSi(CH_3)_3}{\underset{}{C}}HCH_2COOC_2H_5$$

$$\underset{C_6H_5}{\overset{C_6H_5}{\diagdown}}C{=}O \xrightarrow{88\%} \underset{C_6H_5}{\overset{C_6H_5}{\diagdown}}\overset{OSi(CH_3)_3}{\underset{}{C}}CH_2COOC_2H_5$$

Carbonyl compounds with acidic α-protons (*e.g.*, benzylacetone, β-ionone) are converted under these conditions mainly into trimethylsilyl enol ethers.

This Reformatsky-type reaction is also catalyzed by KF in combination with dicyclohexyl-18-crown-6.[3]

[1] R. J. Fessenden and J. S. Fessenden, *J. Org.*, **32**, 3535 (1967).
[2] E. Nakamura, T. Murofushi, M. Shimizu, and I. Kuwajima, *Am. Soc.*, **98**, 2346 (1976).
[3] E. Nakamura, M. Shimizu, and I. Kuwajima, *Tetrahedron Letters*, 1699 (1976).

F

Ferric azide, $Fe(N_3)_3$. Mol. wt. 133.90. Prepared *in situ* from NaN_3 and $Fe_2(SO_4)_3$.

Azidoalkanes. These compounds can be prepared in satisfactory yield from ferric azide and trialkylboranes as shown in equation (I).[1]

(I) $$R_3B + Fe(N_3)_3 + H_2O_2 \longrightarrow RN_3$$

Examples:

$$C_4H_9CH\!=\!CH_2 \xrightarrow[85\%]{} C_4H_9CH_2CH_2N_3$$

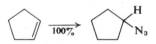

[1] A. Suzuki, M. Ishidoya, and M. Tabata, *Synthesis*, 687 (1976).

Ferric chloride, 1, 390–392; **2,** 199; **3,** 145; **4,** 236; **5,** 307–308; **6,** 260.

Ring expansion of cycloalkanones. 1-Trimethylsilyloxybicyclo[n.1.0]alkanes (1), prepared by Simmons-Smith reaction with silyl enol ethers of cycloalkanones, react with ferric chloride in DMF containing pyridine to form a 3-chlorocyclo-alkanone (2) in fair to high yield. Dehydrochlorination (sodium acetate) yields a 2-cycloalkenone (3) containing one more carbon atom than the starting cycloalkanone.

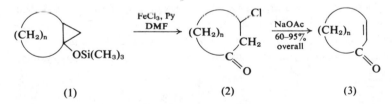

Examples:

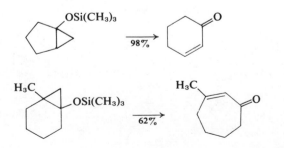

153

This homologation has been applied to bis(trimethylsilyloxy)bicyclo[n.1.0]-alkanes (4) to obtain cycloalkane-1,3-diones (5) in moderate yields (two examples; 68 and 72%).[1]

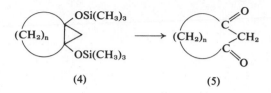

(4)   (5)

Ring expansion induced by $FeCl_3$ has been used in a new synthesis of *dl*-muscone in 23% overall yield from dimethyl tetradecanedioate, which was converted into (6) by the procedure of Ruhlmann (**4**, 537). In this case cyclo-propanation with methylene iodide and diethylzinc (**4**, 153) proved superior to the Simmons-Smith reagent.[2]

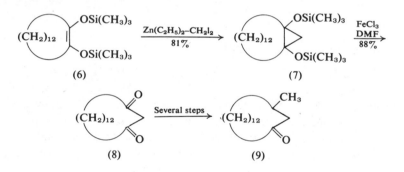

(6)   (7)

(8)   (9)

Oxidative C-demethylation of a phenol. Oxidation of the bridged trione (1) with ferric chloride in ethanol (steam bath) unexpectedly gives the 1,2-naphtho-quinone derivative (2). A quinone methide is a probable intermediate. This unusually mild demethylation is undoubtedly a result of the bridge system.[3]

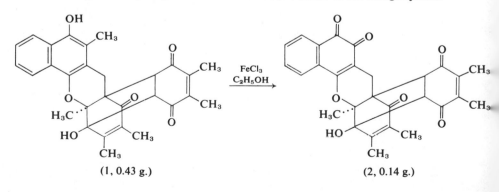

(1, 0.43 g.)   (2, 0.14 g.)

Allenes. Allenes are formed in moderate to high yield by the reaction of propargylic chlorides (terminal and nonterminal) with primary and secondary Grignard reagents (25–50% excess) in the presence of ferric chloride as catalyst.[4]

Examples:

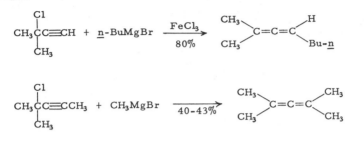

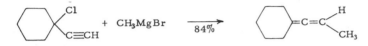

[1] Y. Ito, S. Fujii, and T. Saegusa, *J. Org.*, **41**, 2073 (1976).

[2] Y. Ito and T. Saegusa, *ibid.*, **42**, 2326 (1977).

[3] F. M. Dean, K. B. Hindley, L. E. Houghton, and M. L. Robinson, *J.C.S. Perkin I*, 600 (1976).

[4] D. J. Pasto, G. F. Hennion, R. H. Shults, A. Waterhouse, and S.-K. Chou, *J. Org.*, **41**, 3496 (1976).

Ferric chloride–Sodium hydride.

Reduction of aldehydes and ketones. Addition of ferric chloride to sodium hydride in THF produces a yellow suspension of a reagent that reduces carbonyl compounds to alcohols at room temperature. Best results are obtained with a ratio of 3:1 for NaH and $FeCl_3$. The actual reagent is presumably an iron hydride. Yields of alcohols are in the range of 75–85%. A reagent obtained from $FeCl_2$ and NaH has somewhat different properties.[1]

Note that a reagent obtained from $FeCl_3$ and *n*-butyllithium deoxygenates epoxides to olefins (**6**, 260).

[1] T. Fujisawa, K. Sugimoto, and H. Ohata, *J. Org.*, **41**, 1667 (1976).

Ferric thiocyanate, Fe(SCN)₃, Mol. wt. 230.09; Ferric selenocyanate, Fe(SeCN)₃, mol. wt. 370.78.

Reaction with trialkylboranes. These reagents react with trialkylboranes to form alkyl thiocyanates and alkyl selenocyanates[1]:

$$R_3B + Fe(XCN)_3 \xrightarrow{(H_2O)} 2\,RXCN + (HO)_2BR$$
$$X = S,\ Se$$

The reaction with mixed trialkylboranes is interesting in that secondary and tertiary groups react in preference to primary groups. Thus, 2,3-dimethyl-2-butene can be converted into 1,1,2-trimethylpropyl selenocyanate in 73% yield.[2]

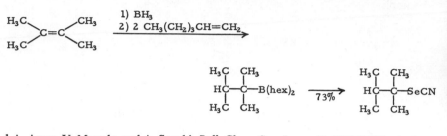

[1] A. Arase, Y. Masuda, and A. Suzuki, *Bull. Chem. Soc. Japan*, **47**, 2511 (1974); A. Arase and Y. Masuda, *Chem. Letters*, 785 (1976).
[2] *Idem, ibid.*, 1115 (1976).

Fluoroxytrifluoromethane, 2, 200; **3**, 146–147; **4**, 237–238; **5**, 312; **6**, 263–264.

Fluorination at saturated carbon. Hesse et al.[1] have observed that adamantane in the presence of radical inhibitors (*e.g.*, nitrobenzene) reacts with CF_3OF to form 1-fluoroadamantane (75% yield). The same reaction can be carried out with F_2 itself. This reaction is characterized by almost exclusive attack at a tertiary position, pronounced tendency to monosubstitution, and a marked polar effect on the rate (adamantane substituted at the 1-position by $NHCOCF_3$ is much less reactive than adamantane). Thus the reaction is considered to involve direct electrophilic fluorination.

This reaction has been extended to direct fluorination at C_9, C_{14}, and C_{17} of steroids.[2]

Examples:

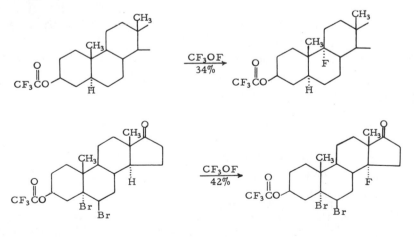

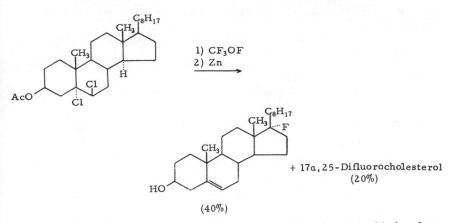

+ 17α,25-Difluorocholesterol
(20%)

(40%)

Addition to double bonds. The reagent reacts with isolated double bonds to afford complex mixtures. Allylic alcohols and acetates, however, react to give only two adducts corresponding to the *cis*-addition of CF_3OF and of F_2.[3]

Example:

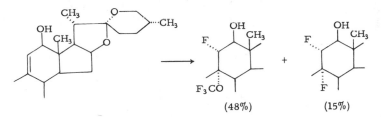

(48%)   (15%)

Reaction with polynuclear arenes. Naphthalene and phenanthrene are unreactive to this reagent but higher arenes such as (1) undergo substitution and oxidation as formulated for one example.[4]

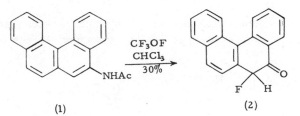

(1)   (2)

Fluorination. α-Diazoketones react with this reagent as shown for α-diazoacetophenone (equation I).[5]

(I) $C_6H_5COCHN_2$ $\xrightarrow[85\%]{\substack{FOCF_3 \\ CFCl_3}}$ $C_6H_5COCHF_2$ + $C_6H_5CO\overset{\displaystyle H}{\underset{\displaystyle F}{C}}-OCF_3$

58:42

[1] D. H. R. Barton, R. H. Hesse, R. E. Markwell, M. M. Pechet, and H. T. Toh, *Am. Soc.*, **98**, 3034 (1976).

[2] D. H. R. Barton, R. H. Hesse, R. E. Markwell, M. M. Pechet, and S. Rozen, *ibid.*, **98**, 3036 (1976).

[3] D. H. R. Barton, L. S. Danks, A. K. Ganguly, R. H. Hesse, G. Tarzia, and M. M. Pechet, *J.C.S. Perkin I*, 101 (1976).

[4] T. B. Patrick, M. H. Le Faivre, and T. E. Koertge, *J. Org.*, **41**, 3413 (1976).

[5] C. Wakselman and J. Leroy, *J.C.S. Chem. Comm.*, 611 (1976).

Formaldehyde, 1, 397–402; **2,** 200–201; **4,** 238–239; **5,** 312–315; **6,** 264–267.

α-Methylenelactones (**5,** 313–314; **6,** 266–267). Grieco *et al.*[1] have used the hydroxymethylation of five- and six-membered lactones with formaldehyde for synthesis of (±)-vernolepin (**2**) and (±)-vernomenin (**3**), elemanolide dilactones of *Vernonia hymenolepis*. The *trans*-decalone (**1**) was a key intermediate to both natural products.

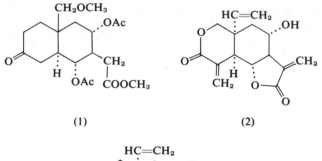

(1)   (2)

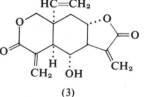

(3)

α-Methylene-γ-thiobutyrolactone (**4**). This lactone has been prepared for the first time by aldol condensation of α-thiobutyrolactone (**1**) with formaldehyde

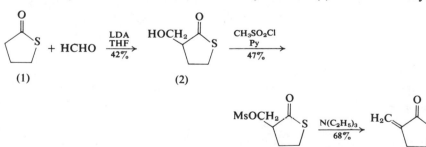

(1)   (2)

(3)   (4)

(LDA, THF, −78°) to give (2). This product was then dehydrated via (3). The thiolactone (4) is unstable at about 55°, hence the usual methods for synthesis of (4) are not suitable.[2]

   Intramolecular amination. The hasubanan skeleton has been constructed by reaction of (1) with HCHO in HCOOH (the yield is lower in CH₃COOH). The product (2) on catalytic hydrogenation is converted into *dl*-3-methoxy-N-methyl-hasubanan (3). This amination may be related to the Sommelet reaction.[3]

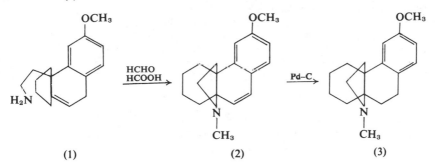

(1)   (2)   (3)

   Prins reaction.[4] The acid-catalyzed addition of formaldehyde to the lactone (1) in acetic acid at 60–80° gives (2) in 75–85% yield.[5] The diol diacetate (2) was transformed into the hydroxy aldehyde (4), readily convertible to an intermediate

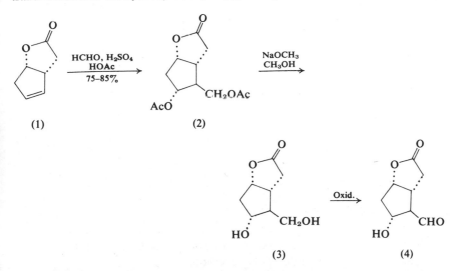

(1)   (2)

(3)   (4)

in the Corey prostaglandin synthesis.[6] Both thioanisole–chlorine (4, 89) and the Pfitzner-Moffatt reagent were satisfactory for the selective oxidation of (3).

   Substituted styrenes. Broos and Anteunis[7] have used a simplified Wittig reaction for synthesis of styrenes (equation I).

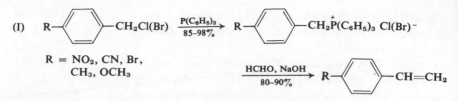

R = NO₂, CN, Br,
    CH₃, OCH₃

[1] P. A. Grieco, M. Nishizawa, S. D. Burke, and N. Marinovic, *Am. Soc.*, **98**, 1612 (1976).
[2] D. H. Lucast and J. Wemple, *Synthesis*, 724 (1976).
[3] S. Shiotani and T. Kometani, *Tetrahedron Letters*, 767 (1976).
[4] E. Arundale and L. A. Mikeska, *Chem. Revs.*, **51**, 505 (1952).
[5] I. Tõmösközi, L. Gruber, G. Kovács, I. Székely, and V. Simonidesz, *Tetrahedron Letters*, 4639 (1976).
[6] E. J. Corey *et al.*, *Am. Soc.*, **91**, 5675 (1969); **92**, 397 (1970).
[7] R. Broos and M. Anteunis, *Syn. Comm.*, **6**, 53 (1976).

Formic acid, 1, 404–407; **2**, 202–203; **3**, 147; **4**, 239–240; **5**, 316–319.

Cleavage of cyclopropylcarbinyl systems (**5**, 317). This useful step in a prostaglandin synthesis has been improved by conversion of the diol (1) into the

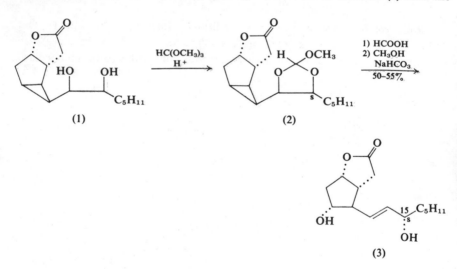

orthoester (2). On solvolysis of (2) the diol (3) is obtained as the major product. In this product the stereochemistry at C_{15} in the precursor is almost completely retained. Only slight epimerization is observed. However, the orthopropionate of (2) shows complete retention of configuration.[1]

Biomimetic polyene cyclizations (**5**, 316). Johnson[2] has reviewed cyclizations of polyenes to polycyclic natural products.

[1] R. C. Kelly and V. VanRheenen, *Tetrahedron Letters*, 1067 (1976).
[2] W. S. Johnson, *Angew. Chem., Int. Ed.*, **15**, 9 (1976).

Furane, Mol. wt. 68.08, b.p. 32°/758 mm.

Diels-Alder reactions. Cycloaddition reactions of furanes are usually only useful in the case of very reactive dienophiles, such as maleic anhydride and dimethyl acetylenedicarboxylate, since the adducts readily undergo a retro Diels-Alder reaction because of the aromatic character of furane. Dauben and Krabbenhoft[1] find, however, that cycloadducts can be formed in fair to high yields by carrying out the reaction at room temperature at 15,000 atm. pressure (15 kbar)[2] for 4–14 hours. The yields in the case of acrylic dienophiles are ~55%, but they are relatively low (0–25%) with crotonic dienophiles. The adduct of furane with dimethyl maleate is obtained in 94% yield as the *endo-cis* isomer.

[1] W. G. Dauben and H. O. Krabbenhoft, *Am. Soc.*, **98**, 1992 (1976).
[2] Apparatus: W. G. Dauben and A. P. Kozikowski, *Am. Soc.*, **96**, 3664 (1974).

G

Glyoxylic acid, 5, 320.

Transthioacetalization. This carbonyl compound and mesoxalic acid (keto-malonic acid) are suitable for transfer of thioacetal groups from various aldehydes and ketones. Acetic acid is a satisfactory medium; a strong mineral acid is also used. Yields, 50–90%.[1]

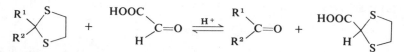

[1] H. Muxfeldt, W.-D. Unterweger, and G. Helmchen, *Synthesis*, 694 (1976).

Gold, 5, 321.

Rearrangement of strained hydrocarbons.[1] Gold salts have been considered to be relatively ineffective for catalysis of rearrangement of strained ring systems. Actually metallic gold and some gold(I) organometallic complexes do catalyze such rearrangements. Surprisingly gold(I) and silver(I) ions often cause different

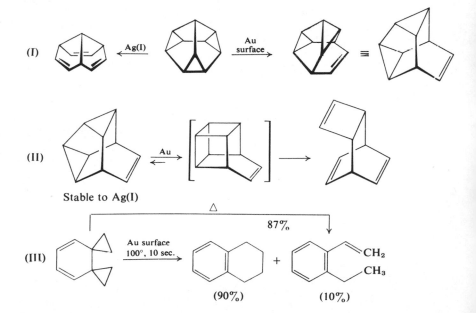

rearrangements as shown in equations (I) and (II). Equation (III) is an example of a difference between thermal isomerization and isomerization on a gold surface.

[1] L.-U. Meyer and A. deMeijere, *Tetrahedron Letters*, 497 (1976).

Grignard reagents, **1**, 415–424; **2**, 205; **5**, 321; **6**, 269–270.

1-*Alkynes*.[1] Grignard reagents react with methoxyallene in THF with catalysis by CuBr to form 1-alkynes (equation I).

(I) $\qquad$ $RMgBr + H_2C{=}C{=}CHOCH_3 \xrightarrow[\text{THF}]{\text{CuBr}} RCH_2C{\equiv}CH + MgBrOCH_3$

***Vinyl ethers*.**[2] If [RCuBr]MgX is used in reaction (I), vinyl ethers are formed (equation II). Both (E)- and (Z)-isomers are formed, but the latter predominate.

(II) $\quad$ $[RCuBr]MgX + H_2C{=}C{=}CHOCH_3 \xrightarrow[-60°]{\text{THF}} $

$$\left[\begin{array}{c} RCH_2C{=}CHOCH_3 \\ | \\ CuBr \end{array}\right] MgX \xrightarrow[80-90\%]{H_3O^+} RCH_2CH{=}CHOCH_3$$

***Ketone synthesis*.** Italian chemists[3] have developed a synthesis of ketones from carboxylic acids. The first step involves reaction of the acid with 2-mercaptophenol (1) to form 1,3-benzoxathiolium perchlorates (2).[4] The products react with Grignard reagents to afford 2,2-disubstituted 1,3-benzoxathioles (3). One of

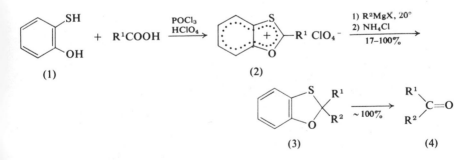

three methods is used to effect the last step: hydrolysis with HCl in DMSO, hydrolysis with $HgCl_2$ in aqueous DMSO, or oxidation with peracetic acid.

***α-Diketone monoketals*.** The reaction of Grignard reagents in HMPT with ketal and thioketal esters, followed by hydrolysis, results in α-diketone monoketals and monothioketals. The latter compounds are attractive for synthetic purposes, since either carbonyl group can be protected (equation I).[5]

Examples:

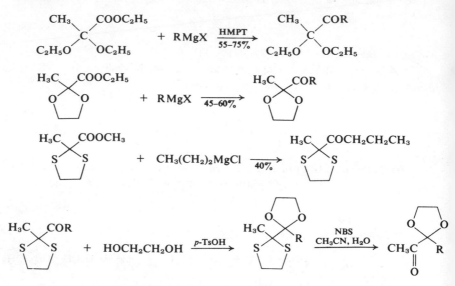

(I)

Trisubstituted allenes. The reaction of aliphatic Grignard reagents with a 1,1-disubstituted propargylic acetate (1) in ether results in formation of a trialkylallene (2). Yields are satisfactory with Grignard reagents of primary halides, but less satisfactory when R^1 is an aryl group.[6]

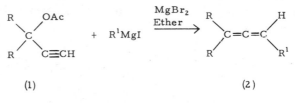

(1)   (2)

[1] J. Meijer and P. Vermeer, *Rec. trav.*, **93**, 183 (1974).

[2] H. Klein, H. Eijsinga, H. Westmijze, J. Meijer, and P. Vermeer, *Tetrahedron Letters*, 947 (1976).

[3] I. Degani, R. Fochi, and P. Tundo, *Gazz. Chim. Ital.*, **105**, 907 (1975).

[4] L. Costa, I. Degani, R. Fochi, and P. Tundo, *J. Heterocyclic Chem.*, **11**, 943 (1974).

[5] F. Huet, M. Pellet and J. M. Conia, *Tetrahedron Letters*, 3579 (1976).

[6] F. Delbecq and J. Gore, *Angew. Chem., Int. Ed. Engl.*, **15**, 496 (1976), and references cited therein.

H

Heptafluoro-1-methylethyl phenyl ketone, $C_6H_5\overset{\overset{\displaystyle O}{\|}}{C}CF(CF_3)_2$. Mol. wt. 274.14, b.p. 80–81°/39 mm.

This ketone is obtained by the reaction of benzoyl chloride with hexafluoropropane in DMF in the presence of KF (66% yield).[1]

Benzoylation.[2] In the presence of TMEDA, this ketone benzoylates alcohols and amines in high yield. The reaction with nitromethane results in α-nitroacetophenone, $C_6H_5COCH_2NO_2$, in 52% yield.

[1] N. Ishisawa and S. Shin-ya, *Bull. Chem. Soc. Japan*, **48**, 1339 (1975).
[2] *Idem, Chem. Letters*, 673 (1976).

Hexa-*n*-butyldistannane, $(C_4H_9)_3Sn{-}Sn(C_4H_9)_3$ (1). Mol. wt. 580.4, b.p. 160–162°/ 0.3 mm.

The reagent is prepared in 70% yield by reaction of tri-*n*-butyltin chloride with Mg in THF.[1]

Photodesulfurization.[2] 1,3-Dithiole-2-thiones (2) on irradiation (300 W high pressure mercury lamp) with an equimolar amount of (1) are converted into tetrahydrofulvalenes (3) in 50–75% yield. These products are of current interest because they form highly conductive charge-transfer complexes with tetracyano-*p*-quinodimethane.[3]

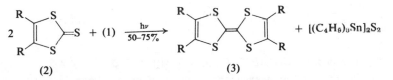

[1] H. Shirai, Y. Sato, and M. Niwa, *Yakugaku Zasshi* (*J. Pharm. Soc. Japan*), **90**, 59 (1970) [*C.A.* **72**, 90593 (1970)].
[2] Y. Ueno, A. Nakayama, and M. Okawara, *Am. Soc.*, **98**, 7440 (1976).
[3] A. F. Garito and A. J. Heeger, *Accts. Chem. Res.*, **7**, 232 (1974).

Hexacarbonylmolybdenum (Molybdenum hexacarbonyl), $Mo(CO)_6$. Mol. wt. 264.01. Suppliers: Alfa ROC/RIC.

Thioacetalization. This metal carbonyl reacts with α-bromo sulfoxides (2:1 mole ratio) in DMF to form thioacetals (equation I). $Mo(CO)_6$ is evidently functioning as both a reagent and a catalyst. $W(CO)_6$ and $Cr(CO)_6$ are essentially inert in this reaction. Deoxygenation of the sulfoxide to a sulfide, which requires

(I) R^1SCH (O, R^2, Br) $\xrightarrow[\text{DME}]{\text{Mo(CO)}_6}$ $\left[R^1SCH \genfrac{}{}{0pt}{}{R^2}{Br} \right]$ $\xrightarrow[\text{40–80\%}]{\text{Mo(CO)}_6}$ $R^2CH \genfrac{}{}{0pt}{}{SR^1}{SR^1}$

an equimolar quantity of $Mo(CO)_6$, is one step in this thioacetalization. The paper suggests a reasonable mechanism.[1]

[1] H. Alper and G. Wall, *J.C.S. Chem. Comm.*, 263 (1976).

Hexadecyltributylphosphonium bromide, 5, 322–323; 6, 271–272.

N-Alkylphthalimides. The phosphonium salt serves as catalyst in a solid–liquid two-phase condensation of alkyl halides or methanesulfonates with potassium phthalimide to form N-alkylphthalimides (equation I).[1] This reaction

(I)   R—X +

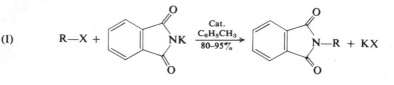

is one step in the Gabriel synthesis of primary amines.[2] The reaction is carried out by adding a solution of RX in toluene to the solid potassium salt. The alkylation is accompanied by almost complete inversion.

[1] D. Landini and F. Rolla, *Synthesis*, 389 (1976).
[2] M. S. Gibson and R. W. Bradshaw, *Angew. Chem., Int. Ed.*, 7, 919 (1968).

Hexafluoroantimonic acid, HF–SbF₅, 5, 309–310; 6, 272–273.

Cyclization of 1,2-diphenylethanes. French chemists[1] have reported a useful synthesis of tricyclic ketones bearing an angular methyl group by reaction of some 1,2-diphenylethanes with this "super acid." The primary products isomerize on long contact with the acid, as shown in the second example.
    Examples:

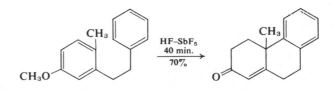

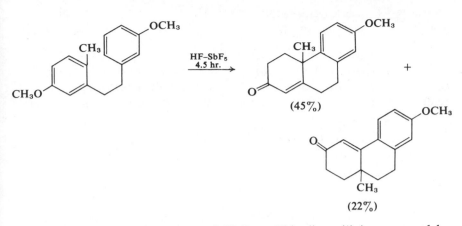

(45%)

+

(22%)

Isomerization of Δ⁴-androstene-3,17-dione. This dione (1) is rearranged by HF–SbF₅ mainly to the dione (2).[2] Generally estrane derivatives are rearranged by super acids to anthrasteroids.[3]

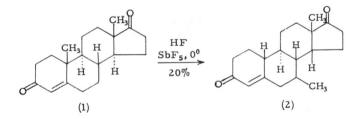

(1)   (2)

[1] J.-P. Gesson and J.-C. Jacquesy, *J.C.S. Chem. Comm.*, 652 (1976).
[2] J. C. Jacquesy, R. Jacquesy, and C. Narbonne, *Bull. soc.*, 1240 (1976).
[3] J. C. Jacquesy, R. Jacquesy, and G. Joly, *Tetrahedron Letters*, 4433 (1974); *idem*, *Tetrahedron*, **31**, 2237 (1975).

Hexamethyldisilazane, 1, 427; 2, 207–208; 5, 323; 6, 273.

Conversion of aldehydes into sulfides. The two-step procedure shown in equation (I) has been used to carry out this transformation.[1]

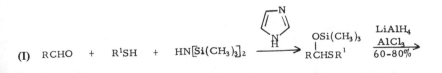

[1] R. S. Glass, *Syn. Comm.*, **6**, 47 (1976).

Hexamethylphosphoric triamide (HMPT), 1, 430–431; **2,** 208–210; **3,** 149–153; **4,** 244–247; **5,** 323–325; **6,** 273–274.

Solvent Effects

Photochemistry of esters. The main reaction observed on photolysis of an ester in various solvents is a Norrish II fragmentation (equation I).[1] When HMPT is used as solvent the main reaction is reduction (equation II).[2] The

(I)

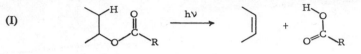

(II)

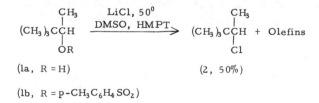

addition of water increases the yield of R′H with respect to R′OH. The HMPT is decomposed.

Chlorodeoxynucleosides. Nucleosides can be chlorinated by thionyl chloride (neat or in pyridine) but in low yields. The reaction is improved considerably by use of HMPT or trimethyl phosphate as solvent. Three of the chlorodeoxy-nucleosides prepared in this way are formulated and the yields are given.[3]

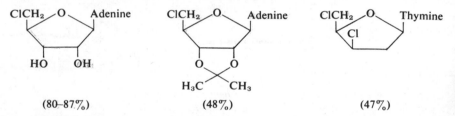

$\qquad$ (80–87%) $\qquad\qquad$ (48%) $\qquad\qquad$ (47%)

Pinacolyl chloride. The conversion of pinacolyl alcohol (1a) into the chloride (2) has presented difficulties because (2) is subject to Wagner-Meerwein re-arrangement to 2,3-dimethyl-2-chlorobutane, which has practically the same

$$(CH_3)_3CCH \quad \xrightarrow[\text{DMSO, HMPT}]{\text{LiCl, }50^0} \quad (CH_3)_3CCH \;+\; \text{Olefins}$$

(1a, R = H) (2, 50%)

(1b, R = p-CH₃C₆H₄SO₂)

boiling point as (2). Even triphenylphosphine dichloride (**1,** 1247–1294) or triphenylphosphine–carbon tetrachloride (**1,** 1247) are ineffective in this case because of rearrangement. However, (2) can be obtained in moderate yield without rearrangement by reaction of the tosylate (1b) with anhydrous lithium chloride in DMSO–HMPT. In the absence of HMPT temperatures $> 115°$ are necessary and large amounts of olefins are obtained.[4]

Substitution of nitrobenzenes. The nitro group of nitro aromatics substituted in the *ortho-* or *para*-position by election-withdrawing groups (CN, NO_2, COC_6H_5, $COOC_2H_5$, $SO_2C_6H_5$, SC_6H_5) is displaced by nucleophiles at $25°$ when the reaction is conducted in HMPT. DMF is less satisfactory as solvent.[5]

Cleavage of alkyl aryl ethers. Ethers such as *p*-nitroanisole, *p*-cyanoanisole, *p*-methoxyacetophenone are cleaved to phenols by $NaNO_2$ in HMPT, often in high yield.[6]

Reaction

Glycidonitriles. White[7] reasoned that α-halonitriles and phosphines should form ylides, which would then undergo Wittig condensation. And, indeed, the lactone aldehyde (1) reacts with dibromoheptanenitrile (2) and hexamethyl-phosphorus triamide to form the glycidonitrile (3) in 97% yield. The actual reagent is considered to be $[(CH_3)_2N]_3P{=}C(CN)(CH_2)_4CH_3$ (a). The product (3) was transformed into the prostaglandin intermediate (5) by known reactions (5, 317).

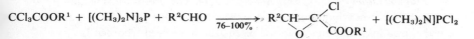

(II)

$$CCl_3COOR^1 + [(CH_3)_2N]_3P + R^2CHO \xrightarrow[76-100\%]{} R^2CH{-}C\overset{Cl}{\underset{COOR^1}{\diagdown}} + [(CH_3)_2N]PCl_2$$

A somewhat similar reaction had been used earlier by Villieras *et al.*[8] for a synthesis of α-haloglycidic esters (equation II).

[1] See, *inter alia*, W. H. Saunders, Jr., *et al.*, *Am. Soc.*, **95**, 5224, 5228 (1973).
[2] H. Deshayes, J. P. Pète, and C. Portella, *Tetrahedron Letters*, 2019 (1976).
[3] D. E. Gibbs and J. G. Verkade, *Syn. Comm.*, **6**, 563 (1976).
[4] M. E. Alonso, *J. Org.*, **41**, 1410 (1976).
[5] N. Kornblum, L. Cheng, R. C. Kerber, M. M. Kestner, B. N. Newton, H. W. Pinnick, R. G. Smith, and P. A. Wade, *ibid.*, **41**, 1560 (1976).
[6] T. Sakai, N. Yasuoka, H. Minato, and M. Kobayashi, *Chem. Letters*, 1203 (1976).
[7] D. R. White, *Tetrahedron Letters*, 1753 (1976).
[8] J. Villieras, G. Lavielle, and J.-C. Combret, *Bull. soc.*, 898 (1971).

***n*-Hexylamine,** $CH_3(CH_2)_5NH_2$. Mol. wt. 101.19, b.p. 129°.

Alkyl nitriles. Primary alkyl bromides are converted into nitriles by reaction with 33% aqueous NaCN in the presence of an amine that is soluble in both the aqueous and organic phases. The minimum chain length for efficient catalysis is about six carbon atoms. Reaction of secondary bromides is sluggish.[1]

[1] W. P. Reeves and M. R. White, *Syn. Comm.*, **6**, 193 (1976).

Hydrazine, 1, 434–445; **2,** 211, **3,** 153; **4,** 248; **5,** 327–329; **6,** 280–281.

Regioselective deacylation. Perbenzoyl derivatives of ribonucleosides are selectively debenzoylated at the 2′-position by hydrazine hydrate. This deacylation is also applicable to peracetates.[1]

Example:

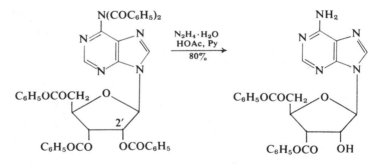

Reduction of nitroarenes to aminoarenes (**1,** 440–441). Du Pont chemists[2] have reported an explosion during attempted reduction of 2-chloro-5-methyl-nitrobenzene to the amine by hydrazine and Pd–C. The main product was shown to be 2-chloro-5-methylphenylhydroxylamine; arylhydroxylamines are known to decompose violently when heated over 90–100°.

Fragmentation of α,β-*epoxyketones* (**5,** 328). The use of this reaction for the synthesis of 5-hexynal from 2,3-epoxycyclohexanone has been published.[3] Further examples and a mechanistic discussion have been recorded.[4]

[1] Y. Ishido, N. Nakazaki, and N. Sakairi, *J.C.S. Chem. Comm.*, 832 (1976).
[2] C. S. Rondestvedt, Jr., and T. A. Johnson, *Chem. Eng. News*, July 4, 38 (1977).
[3] D. Felix, C. Wintner, and A. Eschenmoser, *Org. Syn.*, **55**, 52 (1976).
[4] D. Felix, R. K. Müller, U. Horn, R. Joos, J. Schrieber, and A. Eschenmoser, *Helv.*,
55, 1276 (1972).

Hydriodic acid–Red phosphorus, 1, 449.

Deoxygenation of benzylic ketones. Eisenbraun and co-workers[1] report that this combination is the most satisfactory method for reduction of some benzylic ketones.

Examples:

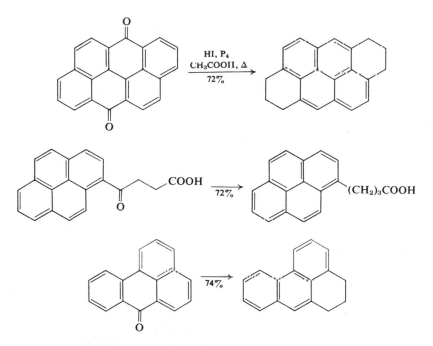

[1] L. L. Ansell, T. Rangarajan, W. M. Burgess, and E. J. Eisenbraun, *Org. Prep. Proc. Int.*,
8, 133 (1976).

Hydrogen bromide–Acetic acid, 5, 335.

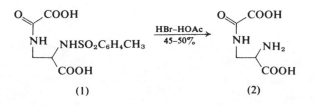

(1)   (2)

Cleavage of a p-*toluenesulfonyl group.* The last step in a recent synthesis of (2), a neurotoxin from *Lathyrys sativus*, involved cleavage of the blocking tosyl group in (1). Reduction with sodium in liquid ammonia and with sodium naphthalenide proved useless. The cleavage was effected in 45–50% yield with hydrogen bromide in acetic acid at 70°.[1]

[1] B. E. Haskell and S. B. Bowlus, *J. Org.*, **41**, 159 (1976).

Hydrogen chloride, 2, 215; **4,** 252; **5,** 335–336; **6,** 285.

Spiro ring systems. Ruppert and White have extended[1] the study of synthesis of spiro carbocyclic systems by cleavage of cyclopropyl ketones (**5,** 41–42). In

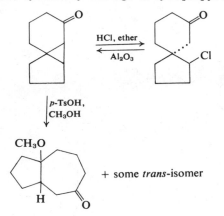

(I)

addition to cleavage by Li–NH₃, two additional methods have been found (equation I).

The cleavage with hydrogen chloride was used in a synthesis of (−)-acorenone

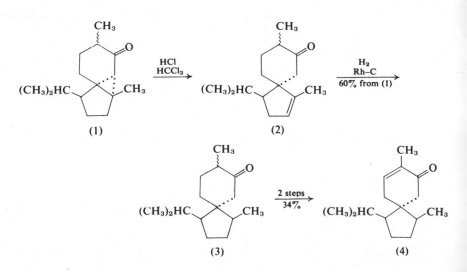

(1)   (2)

(3)   (4)

B (4), which contains three asymmetric centers, from (R)-(+)-limonene.[2] The key intermediate is the cyclopropyl ketone (1). In this case Li–NH₃ cleavage occurred with an undesired inversion of the methyl group at the cyclopropyl ring carbon atom. Cleavage with HCl resulted in the desired cleavage to give (2). The product was converted into (4) by standard reactions.

[1] J. F. Ruppert and J. D. White, *J.C.S. Chem. Comm.*, 976 (1976).
[2] J. F. Ruppert, M. A. Avery, and J. D. White, *ibid.*, 978 (1976).

Hydrogen fluoride, 1, 455–456; **2,** 215–216; **4,** 252; **5,** 336–337; **6,** 285.
Treatment of HF burns has been described.[1]

[1] A. J. Finkel, *Adv. Flourine Chem.*, **7,** 199 (1973).

Hydrogen iodide, 1, 449–450; **2,** 213–214.
Stereocontrolled cleavage of epoxides. The *endo*-epoxide (1) is opened by HI specifically to give, after acetylation, the iodo acetate (2) in nearly quantitative yield. Double dehydrohalogenation leads to the α-methylene-γ-lactone (4).

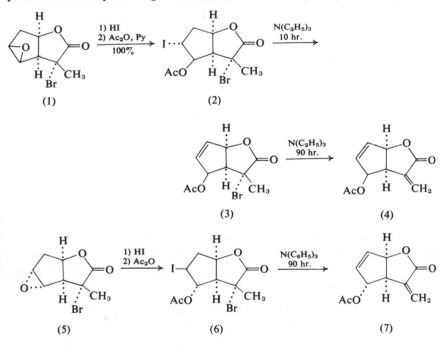

The isomeric *exo*-epoxide (5) is also opened specifically to give (6).[1]

Cleavage of ethers and ketals. Hydrogen iodide in acetonitrile, generated from NaI and diiodomethyl methyl ether,[2] is an effective reagent for cleavage of primary and secondary alkyl methyl ethers (20°, 30 minutes). Hydrogen iodide generated from NaI and TsOH is somewhat less reactive. In other solvents, HI is less effective or even inactive (CH_2Cl_2, DMF). Ketals are also cleaved readily by HI under these conditions.[3]

[1] S. M. Ali and S. M. Roberts, *J.C.S. Chem. Comm.*, 584 (1976).
[2] A. Rieche, H. Gross, and E. Höft, *Ber.*, **93**, 88 (1960).
[3] C. A. Smith and J. B. Grutzner, *J. Org.*, **41**, 367 (1976).

Hydrogen peroxide, 1, 457–471; **2,** 216–217; **3,** 154–155; **4,** 253–255; **5,** 337–339; **6,** 286.

Peroxycarboxylic acids. Peracids are generally prepared by reaction of carboxylic acids with H_2O_2 in a strongly acidic medium (H_2SO_4, CH_3SO_2OH, **1,** 458). Lefort *et al.* now report that they can be made in a neutral medium by reaction of acyl chlorides with 85% H_2O_2 in THF, usually with added pyridine, which can markedly improve the yields in some instances. Diacyl peroxides are sometimes obtained as by-products. The peracids are obtained in 55–70% yield.[1]

β-Hydroxy hydroperoxides. The reaction of some epoxides (1a–1c) with 98% H_2O_2 catalyzed by $HClO_4$ (ether, 0–5°) yields β-hydroxy hydroperoxides (2a–2c). In these cases, only one hydroxy hydroperoxide is formed, that formed

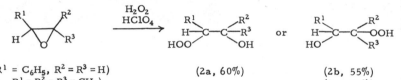

(1a, $R^1 = C_6H_5$, $R^2 = R^3 = H$)
(1b, $R^1 = R^2 = R^3 = CH_3$)
(1c, $R^1 = C_6H_5(CH_2)_2$, $R^2 = R^3 = CH_3$)

(2a, 60%)

(2b, 55%)
(2c, 95%)

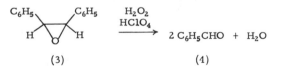

(3)　　　　　　　(4)

by placing the OOH group on the most stable carbonium ion intermediate. Stilbene oxide (3) under these conditions is converted into benzaldehyde (4).[2]

β-Hydroxy hydroperoxides are believed to be involved in biological oxygenations.

[1] J. Y. Nedelec, J. Sorba, and D. Lefort, *Synthesis*, 821 (1976).
[2] V. Subramanyam, C. L. Brizuela, and A. H. Soloway, *J.C.S. Chem. Comm.*, 508 (1976).

Hydrogen peroxide–Hydrochloric acid.

Oxidative chlorination. p-Chloranil is made commercially by reaction of phenol with chlorine. The reaction can also be carried out with 30% H_2O_2 and conc. HCl. Yields are improved by addition of magnesium chloride (equation I). The reaction of α-naphthol with H_2O_2–HCl is also formulated (equation II).[1]

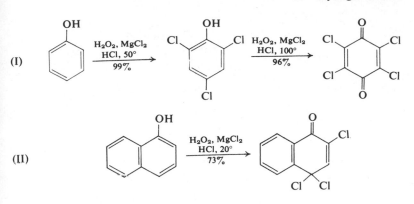

[1] H. Lübbecke and P. Boldt, *Angew. Chem., Int. Ed.*, **15**, 608 (1976).

Hydrogen peroxide–Silver trifluoroacetate.

Alkyl hydroperoxides. These substances can be prepared by gradual addition of the silver salt to an ice-cold solution of an alkyl bromide or iodide and hydrogen peroxide (excess) in ether (equation I). Dialkyl peroxides can be prepared similarly by reaction of alkyl halides and hydroperoxides in pentane (equation II). These reactions are apparently safe.[1]

(I) $RI(Br) + H_2O_2 + CF_3COOAg \xrightarrow[30-60\%]{} ROOH + AgI(Br) + CF_3COOH$

(II) $RI(Br) + R'OOH + CF_3COOAg \xrightarrow[30-90\%]{} ROOR' + AgI(Br) + CF_3COOH$

[1] P. G. Cookson, A. G. Davies, and B. P. Roberts, *J.C.S. Chem. Comm.*, 1022 (1976).

Hydrogen peroxide–Sodium peroxide.

Hydroperoxides.[1] Hydroperoxides (ROOH) can be obtained by oxidation of N-alkyl-N'-tosylhydrazines (RNHNHTs)[2] with $H_2O_2-Na_2O_2$ in THF; yields are in the range of 87–95%.

[1] L. Caglioti, F. Gasparrini, and G. Palmieri, *Tetrahedron Letters*, 3987 (1976).
[2] A. Attanasi, L. Caglioti, F. Gasparrini, and D. Misiti, *Tetrahedron*, **31**, 341 (1975).

Hydrogen selenide, 6, 288.

Conversion of $\diagdown C{=}S$ *into* $\diagdown C{=}Se$. The thione (1) has been converted into

(3) as shown. The selenium compound (3), unlike (1), undergoes phosphate coupling to a mixture of (4) and (5), probably because the $C{=}Se$ bond is weaker than the $C{=}S$ bond.[1] Charge-transfer salts of substances of type (4) and (5) with tetracyanoquinonedimethane are strong electrical conductors.

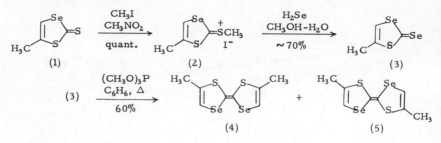

(1)   (2)   (3)

(3) $\xrightarrow[\text{60\%}]{\substack{(CH_3O)_3P \\ C_6H_6, \ \Delta}}$ (4)   +   (5)

[1] E. M. Engler and V. V. Patel, *Tetrahedron Letters*, 423 (1976).

Hydroxylamine hydrochloride, 1, 478–481.

Nitriles. Aldehydes are converted in one step into nitriles when refluxed in DMF with hydroxylamine hydrochloride. Yields are 65–97% (six examples).[1]

$$RCHO + H_2NOH \cdot HCl \xrightarrow[\Delta]{DMF} \left[RC\underset{H}{\overset{NOH}{\diagdown}} \right] \xrightarrow{65-97\%} RC\equiv N$$

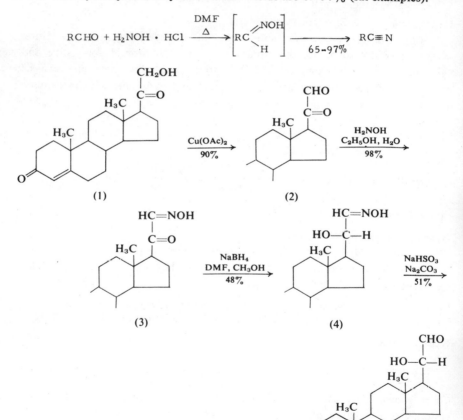

(1)   (2)

(3)   (4)

(5)

20β-OH-21-al steroids. Steroid 20-keto-21-aldehydes (2) can be converted into 20β-hydroxy-21-aldehydes (5) in fair yields, as shown below, by protection of the aldehyde group as the oxime (3). The 20-keto group is reduced more readily by sodium borohydride than the 3-keto group. A 17α-hydroxyl group increases the rate of reduction of the 20-keto group even more.[2]

[1] J. Liebscher and H. Hartmann, *Z. Chem.*, **15**, 302 (1975).
[2] S. Oh and C. Monder, *J. Org.*, **41**, 2477 (1976).

N-Hydroxyphthalimide, 1, 485–486.

Primary alkoxy amines, $RONH_2$. Direct alkylation of hydroxylamine results in attack on nitrogen to give RNHOH. Substances of the type $RONH_2$ can be prepared in two steps as shown.[1]

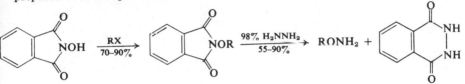

Somewhat higher yields can be obtained by use of N-hydroxynorbornene-5-dicarboximide-2,3 (1).

(1)

[1] A. Rougny and M. Daudon, *Bull. soc.*, 833 (1976).

(1S,2S,5S)-2-Hydroxypinan-3-one, 31–32°, αD −38.9°.

(1). Mol. wt. 168.23, m.p.

Preparation[1]:

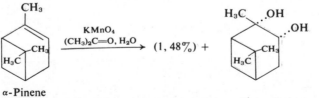

α-Pinene   (1, 48%) +

D-α-Amino acids. Japanese chemists[2] have developed an efficient asymmetric synthesis of D-α-amino acids (4) by alkylation of the chiral Schiff base (2) formed from (1) and the *t*-butyl ester of glycine (equation I). The overall yields of (4) are 50–80%. The optical yields are 65–85%.

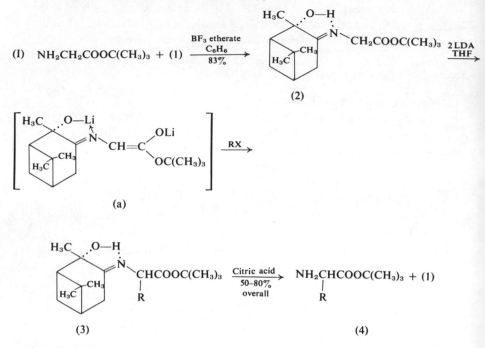

(I) $NH_2CH_2COOC(CH_3)_3$ + (1) $\xrightarrow[\;83\%\;]{\substack{BF_3 \text{ etherate}\\ C_6H_6}}$

(2) $\xrightarrow[\;THF\;]{2\,LDA}$

(a)

(3) $\xrightarrow[\substack{50-80\%\\ overall}]{Citric\ acid}$ $NH_2CHCOOC(CH_3)_3$ + (1)   (4)

[1] R. G. Carlson and J. K. Pierce, *J. Org.*, **36**, 2319 (1971).
[2] S. Yamada, T. Oguri, and T. Shioiri, *J.C.S. Chem. Comm.*, 136 (1976).

I

Alkene synthesis (**4,** 37, 110). Evans et al.[1] have developed variations of the Zweifel[2] boron-mediated olefination reaction that lead to improved results. Boronic ester ate complexes (a) and (b) were used for the rearrangement, which

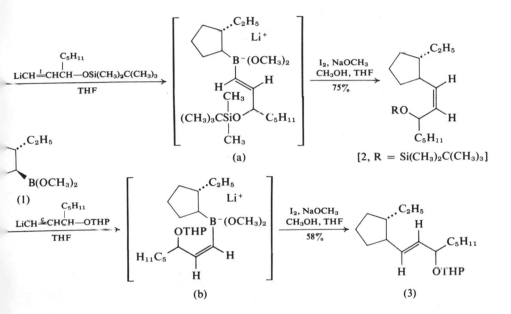

was carried out with I_2 (5 equiv.) and sodium methoxide (3 equiv.) in CH_3OH–THF (2:1).

Epoxidation. Italian chemists[3] have found that the epoxidation method of Cornforth and Green (**3,** 159) when applied to Δ^2-cholestene (1) leads to the *trans, diaxial* diol (2). They reasoned that the epoxide is formed but hydrolyzes in the acidic medium. They then carried out the reaction in the presence of silver oxide to remove hydrogen iodide and obtained the oxides (3) and (4) in high yield. The reaction was then shown to be applicable to other alkenes (cyclohexene oxide, 80%).

Optically active iodides. Trialkylboranes react with iodine in the presence

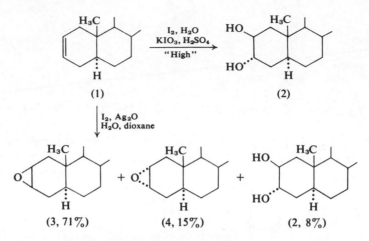

(1)   (2)

(3, 71%)   (4, 15%)   (2, 8%)

of base to form iodides. For example, tri-*exo*-norbornylborane (1), obtained by hydroboration of norbornene, is converted into *endo*-2-iodonorbornane (2) in 70% yield (based on conversion of two of the alkyl groups to the iodide). This reaction is unexpected since usually the carbon–boron bond is broken with retention of configuration.[4,5]

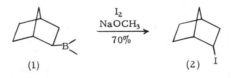

(1)   (2)

This new reaction has been used to prepare optically active iodides such as (R)-2-iodobutane (equation I) with inversion of the C—B bond. Note that reaction of this borane with alkaline hydrogen peroxide gives (S)-2-butanol.

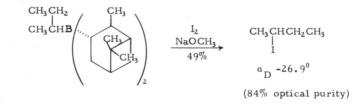

(I)

$$\begin{array}{c} CH_3CHCH_2CH_3 \\ | \\ I \end{array}$$

$^\alpha_D$ -26.9°

(84% optical purity)

Conjugated diynes (*cf.* **5**, 346). A convenient route to these substances, both symmetrical and unsymmetrical, has been developed by Sinclair and Brown.[6] Treatment of 1-alkynyldisiamylboranes with lithium alkynes forms an ate complex (1), which on reaction with iodine at −78° in THF gives a diyne (2). It is also possible to carry out this synthesis without isolation of the 1-alkynyl-

$$Sia_2BC{\equiv}CR \; + \; LiC{\equiv}CR^I \quad \xrightarrow{\text{THF, } -78^0} \quad \left[Li^+ \left[Sia_2\bar{B}(C{\equiv}CR)(C{\equiv}CR^I) \right] \right]$$

$$(1)$$

$$\xrightarrow[80-95\%]{\underset{THF, -78^0}{I_2}} \quad RC{\equiv}C-C{\equiv}CR^I$$

$$(2)$$

disiamylborane by reaction of borane–dimethyl sulfide with alkynes, but yields in this one-pot procedure are significantly lower (60–70%).

[1] D. A. Evans, R. C. Thomas, and J. A. Walker, *Tetrahedron Letters*, 1427 (1976).
[2] G. Zweifel, R. P. Fisher, J. T. Snow, and C. C. Whitney, *Am. Soc.*, **93**, 6309 (1971); *ibid.*, **94**, 6560 (1972).
[3] M. Parrilli, G. Barone, M. Adinolfi, and L. Mangoni, *Tetrahedron Letters*, 207 (1976).
[4] H. C. Brown, N. R. DeLue, G. W. Kabalka, and H. C. Hedgecock, Jr., *Am. Soc.*, **98**, 1290 (1976).
[5] N. R. DeLue and H. C. Brown, *Synthesis*, 114 (1976).
[6] J. A. Sinclair and H. C. Brown, *J. Org.*, **41**, 1078 (1976).

Iodine tris(trifluoroacetate), 4, 263–264.

Oxidation of alkanes and ethers. Tertiary CH groups of alkanes are readily oxidized by $I(OCOCF_3)_3$; secondary CH_2 groups require days; neopentanes are not oxidized.[1]

Examples:

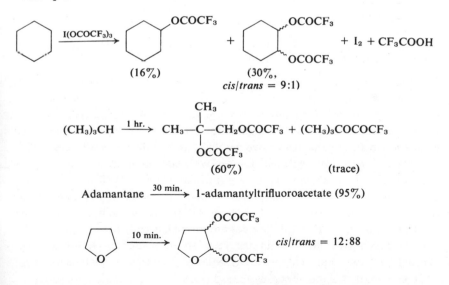

$$Adamantane \xrightarrow{\text{30 min.}} \text{1-adamantyltrifluoroacetate (95\%)}$$

[1] J. Buddrus and H. Plettenberg, *Angew. Chem., Int. Ed.*, **15**, 436 (1976).

N-Iodosuccinimide–Carboxylic acids.

trans-1,2-Iodocarboxylates. The reaction of a double bond with NIS and a carboxylic acid in CHCl$_3$ at 60° (3 hours) results in formation of a *trans*-1,2-iodocarboxylate (equation I).[1]

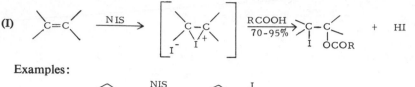

Examples:

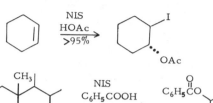

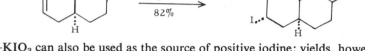

I$_2$–KIO$_3$ can also be used as the source of positive iodine; yields, however, are lower.

French chemists[2] have used NBS and AgOAc in HOAc to obtain *trans*-1,2-bromoacetates.

Caution: Some *vic*-iodo alkoxy compounds, particularly those in the terpene field, have been found to explode spontaneously on standing.[3]

[1] M. Adinofi, M. Parrilli, G. Barone, G. Laonigro, and L. Mangoni, *Tetrahedron Letters*, 3661 (1976).

[2] D. Jasserand, J. P. Girard, J. C. Rossi, and R. Granger, *ibid.*, 1581 (1976).

[3] D. R. Dimmel, *Chem. Eng. News*, July 4, 38 (1977).

Ion-exchange resins, 1, 511–519; **2,** 227–228; **4,** 266–268; **5,** 355–356; **6,** 302–304.

Quaternary ammonium hydroxides (4, 267–268). The use of an anion-exchange resin for conversion of quaternary ammonium halides into the hydroxide has been published.[1] The method is suitable for even very sensitive compounds and is superior to silver oxide (expensive) or thallous ethoxide (expensive and toxic).

Alkyl fluorides by exchange from alkyl halides or methanesulfonates. The resin used for the reaction is the F$^-$ form of Amberlyst-A26 (Rohm and Haas), a macroreticular anion-exchange resin containing ammonium groups. When this material and primary alkyl halides or sulfonates are refluxed in a solvent (pentane, hexane, ether), alkyl fluorides are formed, usually in satisfactory yields. Alkenes accompany fluorides in the reaction of secondary substrates. This reaction has been conducted previously under phase-transfer catalysis (5, 322).[2]

[1] C. Kaiser and J. Weinstock, *Org. Syn.*, **55**, 3 (1976).

[2] G. Cainelli, F. Manescalchi, and M. Panunzio, *Synthesis*, 472 (1976).

Iron pentacarbonyl (Pentacarbonyliron), 1, 519; **3,** 167; **5,** 357–358; **6,** 304–305.

N,N'-Disubstituted ureas. These compounds can be prepared in about 50–95% yield by reaction of aryl or alkyl nitro compounds with bromomagnesium alkyl or aryl amides in the presence of iron pentacarbonyl (equation I).[1]

(I)
$$R^1NHMgBr \ + \ R^2NO_2 \ \xrightarrow[50-95\%]{\begin{array}{c}1)\ Fe(CO)_5\\2)\ H_2SO_4\end{array}} \ R^1NHCONHR^2$$

[1] M. Yamashita, K. Mizushima, Y. Watanabe, T. Mitsudo, and Y. Takegami, *J.C.S. Chem. Comm.,* 670 (1976).

Isopropylidenetriphenylphosphorane, 5, 361.

Chrysanthemic esters. Belgian chemists[1] noted that chrysanthemic acid could be formed from two isopropylidene units and, indeed, they obtained *trans*-methyl chrysanthemate (2) in 60% yield by reaction of methyl *trans*-4-oxobutenoate (1)[2] with isopropylidenetriphenylphosphorane (−78 → 20°). Cyclopentylidenetriphenylphosphorane reacts with (1) in the same way, but other dialkylmethylenetriphenylphosphoranes react to form dienoic esters.

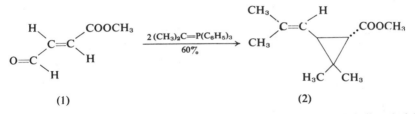

(1)   (2)

This synthesis was modified as shown to prepare methyl 2,2-dimethyl-3-formylcyclopropanecarboxylate (3).

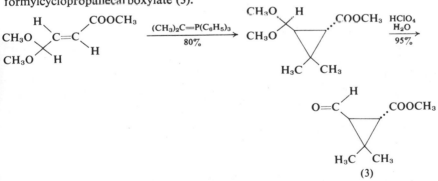

(3)

Krief *et al.*[3] have also reported a stereospecific synthesis of methyl *cis*-chrysanthemate (8) from (4), available in three steps from propargyl alcohol. Isopropylidenetriphenylphosphorane does not react with (4), but *cis*-cyclopropanation can be accomplished with isopropylidenediphenylsulfurane (2, 180–181). The

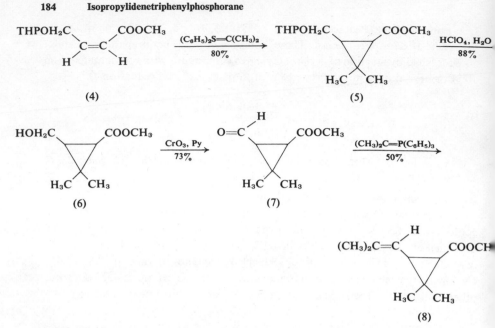

(4)   (5)

(6)   (7)

(8)

protecting group is removed and the resulting alcohol (6) is oxidized with Collins reagent. The final step is reaction with isopropylidenetriphenylphosphorane.

Earlier syntheses of chrysanthemic acids have been discussed by Thomas.[4]

[1] M. J. Devos, L. Hevesi, P. Bayet, and A. Krief, *Tetrahedron Letters*, 3911 (1976).
[2] Available by ozonolysis of methyl sorbate.
[3] M. Sevrin, L. Hevesi, and A. Krief, *Tetrahedron Letters*, 3915 (1976).
[4] A. F. Thomas, in *Total Synthesis of Natural Products*, Vol. II, J. ApSimon, Ed., Wiley, New York, 1973, pp. 49–58.

L

Lead tetraacetate (LTA), 1, 537–563; 2, 234–238; 3, 168–171; 4, 278–282; 5, 365–370; 6, 313–317.

Oxidative decarboxylation. Benzo[1,2;3,4]dicyclobutene has been prepared for the first time by the route formulated in equation (I).[1]

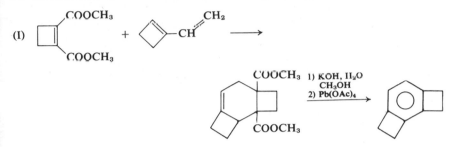

(I)

Oxidative decarboxylation of dihydroaromatic acids. Some years ago Birch[2] mentioned that 1-substituted 1,4-dihydrobenzoic acids are decarboxylated to arenes by LTA. The starting materials are readily available by a one-step Birch reduction (Li–NH$_3$) and alkylation from benzoic acids.[3] Birch and Slobbe[4] have now extended the early work and shown that this process is useful synthetically. One example is the synthesis of olivetol dimethyl ether (equation I). Equation II formulates the synthesis of a useful intermediate to dihydrojasmone.

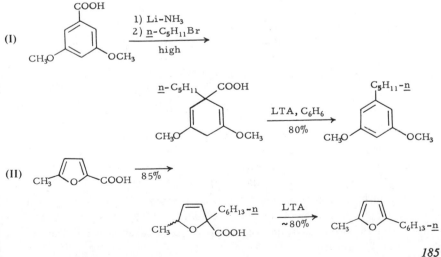

185

Catalytic amounts of cupric acetate were added in these decarboxylations; this variant results in somewhat purer products.

Oxidative decarboxylation of C₂₀-gibberellins. Oxidative decarboxylation of the gibberellin (1) gives a mixture of the γ- and δ-lactones (2) and (3). γ-Lactones (2) are natural products and biologically active; lactones of type (3) are inactive.[5]

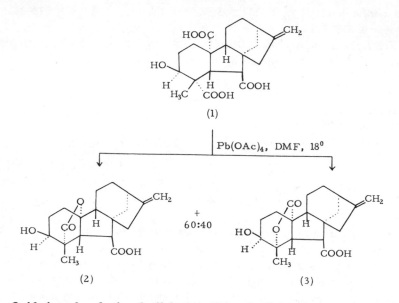

$Pb(OAc)_4$, DMF, 18°

+

60:40

(2)   (3)

Oxidation of enol trimethysilyl ethers. Trimethysilyl enol ethers of aryl methyl ketones are converted by LTA in benzene at 20° into aryl-substituted phenacyl acetates. An example is formulated in equation (I). The same reaction is observed with lead tetrabenzoate.[6]

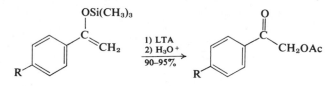

1) LTA
2) H_3O^+
90-95%

Oxidative seco-rearrangement. Trost and Hiroi[7] have devised a method for ring cleavage of a cycloalkanone, as illustrated for the dithiane (1) derived from cyclohexanone and $TsS(CH_2)_3STs$ (**4, 539**). This can be reduced to the alcohol or converted into (2) as shown. Oxidation of (2) with 2.6 equiv. of lead tetracetate proceeds in high yield to give the rearranged α-thio thioester (3). This can be cleaved to the keto ester (4) or converted into the keto *trans*-enoate (5). The paper cites the cleavage of an unsymmetrical cyclohexanone and cleavages of

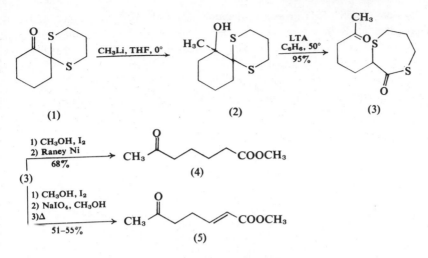

(1)   (2)   (3)

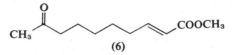

1) CH₃OH, I₂
2) Raney Ni
   68%

(3)

CH₃ ...COOCH₃
(4)

1) CH₃OH, I₂
2) NaIO₄, CH₃OH
3) Δ
   51–55%

CH₃ ...COOCH₃
(5)

cyclononanone and cyclodecanone. In fact this cleavage of cyclononanone was used for a synthesis of the methyl ester of queen's substance (6).

CH₃ ...COOCH₃
(6)

α-Acetoxy aldehydes. These compounds can be prepared in moderate yield by oxidation of sodium glycidates with 2 equiv. of LTA and pyridine, with benzene as solvent (equation I).[8]

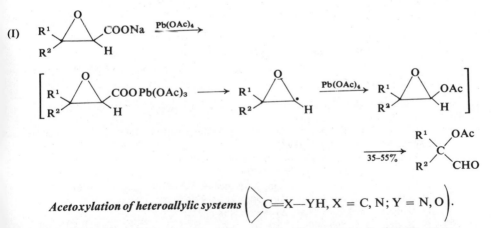

Acetoxylation of heteroallylic systems $\left(C{=}X{-}YH, X = C, N; Y = N, O \right)$.

Acetoxylation of enamines, enols, hydrazones, and oximes by lead tetraacetate has been reviewed.[9]

Oxidation of phenols. Oxidation of mesitol (1) under standard conditions of Wessely (**1**, 550–551) gives the dienones (**2**) and (**3**).[10] Yates has used α,β-unsaturated acids in this oxidation for a general synthesis of bicyclo[2.2.2]-octenones. Thus oxidation of (**1**) with LTA in acrylic acid gives the dienones

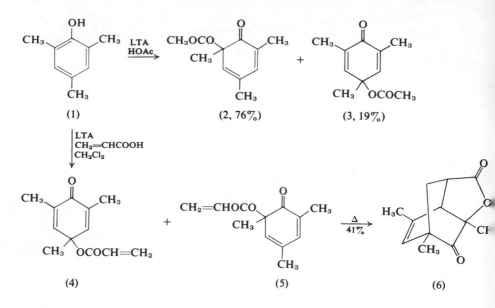

(4) and (5); when heated the latter product undergoes an intramolecular Diels-Alder reaction to form (6).[11]

The reaction is general for *ortho*-substituted phenols and for α,β-unsaturated acids, but yields are somewhat low (*ca.* 20%) in the case of cresol. One useful result of this route is that the bicyclo[2.2.2]octenones are regioisomers of the adducts formed by intermolecular Diels-Alder reactions of the 2,4-dienones obtained by usual Wessely oxidation.[12]

[1] R. P. Thummel, *Am. Soc.*, **98**, 628 (1976).
[2] A. J. Birch, *J. Chem. Soc.*, 1551 (1950).
[3] H. van Bekkum, C. B. van den Bosch, G. van Minnen-Pathuis, J. C. de Mos, and A. M. van Wijk, *Rec. trav.*, **90**, 137 (1971).
[4] A. J. Birch and J. Slobbe, *Tetrahedron Letters*, 2079 (1976).
[5] S. Hunig and G. Wehner, *Synthesis*, 180 (1975).
[6] G. M. Rubottom, J. M. Gruber, and K. Kincaid, *Syn. Comm.*, **6**, 59 (1976).
[7] B. M. Trost and K. Hiroi, *Am. Soc.*, **98**, 4313 (1976).
[8] B. D. Kulkarni, and A. S. Rao, *Synthesis*, 454 (1976).
[9] R. N. Butler, *Chem. Ind.*, 499 (1976).
[10] W. A. Bubb and S. Sternhell, *Tetrahedron Letters*, 4499 (1970).
[11] D. J. Bichan and P. Yates, *Canad. J. Chem.*, **53**, 2054 (1975).
[12] P. Yates and H. Auksi, *J.C.S. Chem. Comm.*, 1016 (1976).

Lead tetrabenzoate, 1, 563.

Oxidation of trimethylsilyl enol ethers.[1] The reaction of these enol ethers with lead tetrabenzoate in methylene chloride (20°) followed by treatment with triethylammonium fluoride[2] affords α-benzoyloxy carbonyl compounds in high yield (equation I).

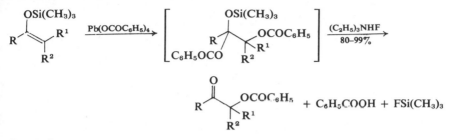

Examples:

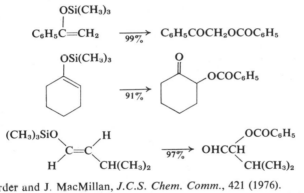

[1] J. R. Bearder and J. MacMillan, *J.C.S. Chem. Comm.*, 421 (1976).
[2] G. M. Rubottom, J. M. Gruber, and G. M. Mong, *J. Org.*, **41**, 1673 (1976).

D- and L-*t*-Leucine *t*-butyl esters, $(CH_3)_3CCH(NH_2)COOC(CH_3)_3$ (1). Mol. wt. 185.26; α_D + 54.6°, L-(1); α_D − 55.2°, D-(1).

Asymmetric synthesis of aldehydes. The conjugate addition of Grignard reagents to α,β-unsaturated aldimines, prepared from L-(1) or D-(1), results, after hydrolysis, in optically active aldehydes (40–55% yield) in high optical yields, 91–98%. The configuration of the resulting aldehyde is determined by that of the amino acid ester. (R)-Aldehydes are obtained when D-(1) is used. Yields of aldehydes are low when $(CH_3)_2CHCH(NH_2)COOC(CH_3)_3$ is used as the chiral reagent, probably because 1,2-addition becomes competitive.

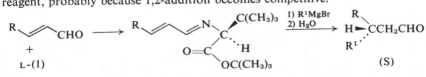

[1] S. Hashimoto, S. Yamada, and K. Koga, *Am. Soc.*, **98**, 7450 (1976).

1-Lithiocyclopropyl phenyl sulfide, 5, 372–373; **6**, 319–320.

Cyclopentane annelation. Trost and Keeley[1] have described a new method for cyclopentane annelation (equation I). Yields are high in all steps, and the

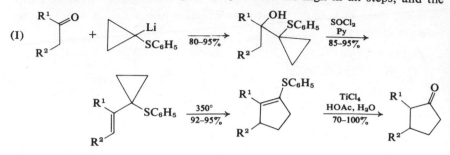

vinylcyclopropane rearrangement appears to be highly stereoselective. The paper reports five examples.

[1] B. M. Trost and D. E. Keeley, *Am. Soc.*, **98**, 248 (1976).

Lithio-N,N-dimethylthiopivalamide, stable up to −50°.   (1). Mol. wt. 151.20,

The anion is prepared by reaction of the thioamide with *sec*-butyllithium in THF–TMEDA at −78°.

Aminoalkylation. This organolithium compound (1) reacts with alkyl halides and with ketones as formulated (equations I and II). Further transformations of the products are also indicated.[1]

(I)   (1) + RX $\xrightarrow{45-82\%}$

(II)   (1) + R¹R²C=O $\xrightarrow{15-70\%}$ $\xrightarrow{H_2O_2,\ OH^-}$

LiAlH₄

[1] D. Seebach and W. Lubosch, *Angew. Chem., Int. Ed.*, **15**, 313 (1976).

2-Lithio-2-methylthio-1,3-dithiane, (1), 4, 216–217.
    Preparation:

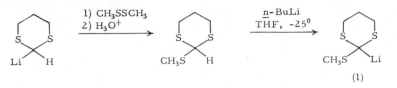

Ester synthesis. Use of this reagent permits the formal oxidation of α-hydroxyaldehydes and γ-ketoaldehydes to the corresponding esters.[1]
    Examples:

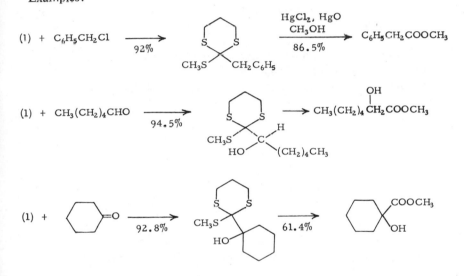

[1] W. Woessner, *Chem. Letters*, 43 (1976).

1-Lithio-3,3,6,6-tetramethoxy-1,4-cyclohexadiene (1). Mol. wt. 206.16.
    Preparation by anodic oxidation of 2-bromo-1,4-dimethoxybenzene followed by metallation:

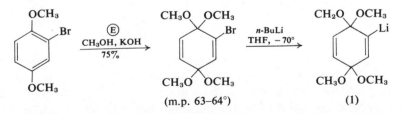

(m.p. 63–64°)   (1)

Reactions. This metallated quinone ketal is a versatile quinone carbanion synthon. For example, it reacts with cyclohexanone to give (2) in high yield; the

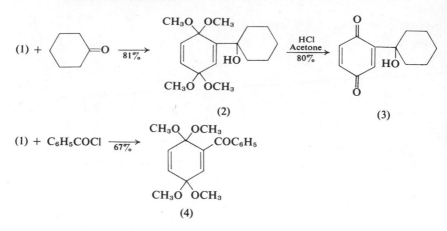

(2)   (3)

(4)

product can be hydrolyzed to the functionalized quinone (3). A similar reaction with benzoyl chloride is formulated.

Thus the polarity of *p*-benzoquinone is reversed in reactions of (1).[1]

[1] M. J. Manning, P. W. Raynolds, and J. S. Swenton, *Am. Soc.*, **98**, 5008 (1976).

1-Lithio-2-vinylcyclopropane, *cis*- and *trans*-, 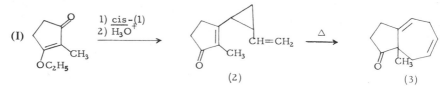 (1). Mol. wt. 74.05.

The reagents are prepared by reaction of *t*-butyllithium in ether–pentane ($-78°$) with the 1-bromo-2-vinylcyclopropanes.[1]

Cycloheptane derivatives.[2,3] Annelated cycloheptane derivatives have been prepared by reaction of (1) with 3-alkoxyenones to form 1,2-divinylcyclopropanes (2), which rearrange thermally to annelated cycloheptadiene derivatives (3) (equation I). Actually, either *trans*- or *cis*-(1) or a mixture of the two can be used since both *cis*- or *trans*-(2) rearrange on thermolysis to the same product (3), although *trans*-(2) requires somewhat more drastic conditions. The overall yield of (3) using a 7:3 mixture of *cis*- and *trans*-(1) is 72%.

(I)

(2)   (3)

The method has been extended to the preparation of 4-cyclohepteneones, such as karahanaenone (8) (equation II). In this case the reaction involves reaction of isobutyraldehyde with a mixture of *cis*- and *trans*-1-lithio-2-methyl-2-vinylcyclopropane (4) followed by oxidation with pyridinium chlorochromate to give a mixture of ketones (5). Silylation and thermal rearrangement as before lead to (7); the ketone (8) is obtained on desilylation.

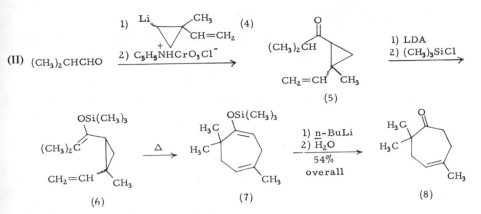

(II) $(CH_3)_2CHCHO$ 1) ... 2) $C_5H_5NHCrO_3Cl^-$

(4) (5) 1) LDA 2) $(CH_3)_3SiCl$

(6) Δ (7)   1) $\underline{n}$-BuLi 2) H_2O 54% overall   (8)

[1] Prepared by the reaction of butadiene with dibromocarbene to form 1,1-dibromo-2-vinylcyclopropane (60% yield) followed by reduction with tri-n-butyltin hydride (75–80% yield). L. Skattebøl, *J. Org.*, **29**, 2951 (1964); D. Seyferth, H. Yamazaki, and D. L. Alleston, *ibid.*, **28**, 703 (1963); J. Landgrebe and L. Becker, *ibid.*, **33**, 1173 (1968).

[2] P. A. Wender and M. P. Filosa, *J. Org.*, **41**, 3490 (1976).

[3] J. P. Marino and L. J. Browne, *Tetrahedron Letters*, 3245 (1976).

(1-Lithiovinyl)trimethylsilane, $(CH_3)_3Si$ \ C=CH_2 / Li (1). Mol. wt. 106.17.

This reagent is prepared from (1-bromovinyl)trimethylsilane[1] by metal–halogen exchange with t-butyllithium.

Alkene synthesis. The conversion of RCHO to (E)-RCH=CHCH$_2$Cl (5) can be conducted as formulated in equation (I).[2] The conversion of (3) to (4) is

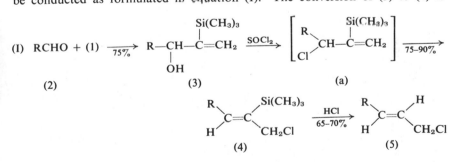

(I) RCHO + (1) $\xrightarrow{75\%}$ R—CH—C=CH$_2$ $\xrightarrow{SOCl_2}$ [...] $\xrightarrow{75-90\%}$

(2)   (3)   (a)

... $\xrightarrow[65-70\%]{HCl}$ ...

(4)   (5)

highly stereoselective, only traces of the (E)-isomer being formed. The final step involves replacement of the trimethylsilyl group by a proton with retention of configuration.

The reaction formulated in (I) has been modified to provide a route to both (E)- and (Z)-disubstituted alkenes.[3] Thus reaction of (4) with lithium dialkyl cuprates gives the vinylsilane (6), which on desilylation gives the (E)-alkene

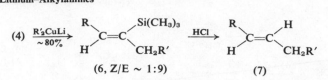

(4) $\xrightarrow[\sim 80\%]{R'_2CuLi}$ (6, Z/E ~ 1:9) $\xrightarrow{HCl}$ (7)

predominantly. The isomeric vinylsilanes are obtained by acetylation of the allylic alcohol (3) and reaction of the acetate (8) with lithium dialkyl cuprates.

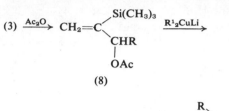

(3) $\xrightarrow{Ac_2O}$ $CH_2=C\begin{smallmatrix}Si(CH_3)_3\\\\CHR\\|\\OAc\end{smallmatrix}$ $\xrightarrow{R^1{}_2CuLi}$

(8)

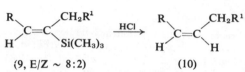

(9, E/Z ~ 8:2) $\xrightarrow{HCl}$ (10)

The stereochemistry of this reaction is sensitive to the copper reagent, the size of R^1, and the temperature, but in any case the (E)-isomer predominates; on desilylation the (Z)-alkene is formed.

The new reaction was used for a stereoselective synthesis of the sex pheromone of the gypsy moth (disparlure) (equation II).

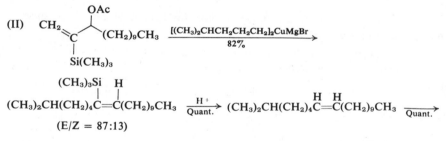

(II) $CH_2=C\begin{smallmatrix}OAc\\|\\ \\|\\Si(CH_3)_3\end{smallmatrix}(CH_2)_9CH_3$ $\xrightarrow[82\%]{[(CH_3)_2CHCH_2CH_2CH_2]_2CuMgBr}$

$(CH_3)_2CH(CH_2)_4\overset{(CH_3)_3Si\ \ \ H}{\underset{}{C}}=\overset{}{C}(CH_2)_9CH_3$ $\xrightarrow[Quant.]{H^+}$ $(CH_3)_2CH(CH_2)_4\overset{H\ \ H}{C}=C(CH_2)_9CH_3$ $\xrightarrow{Quant.}$

(E/Z = 87:13)

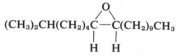

$(CH_3)_2CH(CH_2)_4\underset{H\ \ H}{\overset{O}{C-C}}(CH_2)_9CH_3$

[1] A. G. Brook, J. M. Duff, and D. G. Anderson, *Canad. J. Chem.*, **48**, 561 (1970); A. G. Brook and J. M. Duff, *ibid.*, **51**, 2024 (1973).

[2] T. H. Chan, W. Mychajlowskij, B. S. Ong, and D. N. Harpp, *J. Organometal. Chem.*, **107**, C1 (1976).

[3] W. Mychajlowskij and T. H. Chan, *Tetrahedron Letters*, 4439 (1976).

Lithium–Alkylamines, 1, 574–581; **2**, 245; **3**, 175; **4**, 287–288; **5**, 378–379; **6**, 322.

Reduction of selenides. The carbon–selenium bond of methyl or phenyl alkyl selenides (1) is cleaved easily ($-10°$) by lithium in ethylamine or by Raney nickel; the former method usually gives higher yields of alkanes (2). Both these methods have been used with sulfur compounds. Selenoacetals are also reduced to alkanes by either method.[1]

$$R^2-\underset{\underset{R^3}{|}}{\overset{\overset{R^1}{|}}{C}}-SeR \quad \xrightarrow[\underset{75-90\%}{}]{\overset{Li}{C_2H_5NH_2}} \quad R^2-\underset{\underset{R^3}{|}}{\overset{\overset{R^1}{|}}{C}}-H$$

$$(1,\ R = CH_3,\ C_6H_5) \qquad\qquad (2)$$

Thus a carbonyl group can be converted into a CH_2 group by conversion into a selenoacetal followed by reduction or into CHR by alkylation of the α-seleno-carbanion followed by reduction.

[1] M. Sevrin, D. Van Ende, and A. Krief, *Tetrahedron Letters*, 2643 (1976).

Lithium–Ammonia, 1, 601–603; 2, 205; 3, 179–182; 4, 288–290; 5, 379–381; 6 322–323.

Unusual reduction of cyclopropyl phenyl ketone. Cyclopropyl ketones are reduced by Li–NH_3 with cleavage of the cyclopropane ring; for example, cyclopropyl methyl ketone is reduced to pentanone-2 and pentanol-2.[1] Unexpectedly cyclopropyl phenyl ketone is reduced without cleavage (equation I).[2]

(I)

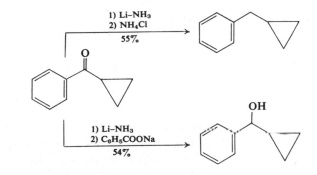

[1] R. V. Volkenburgh *et al.*, *Am. Soc.*, **71**, 3595 (1949).
[2] S. S. Hall and C.-K. Sha, *Chem. Ind.*, 216 (1976).

Lithium acetylide–Ethylenediamine, 1, 574; 5, 382.

Reaction with a ketone. Lithium acetylide alone reacts with the ketone (1)

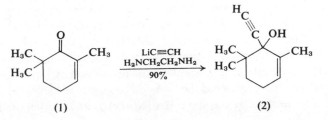

to form the desired carbinol (2) in low yield, but the yield of (2) is 90% if the complex with ethylenediamine is used.

[1] C. Schmidt, *Canad. J. Chem.*, **54**, 2310 (1976).

Lithium aluminum hydride, 1, 581–595; **2,** 242; **3,** 176–177; **4,** 291–293; **5,** 382–389; **6,** 325–326.

Reduction of 1,2-epoxides. Mihailović *et al.* have discussed factors influencing the ease of reduction of 1,2-epoxycycloalkanes and 1,2-epoxyalkanes by LAH. In the case of the former compounds, *cis*-epoxides are reduced more readily than the *trans*-isomers, the differences being more marked in C_{11}–C_{15} rings. The ease of reduction is about the same for C_5–C_7 rings and then increases for C_{11}–C_{15} rings. These differences can be explained in terms of hindrance to backside attack. In the case of 1,2-epoxyalkanes, primary-secondary epoxides are reduced readily, but disecondary epoxides are reduced slowly, particularly if the alkyl group is branched.[1]

Reduction of an α,β-unsaturated epoxide. Reduction of 3β-benzoyloxy-14α, 15α-epoxy-5α-cholest-7-ene (1) with lithium aluminum hydride (or deuteride) or

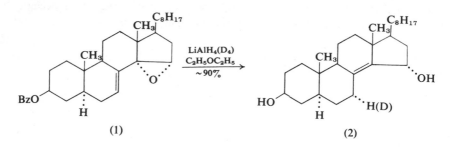

(1)   (2)

with lithium triethylborohydride (or deuteride) gives $\Delta^{8(14)}$-5α-cholestene-3β-15α-diol (2) in high yield; the hydride introduced at the 7-position has the α-configuration.[2] A similar reductive rearrangement has been observed with carbohydrate epoxides.[3]

[1] M. Lj. Mihailović, V. Andrejević, J. Milovanoić, and J. Janković, *Helv.*, **59**, 2305 (1976).
[2] E. J. Parish and G. J. Schroepfer, Jr., *Tetrahedron Letters*, 3775 (1976).
[3] B. Fraser-Reid, S. Y.-K. Tam, and B. Radatus, *Canad. J. Chem.*, **53**, 2005 (1975).

Lithium aluminum hydride–Cuprous iodide, 6, 326–327.

Conjugate reduction of enones. Enones are reduced in high yield to saturated ketones by LiAlH₄–CuI (4:1). The reagent is superior to TiCl₃–LiAlH₄ in respect to yield but not in regioselectivity. LiAlH₄–FeCl₃ does not reduce enones at 0 or 20°. The effective reagent is considered to be H₂AlI, which was synthesized independently and shown to produce the same results as LiAlH₄–CuI.[1]

Ashby and Lin[2] have continued the study on conjugate reduction of enones by H₂AlI and have found that the alane HAl[N(*i*-Pr)₂]₂ and the borane HBI₂ are particularly useful for this purpose. The former reagent was prepared by addition of diisopropylamine to aluminum hydride in THF. HBI_2 was prepared by addition of iodine in THF to BH_3 in THF at 0°.

[1] E. C. Ashby, J. J. Lin, and R. Kovar, *J. Org.*, **41**, 1939 (1976).
[2] E. C. Ashby and J. J. Lin, *Tetrahedron Letters*, 3865 (1976).

Lithium bis(trimethylsilyl)amide, 4, 296; **5,** 393–394.

Directed aldol condensation. A recent synthesis of the gingerols (2), the pungent principles of ginger, involves aldol reactions with the O-trimethylsilyl ether of gingerone (1). Reaction with this base affords the anion (a), which undergoes directed aldol condensation to form gingerols. The yield of [6]-gingerol

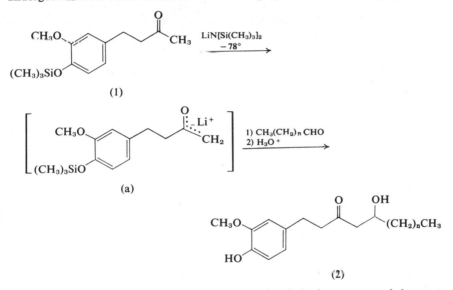

(2, n = 4) was 57%. The base does not cleave the silyl ether group and does not deprotonate the α-methylene group of (1).

The anion (a) also undergoes directed Claisen condensation with N-hexanoyl-imidazole to form [6]-gingerdione (3).[1]

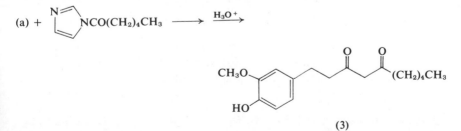

[1] P. Denniff and D. A. Whiting, *J.C.S. Chem. Comm.*, 712 (1976).

Lithium 1,1-bis(trimethylsilyl)-3-methylbutoxide, $\underset{\underset{Si(CH_3)_3}{|}}{\overset{\overset{Si(CH_3)_3}{|}}{LiOCCH_2CH(CH_3)_2}}$ (1). Mol.
wt. 238.44.

This hindered base is prepared[1] by treatment of the corresponding alcohol[2] with *n*-butyllithium in THF.

Regiospecific directed aldol reactions.[1] The kinetic enolate of a methyl ketone can be generated by this base even in the presence of an aldehyde. (Stork and co-workers[3] have previously used LDA.) Examples are shown in equation

(I) $n\text{-}C_5H_{11}CHO + CH_3COCH_2CH(CH_3)_2 \xrightarrow[84\%]{\overset{(1),\ THF}{-40°}}$

$$\overset{\overset{OH}{|}}{n\text{-}C_5H_{11}CHCH_2COCH_2CH(CH_3)_2}$$

(II) $(CH_3)_2CHCHO + CH_3COCH_2CH_2C_6H_5 \xrightarrow{71\%}$

$$\overset{\overset{OH}{|}}{(CH_3)_2CHCHCH_2COCH_2CH_2C_6H_5}$$

(III) $C_6H_5CH_2CH_2CHO + CH_3COCH_2CH(CH_3)_2 \xrightarrow{86\%}$

$$\overset{\overset{OH}{|}}{C_6H_5CH_2CH_2CHCH_2COCH_2CH(CH_3)_2}$$

(I)–(III). None of the regioisomer was detected. The base (2) is almost as satisfactory as (1).

$$\underset{\underset{Si(CH_3)_3}{|}}{\overset{\overset{Si(CH_3)_3}{|}}{LiOCCH_3}}$$

(2)

β-Hydroxy esters.[4] The enolate of ethyl acetate can be generated selectively with (1), even in the presence of aldehydes (equation I). Ethyl acetoacetate is not formed. The base is useful therefore for aldol condensations.

(I) $CH_3COOC_2H_5 + C_6H_5CH_2CH_2CHO \xrightarrow[77\%]{(1)} \underset{\underset{OH}{|}}{C_6H_5CH_2CH_2CHCH_2COOC_2H_5}$

Examples:

$$CH_3COOC_2H_5 + (CH_3)_2CHCHO \xrightarrow{74\%} (CH_3)_2CHCHCH_2COOC_2H_5$$
$$|$$
$$OH$$

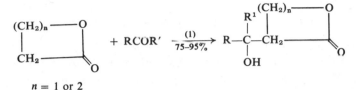

$$CH_3COOC_2H_5 +$$

In a related reaction, ethyl acetate is replaced by γ-butyrolactone or δ-valerolactone (equation II).

(II)

$$+ RCOR' \xrightarrow[75-95\%]{(1)}$$

$$n = 1 \text{ or } 2$$

[1] I. Kuwajima, T. Sato, M. Arai, and N. Minami, *Tetrahedron Letters*, 1817 (1976).
[2] J. P. Picard, R. Calas, J. Dunoguès, and N. Duffaut, *J. Organometal. Chem.*, **26**, 183 (1971).
[3] G. Stork, G. A. Kraus, and G. A. Garcia, *J. Org.*, **39**, 3459 (1974).
[4] I. Kuwajima, N. Minami, and T. Sato, *Tetrahedron Letters*, 2253 (1976).

Lithium bis(2-vinylcyclopropyl) cuprate, $\left(CH_2{=}CH{-}\triangledown \right)_2 CuLi$ (1). Mol. wt. 204.69.

This reagent is prepared from 1-lithio-2-vinylcyclopropane (this volume).

Conjugate addition. The cuprate (1) reacts with methyl propiolate at $-78°$ to give (2) and the rearrangement product (3) (equation I). A further example of conjugate addition is shown in equation (II).[1]

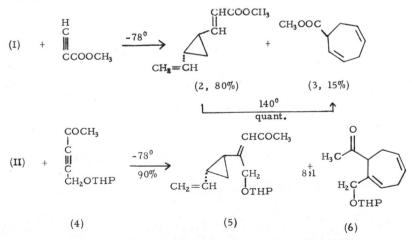

(2, 80%)   (3, 15%)

$$\underset{\text{quant.}}{\overset{140°}{\longrightarrow}}$$

(4)   (5)   8:1   (6)

[1] J. P. Marino and L. J. Browne, *Tetrahedron Letters*, 3241, 3245 (1976).

Lithium bromide–Hexamethylphosphoric triamide, 5, 395.

Decarbomethoxylation. Decarbomethoxylation with a lithium halide in HMPT, used previously only for β-keto esters (5, 323), has been used for conversion of (1) into tetraselenofulvalene (2).[1]

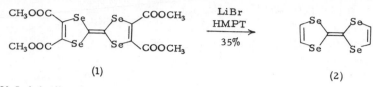

(1)   (2)

[1] M. V. Lakshmikantham and M. P. Cava, *J. Org.*, **41**, 882 (1976).

Lithium carbonate–Lithium halide, 1, 606–608; **2,** 245–246; **4,** 298; **5,** 298.

Dehydrohalogenation. White and co-workers[1] have reported a very efficient dehydrobromination of (1) to (2) by use of lithium chloride and lithium carbonate in HMPT (105°). The same conditions have been used for conversion of (3) to

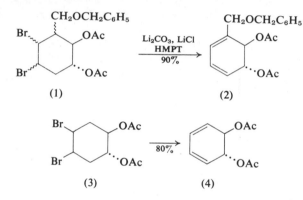

(4).[2] In this case use of numerous bases, including DBN, proved unsuccessful.

[1] M. R. Demuth, P. E. Garrett, and J. D. White, *Am. Soc.*, **98**, 634 (1976).
[2] K. L. Platt and F. Oesch, *Synthesis*, 449 (1977).

Lithium di-*t*-butylbiphenyl, Li^+ [$(CH_3)_3C$—⟨benzene⟩—⟨benzene⟩—$C(CH_3)_3$]$^{\cdot -}$ (1).

This blue radical anion is formed by addition of slivers of lithium to the biphenyl in THF at low temperatures (argon). The reagent decomposes slowly at 20°.

Reaction with alkyl chlorides. The radical anion reacts with alkyl chlorides

(1) $RCl + (1) \xrightarrow[>90\%]{-78°} [RLi] \xrightarrow{H_2O} RH + (CH_3)_3C$—⟨benzene⟩—⟨benzene⟩—$C(CH_3)_3$

to form alkyllithiums in high yield (equation I). It is superior to sodium naphthalenide for this purpose, since the latter reagent is mainly alkylated by alkyl halides.[1]

[1] P. K. Freeman and L. L. Hutchinson, *Tetrahedron Letters*, 1849 (1976).

Lithium 9,9-di-*n*-butyl-9-borabicyclo[3.3.1]nonate (1), **6**, 62–63.

Reduction of epoxides. Epoxides substituted by aryl groups are reduced by this ate complex of 9-BBN at the more substituted carbon atom, whereas aliphatic epoxides are reduced less readily and at the less hindered position.[1] The behavior of (1) generally differs from that of complex metal hydrides.

Examples:

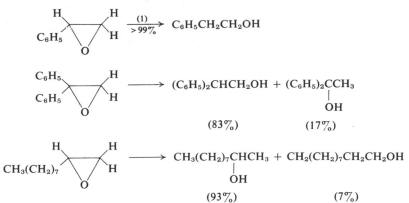

[1] Y. Yamamoto, H. Toi, A. Sonoda, and S.-I. Murahashi, *J.C.S. Chem. Comm.*, 672 (1976).

Lithium diethylamide–Hexamethylphosphoric triamide.

Activated lithium diethylamide. Lithium diethylamide generated *in situ* from lithium and the amine in HMPT–benzene behaves as a superbase. Thus it can metallate even aldimines in the α-position (equation I).[1] In a similar reaction,

hydrazones of straight-chain aldehydes are metallated at the positions α to the azomethine group (equation II). The reaction takes a different course with

(II) $RCH_2CH{=}N{-}N(CH_3)_2$ $\xrightarrow{\begin{array}{c}\text{Li, } HN(C_2H_5)_2\\ \text{HMPT, } C_6H_6\end{array}}$

$$\overset{-}{R}CHCH{=}N{-}N(CH_3)_2 \quad Li^+ \quad \xrightarrow[75\text{-}80\%]{R'Br} \quad \underset{R'}{\overset{R}{>}}CHCH{=}N{-}N(CH_3)_2$$

α-branched hydrazones to give nitriles (equation III). On the other hand, β-

(III)
$$R^1\underset{R^2}{\overset{R}{\diagdown}}C-CH=N-N(CH_3)_2 \xrightarrow[70-95\%]{\begin{array}{c}1)\ \text{Base}\\2)\ H_2O\end{array}} R^1\underset{R^2}{\overset{R}{\diagdown}}C-C\equiv N$$

branched hydrazones are converted into a mixture of the initial hydrazone and the nitrile in which the former predominates.[2]

Actually the reaction formulated in equation (I) involves formation of lithium N,N-dimethylamide, which is a very strong base in HMPT.[3] Electrophiles other than H^+ can be used. Alkylating reagents can be used to form substituted nitriles (equation IV). β-Hydroxy nitriles are obtained by addition of aldehydes

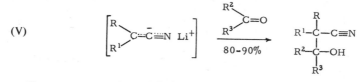

(IV)
$$\underset{R^1}{\overset{R}{\diagdown}}CHCH=NN(CH_3)_2 \xrightarrow{\text{Base}} \left[\underset{R^1}{\overset{R}{\diagdown}}CH-C\equiv N + LiN(CH_3)_2\right] \longrightarrow$$

$$\left[\underset{R^1}{\overset{R}{\diagdown}}\overset{Li^+}{\underset{}{C}}\cdots C\equiv N\right] \xrightarrow[75-95\%]{R^2X} R^1-\overset{R}{\underset{R^2}{\overset{|}{C}}}-C\equiv N$$

or ketones (equation V). When oxiranes are used the γ-hydroxy nitriles formed

(V)
$$\left[\underset{R^1}{\overset{R}{\diagdown}}\overset{-}{C}\cdots C\equiv N\ Li^+\right] \xrightarrow[80-90\%]{\underset{R^3}{\overset{R^2}{\diagdown}}C=O} R^1-\overset{R}{\underset{}{\overset{|}{C}}}-C\equiv N \\ R^2-\overset{|}{\underset{R^3}{\overset{|}{C}}}-OH$$

cyclize spontaneously to 2-iminotetrahydrofuranes (equation VI). The products are readily hydrolyzed to γ-butyrolactones.

(VI)
$$\left[\underset{R^1}{\overset{R}{\diagdown}}C\cdots C\equiv N\ Li^+\right] \xrightarrow{\underset{R^3}{\overset{R^2}{\diagup}}\triangle}$$

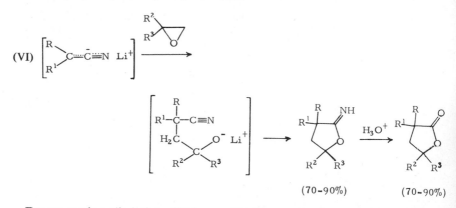

(70-90%)   (70-90%)

Deprotonation-alkylation of imines and hydrazones has been used for synthesis of substituted pyruvaldehydes (equation VII).[4]

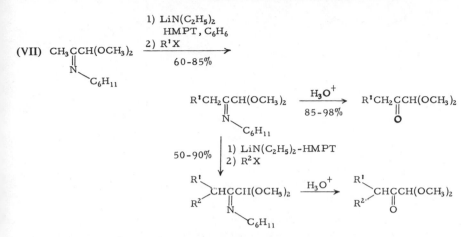

(VII) $CH_3CCH(OCH_3)_2$

This activated base also converts N,N-disubstituted amides into the α-anions, which react with various electrophiles (equation VIII).[5]

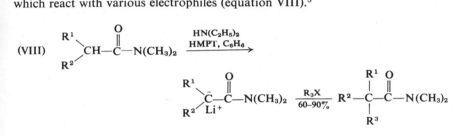

This metallation was used in a synthesis of *dl*-piperitone (1).

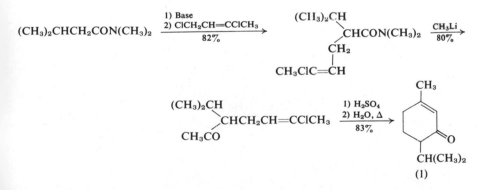

Cyclopropanes. The reaction of γ,δ-unsaturated epoxides with lithium diethylamide and HMPT results in formation of disubstituted cyclopropanes.[6] A typical example is shown in equation (IX), which also indicates the importance of a polar solvent. The reaction involves formation of the allylic carbanion

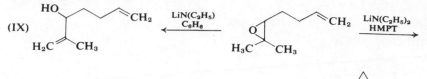

(IX)

(two isomers)

followed by γ-elimination. In benzene or ether β-elimination is favored. This isomerization is comparable to that of γ,δ-epoxy ketones (Gaoni, **5**, 451–452) and of γ,δ-epoxy nitriles (Stork *et al.*[7]).

[1] T. Cuvigny and H. Normant, *Organometal. Chem. Syn.*, **1**, 237 (1971).
[2] T. Cuvigny, J. F. Le Borgne, M. Larchevêque, and H. Normant, *Synthesis*, 237 (1976).
[3] J. F. Le Borgne, T. Cuvigny, M. Larchevêque, and H. Normant, *ibid.*, 238 (1976).
[4] T. Cuvigny and H. Normant, *ibid.*, 198 (1977).
[5] P. Hullot, T. Cuvigny, M. Larchevêque, and H. Normant, *Canad. J. Chem.*, **54**, 1098 (1976).
[6] M. Apparu and M. Barrelle, *Tetrahedron Letters*, 2837 (1976).
[7] G. Stork *et al.*, *Am. Soc.*, **96**, 5270 (1974).

Lithium diisopropylamide (LDA), **1**, 611; **2**, 249; **3**, 184–185; **4**, 298–302; **5**, 400–406; **6**, 334–339.

Pyrroles; tetrahydropyridazines. LDA in THF converts aromatic ketazines into dianions, which on hydrolytic work-up are converted into either pyrroles or tetrahydropyridazines, depending on the substrate.[1] In the Piloty pyrrole synthesis[2] ketazines are converted into pyrroles by treatment with $ZnCl_2$ or HCl. Examples:

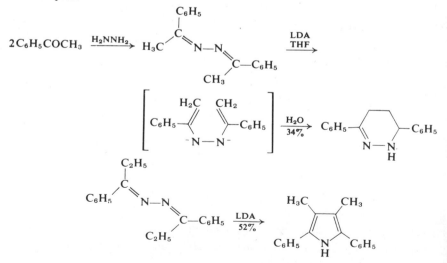

2,5-Dialkylpyrrolidines. Fraser and Passannanti[3] have developed a general route to these compounds. The method is illustrated by the synthesis of 2-ethyl-5-heptylpyrrolidine, a constituent of the venom of the fire ant.

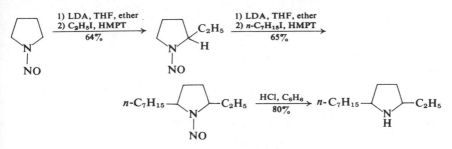

α-Methylenelactones. Attempts to generate the trianion of itaconic acid (LDA, −78°, THF) are only moderately successful, but dianions of monoesters of the acid are readily generated. These react with aldehydes and ketones to give adducts that cyclize to α-methylenelactones in the presence of acids.[4]

Examples:

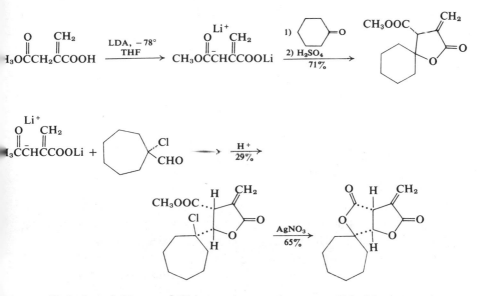

Alkylation of thioesters.[5] This reaction can be accomplished by conversion to the lithium enolate in an aprotic solvent followed by alkylation; both steps are carried out at −78° (equation I). Yields are often improved by addition of

(I) $$CH_3CSR \xrightarrow[THF, -78°]{LDA} CH_2=CSR \xrightarrow[65-80\%]{R'X \atop -78 \to 25°} R'CH_2CSR$$

HMPT. The same conditions were also successful for alkylation of γ-thiobutyro-lactone (equation II).

(II)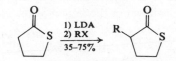

If a base such as potassium t-butoxide is used in place of LDA, thioethers are formed on alkylation.[6]

α-Hydroxycarboxamides. N-Protected formamides (1) are converted by LDA into the anion (2), which reacts with carbonyl compounds to form (3). The protecting group is removed by acid treatment.[7]

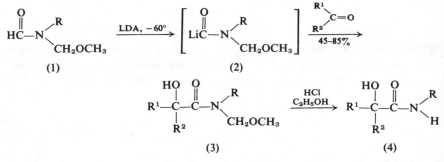

α,β-Unsaturated ketones.[8] 2-Isoxazolines (1)[9] are cleaved to unsaturated oximes (2) by LDA, or by n-butyllithium or ethylmagnesium bromide in lower yields. The products are converted into α,β-unsaturated ketones (3) by the procedure of Timms and Wildsmith (4, 506–507).

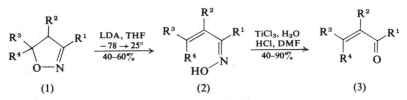

Shapiro-Heath olefin synthesis (2, 418–419; 4, 511; 5, 678–679; 6, 598–599). Vedejs[10] has mentioned that LDA can be used instead of alkyllithiums for conversion of the more typical tosylhydrazones into olefins at 0° in THF. This variation was used in the synthesis of a new (CH)$_{12}$-hydrocarbon (1).

(1)

Trianions of 2,4,6-triketones. Formation of the trianion of 1-phenyl-1,3,5-hexanetrione (1) with $LiNH_2$ fails, but it proceeds rapidly and completely with LDA (3 equiv.) in THF. The trianion reacts preferentially at the methyl terminus;

$$C_6H_5COCH_2COCH_2COCH_3 \xrightarrow[\text{THF}]{\text{LDA}} [C_6H_5CO\bar{C}HCO\bar{C}HCO\bar{C}H_2] \xrightarrow[82\%]{\substack{\text{1) } CO_2 \\ \text{2) HCl, } H_2O}}$$

$$\underset{(1)}{} \qquad \underset{(a)}{}$$

$$C_6H_5COCH_2COCH_2COCH_2COOH$$
$$(2)$$

an example is the synthesis of the 3,5,7-triketo acid (2). Indeed this sequence is general for synthesis of acids of this type.[11]

Conversion of endo-*peroxides into furanes.* This transformation can be accomplished by treatment of an *endo*-peroxide, obtained from a 1,3-diene, with LDA at $-70°$ (1 hour) followed by quenching with *p*-toluenesulfonyl chloride or chromatography on silica gel.[12]

Example:

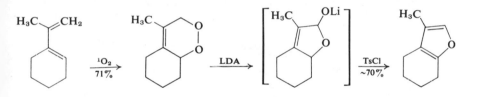

[1] Z. Yoshida, T. Harada, Y. Tamaru, *Tetrahedron Letters*, 3823 (1976).
[2] O. Piloty, *Ber.*, **43**, 489 (1910).
[3] R. R. Fraser and S. Passannanti, *Synthesis*, 540 (1976).
[4] R. M. Carlson and A. R. Oyler, *J. Org.*, **41**, 4065 (1976).
[5] R. A. Gorski, G. J. Wolber, and J. Wemple, *Tetrahedron Letters*, 2577 (1976).
[6] G. E. Wilson, Jr., and J. G. Riley, *ibid.*, 379 (1972).
[7] U. Schöllkopf and H. Beckhaus, *Angew. Chem., Int. Ed.*, **15**, 293 (1976).
[8] V. Jäger and H. Grund, *Angew. Chem., Int. Ed.*, **15**, 50 (1976).
[9] Preparation by cycloaddition of nitrile oxides to alkenes: M. Christl and R. Huisgen, *Ber.*, **106**, 3345 (1973).
[10] E. Vedejs and R. A. Shepherd, *J. Org.*, **41**, 742 (1976).
[11] T. M. Harris, G. P. Murphy, and A. J. Poje, *Am. Soc.*, **98**, 7733 (1976).
[12] B. Harirchian and P. D. Magnus, *Syn. Comm.*, **7**, 119 (1977).

Lithium diisopropylamide–Hexamethylphosphoric triamide.

Cyclopropanes. A new synthesis of *trans*-chrysanthemic acid (3) involves treatment of (1) with LDA (2 equiv.) and HMPT in THF at $-70°$. The only identifiable product (2) was converted by known methods into (3). In this or

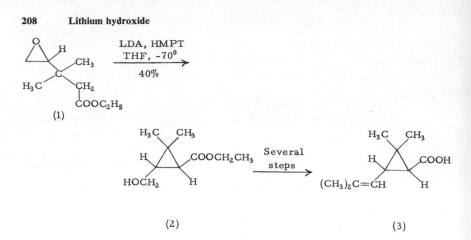

(2)   (3)

related cyclizations of β,γ-epoxy carbanions, no evidence was obtained for formation of a four-membered carbocycle.[1]

[1] J. H. Babler and A. J. Tortorello, *J. Org.*, **41**, 885 (1976).

Lithium hydroxide, LiOH. Mol. wt. 23.95. Suppliers: Alfa, ROC/RIC, others.

Oxetanes. Treatment of 1-(β,γ-epoxypropyl)cyclohexane-1-ol (1) with lithium hydroxide in 75% aqueous DMSO affords the oxetane (2) and the triol (3).

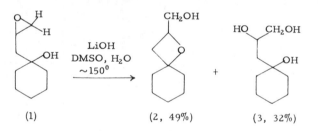

(1)   (2, 49%)   (3, 32%)

KOH and NaOH effect the same reaction but require longer reaction times. The reaction of (1) with a base (NaH, *n*-BuLi) in anhydrous solvents results in formation of the oxolane dimer (4) and oligomers.[1]

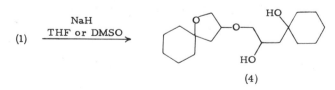

(4)

[1] A. Murai, M. Ono, and T. Masamune, *J.C.S. Chem. Comm.*, 864 (1976).

Lithium iodide, 5, 410–411.

Ester cleavage (**1**, 615–617). A recent review on ester cleavage cites about 30 examples of cleavage with lithium iodide. Direct comparison has established that LiI is superior to other lithium halides, the order being LiI > LiBr > LiCl in pyridine or DMF. Methyl esters are cleaved readily, but ethyl esters can be cleaved but more slowly.[1]

[1] J. E. McMurry, *Org. React.*, **24**, 187 (1976).

Lithium N-isopropylcyclohexylamide, 4, 306–309; 5, 411–412.

Ester enolate Claisen rearrangement (**4**, 307–308). Ireland *et al.*[1] have reported details of this [3,3]sigmatropic rearrangement and also recent improvements. Rearrangement of the *t*-butyldimethylsilylketene acetal of the enolate (equation I) is generally superior to rearrangement of the trimethylsilylketene acetal as used previously. A further advantage is that the silyl esters can be converted into the

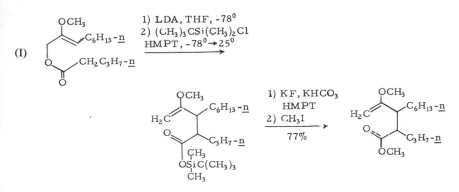

(I)

methyl esters as shown. LDA, as well as LiICA, can be used to generate the enolate (equation I).

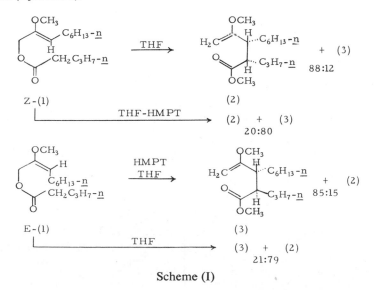

Scheme (I)

A further development in the rearrangement is that the stereochemistry can be controlled to a high degree by the choice of solvent (6, 276–277). Thus the stereo-chemistry of rearrangement of (Z)- and (E)-1 can be markedly controlled by choice of solvent: THF alone or THF–HMPT (scheme I). This stereochemistry appears to be general. Thus similar results were obtained in the enolization of 3-pentanone (equation II).

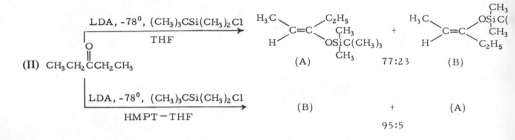

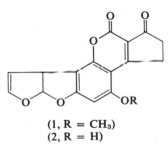

[1] R. E. Ireland, R. H. Mueller, and A. K. Willard, *Am. Soc.*, **98**, 2868 (1976).

Lithium 2-methylpropane-2-thiolate (Lithium *t*-butylmercaptide), 5, 412.

Demethylation. A group at the Massachusetts Institute of Technology[1] used this reagent for demethylation of aflatoxin B_1 (1) to the less toxic metabolite aflatoxin P_1 (2); yield, 45%.

(1, R = CH₃)
(2, R = H)

[1] G. Büchi, D. Spitzner, S. Paglialunga, and G. N. Wogan, *Life Sciences*, **13**, 1143 (1973).

Lithium phenylethynolate, $C_6H_5C \vdots \bar{C} \vdots OLi^+$ (1). Mol. wt. 124.06.

Preparation:

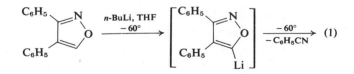

3-Phenyl-β-lactones; alkenes. The reagent reacts with carbonyl compounds to afford, after protonation, β-lactones (2), which lose CO_2 on heating to form alkenes (3).

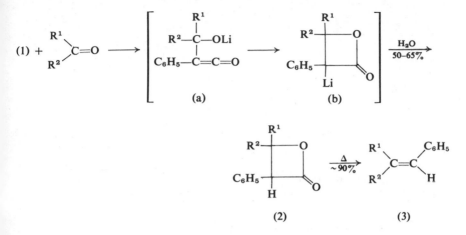

[1] U. Schöllkopf and I. Hoppe, *Angew. Chem., Int. Ed.,* 765 (1975).

Lithium phenylthio(*t*-butyl)cuprate, (1), **5,** 414–415.

t-Butyl phenyl ketone (**5,** 414). The synthesis of this ketone has been published.[1] The reaction is generally applicable to synthesis of *sec-* and *t*-alkyl ketones from carboxylic acid chlorides.

Posner's reagent has also been used for synthesis of bis-*t*-butyl diketones by reaction with dicarboxylic acid chlorides (equation I).[2]

$$\text{(I)} \qquad \text{ClCOCOCl} + (1) \xrightarrow[\underset{100\%}{}]{\text{THF, }-70^{0}} (CH_3)_3CCOCOC(CH_3)_3$$

[1] G. H. Posner and C. E. Whitten, *Org. Syn.,* **55,** 122 (1976).
[2] G. B. Bennett, J. Nadelson, L. Alden, and A. Jani, *Org. Prep. Proc. Int.,* **8,** 13 (1976).

Lithium phenylthio(cyclopropyl)cuprate, $\left(\text{▷}-CuSC_6H_5\right)Li$ (1). Mol. wt. 220.72.

The mixed cuprate is prepared by addition of cyclopropyllithium in ether to a slurry of phenylthiocopper in THF at $-78°$; the reaction is then allowed to warm to $-20°$.

Cyclopentane annelation.[1] This cuprate reacts with β-halo-α,β-unsaturated ketones to form β-cyclopropylenones. The products rearrange on thermolysis to cyclopentane systems. An example of this annelation method is formulated in equation (I).

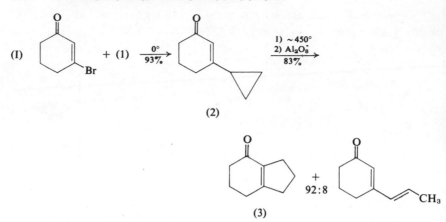

(2)

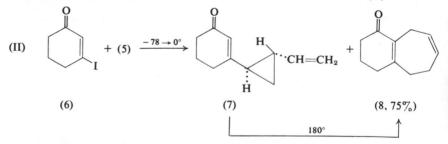

(3)

Cycloheptane annelation.[2] The method cited above has been extended to the synthesis of β-(2-vinylcyclopropyl)enones. The cuprate for this purpose is lithium phenythio(2-vinylcyclopropyl) cuprate, $\left(CH_2=CH \underset{\bigtriangledown}{} CuSC_6H_5 \right) Li$, (5),

prepared by the reaction of 2-vinylcyclopropyllithium with phenylthiocopper. An example of cycloheptane annelation is formulated in equation (II).

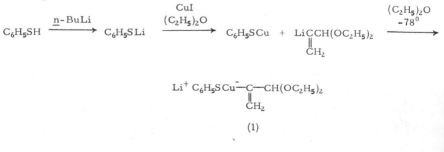

(6)   (7)   (8, 75%)

¹ E. Piers, C. K. Lau, and I. Nagakura, *Tetrahedron Letters*, 3233 (1976).
² E. Piers and I. Nagakura, *ibid.*, 3237 (1976).

Lithium phenylthio[(α-diethoxymethyl)vinyl] cuprate, (1). Mol. wt. 308.84.
    Preparation:

$$C_6H_5SH \xrightarrow{\underline{n}-BuLi} C_6H_5SLi \xrightarrow[\substack{CuI \\ (C_2H_5)_2O}]{} C_6H_5SCu + \underset{\substack{\| \\ CH_2}}{LiCCH(OC_2H_5)_2} \xrightarrow[-78^0]{(C_2H_5)_2O}$$

$$Li^+ C_6H_5S\,Cu^- \underset{\substack{\| \\ CH_2}}{-C-CH(OC_2H_5)_2}$$

(1)

Reactions.[1] This mixed cuprate transfers the three-carbon unit selectively to allylic halides (equations I and II). The cuprate also undergoes conjugate addition to α,β-unsaturated ketones (equation III).

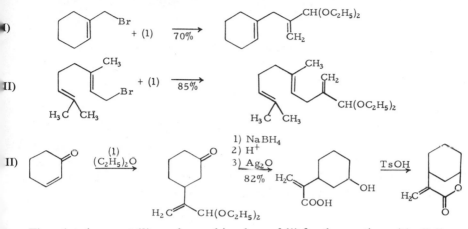

The related reagent (2) can be used in place of (1) for the reaction with allylic halides, but it undergoes only 1,2-addition reactions with α,β-unsaturated ketones.

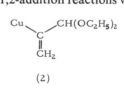

(2)

[1] P. A. Grieco, C.-L. Wang, and G. Majetich, *J. Org.*, **41**, 726 (1976).

Lithium 2,2,6,6-tetramethylpiperidide (LiTMP), 4, 310–311; 5, 417; 6, 345–348.

Cyclopropyl ethers; cyclopropenyl ethers.[1] Cyclopropyl ethers can be prepared in satisfactory yield by reaction of an alkene (excess) with chloromethyl ethers and LiTMP. The base eliminates HCl from the ether with formation of ROCH.

Examples:

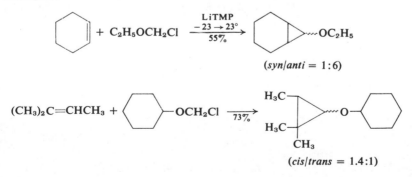

$$CH_3$$
$$(CH_3)_3SiOC{=}CH_2 + (CH_3)_2CHOCH_2Cl \xrightarrow[55\%]{}$$

H₃C— ⟩···OCH(CH₃)₂, OSi(CH₃)₃

Cyclopropenyl ethers are obtained when an alkyne is used as substrate.

Example:

$$C_2H_5C{\equiv}CC_2H_5 + (CH_3)_3CCH_2OCH_2Cl \xrightarrow[57\%]{LiTMP \; 0-5°}$$

with C_2H_5 substituents ···OCH₂C(CH₃)₃

Cyclopropanols. This new carbenoid reaction has been extended to a synthesis of cyclopropanols[2] by use of β-chloroethyl chloromethyl ether as the carbenoid precursor. The final step involves cleavage of a β-chloroethyl ether with *n*-butyllithium, a reaction first used by Schöllkopf.[3]

Examples:

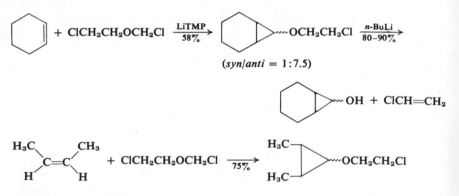

$$+ ClCH_2CH_2OCH_2Cl \xrightarrow[58\%]{LiTMP} \cdots OCH_2CH_2Cl \xrightarrow[80-90\%]{n\text{-BuLi}}$$

(syn/anti = 1:7.5)

···OH + ClCH=CH₂

$$\underset{H}{\overset{H_3C}{>}}C{=}C\underset{H}{\overset{CH_3}{<}} + ClCH_2CH_2OCH_2Cl \xrightarrow[75\%]{}$$

H₃C ⟩···OCH₂CH₂Cl, H₃C

Olofson and co-workers[4] have also prepared cyclopropanols by treatment of a chloromethyl ester (1) with this base to generate an intermediate acyloxycarbene (a). Reaction of the carbene with an alkene gives a cyclopropyl ester (2), which is converted into a cyclopropanol (3) by LiAlH₄ or CH₃Li. Both the *cis*- and

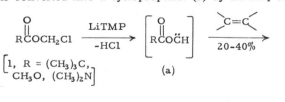

$$\text{RCOCH}_2\text{Cl} \xrightarrow[-HCl]{LiTMP} [\text{RCOCH}] \xrightarrow[20-40\%]{>C=C<}$$

$$\begin{bmatrix}1, \; R = (CH_3)_3C,\\ CH_3O, \; (CH_3)_2N\end{bmatrix}$$

(a)

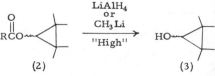

$$\text{RCO}{\cdots} \xrightarrow[\text{"High"}]{LiAlH_4 \; or \; CH_3Li} \text{HO}{\cdots}$$

(2)   (3)

trans-isomers of (2) are obtained. Yields are not high, but the method is convenient for preparation of small quantities of cyclopropanols.

α-*Metallated isocyanides; β-amino alcohols.* Even weakly acidic isocyanides that are not metallated by *n*-butyllithium can be metallated in the α-position by lithium tetramethylpiperidide. This reaction has been used as the first step in a synthesis of 2-oxazolines and of β-amino alcohols obtained on basic or acidic hydrolysis.[5]

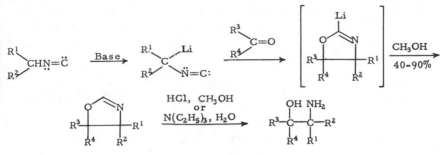

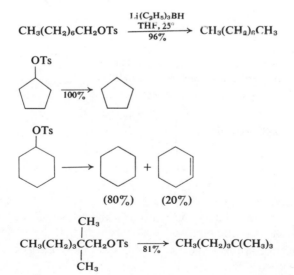

[1] R. A. Olofson, K. D. Lotts, and G. N. Barber, *Tetrahedron Letters*, 3779 (1976).
[2] G. N. Barber and R. A. Olofson, *ibid.*, 3783 (1976).
[3] U. Schöllkopf, *Angew. Chem., Int. Ed.*, 7, 588 (1968).
[4] R. A. Olofson, K. D. Lotts, and G. N. Barber, *Tetrahedron Letters*, 3381 (1976).
[5] U. Schöllkopf, F. Gerhart, I. Hoppe, R. Harms, K. Hantke, K.-H. Scheunemann, E. Eilers, and E. Blume, *Ann.*, 183 (1976).

Lithium triethylborohydride (Super Hydride), 4, 313; **6,** 348–349.

Deoxygenation of alcohols. This hydride is superior to lithium aluminum hydride for reduction of tosylates to hydrocarbons.[1]

Examples:

$$CH_3(CH_2)_6CH_2OTs \xrightarrow[\substack{Li(C_2H_5)_3BH \\ THF, 25° \\ 96\%}]{} CH_3(CH_2)_6CH_3$$

OTs

$\xrightarrow{100\%}$

OTs

$\longrightarrow$ + (80%)   (20%)

$$CH_3(CH_2)_3\overset{\displaystyle CH_3}{\underset{\displaystyle CH_3}{C}}CH_2OTs \xrightarrow{81\%} CH_3(CH_2)_3C(CH_3)_3$$

Holder and Matturro[2] have independently found that this hydride is more effective than $LiAlH_4$ for reduction of mesylates, even hindered ones, to alkanes.

[1] S. Krishnamurthy and H. C. Brown, *J. Org.*, **41**, 3064 (1976).
[2] R. W. Holder and M. G. Matturro, *ibid.*, **42**, 2166 (1977).

Lithium trimethylsilyldiazomethane, $\overset{N_2}{\overset{\|}{Li\overset{}{C}Si(CH_3)_3}}$ (1). Mol. wt. 120.16.

This reagent is prepared by metallation of trimethylsilyldiazomethane with *n*-butyllithium at $-100°$. The preparation of trimethylsilyldiazomethane by Seyferth *et al.* has been mentioned (**4**, 543). Schöllkopf and Scholz[1] have used a variation of this preparation (equation I).

(I) $(CH_3)_3SiCH_2Cl + NaNHCOCH_3 \xrightarrow[90\%]{\overset{Xylene}{\Delta}} (CH_3)_3SiCH_2NHCOCH_3 \xrightarrow[94\%]{NaNO_2}$

$$(CH_3)_3SiCH_2\overset{\overset{NO}{|}}{N}COCH_3 \xrightarrow{C_6H_5CH_2CH_2NH_2} H\overset{\overset{N_2}{\|}}{C}Si(CH_3)_3$$

Homologation of carbonyl compounds to aldehydes. Schöllkopf and Scholz have reported a three-step procedure for this transformation (equation I). The last step was first used by Stork and Ganem.[2]

(I)   (1) + $\overset{R^1}{\underset{R^2}{>}}C{=}O \xrightarrow[60-80\%]{HOAc} R^1\overset{\overset{OH}{|}}{\underset{\underset{R^2}{|}}{C}}\overset{\overset{N_2}{\|}}{C}Si(CH_3)_3 \xrightarrow[\sim 50\%]{30-40°, -N_2}$

$$R^1\overset{\overset{O}{\|}}{\underset{\underset{R^2}{|}}{C}}CHSi(CH_3)_3 \xrightarrow{H_3O^+} \left[\overset{R^1}{\underset{R^2}{>}}C{=}C\overset{OH}{\underset{H}{<}}\right] \longrightarrow R^1\overset{\overset{H}{|}}{\underset{\underset{R^2}{|}}{C}}\overset{}{C}\overset{O}{\underset{H}{<}}$$

[1] U. Schöllkopf and H.-U. Scholz, *Synthesis*, **271** (1976).
[2] G. Stork and B. Ganem, *Am. Soc.*, **95**, 6152 (1973).

Lithium trisiamylborohydride, $Li\left[\overset{CH_3}{\underset{(CH_3)_2CH}{>}}CH\right]_3$ BH, $LiSia_3BH$. Mol. wt. 232.1

This complex borohydride is prepared by addition of *t*-butyllithium in *n*-pentane to trisiamylborane in THF at $-78°$.

Stereoselective reduction of cyclic ketones.[1] This borohydride reduces cyclic ketones to the corresponding thermodynamically less stable alcohols even more stereoselectively than lithium tri-*sec*-butylborohydride (L-Selectride, **4**, 312–313). It is also superior in this respect to lithium tris(*trans*-2-methylcyclopentyl)borohydride.

Examples:

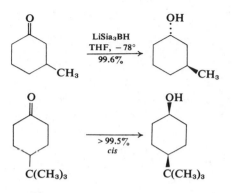

The reagent reduces camphor to the *exo*-alcohol, but only slowly (25°, 72 hours), probably because of the large steric requirements of this complex borohydride.

[1] S. Krishnamurthy and H. C. Brown, *Am. Soc.*, **98**, 3383 (1976).

M

Magnesium, 1, 627–629; **2,** 254; **3,** 189; **4,** 315; **5,** 419; **6,** 351–352.

Highly reactive form (**5,** 419). Details for the preparation of Rieke's[1] activated and pyrophoric magnesium have been submitted to *Organic Synthesis*. The powder should be kept under argon. The preparation includes synthesis of benzoic acid in 70–89% yield from bromobenzene.

Coupling. The interesting strained cyclobutaarene (2;1,2,5,6,9,10-hexahydro-triscyclobuta[*b,h,n*]triphenylene) has been obtained in low yield by treatment of a concentrated solution of (1) with magnesium at 20°. Unlike triphenylene, (2) is readily hydrogenated to a hydrocarbon in which the central benzene ring is intact. The high strain in (2) is also reflected in the electronic spectrum.[2]

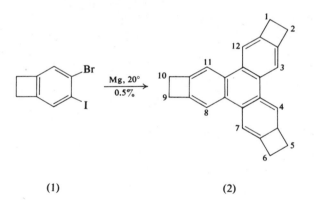

(1)   (2)

[1] R. D. Rieke, S. E. Bales, and P. M. Hudnall, *Org. Syn.*, submitted (1976).
[2] H. R. L. Funk and K. P. C. Vollhardt, *Angew. Chem., Int. Ed.*, **15,** 53 (1976).

Magnesium bromide etherate, 1, 629–630.

Rearrangement of α,β-epoxysilanes. Hudrlik and collaborators[1] have extended the well-known rearrangement of epoxides to ketones[2] with Lewis acids to the rearrangement of α,β-epoxysilanes. In this case, cleavage of seven substrates was observed to result from α-cleavage. In two cases α-bromo-β-hydroxysilanes were obtained; in the others, β-ketosilanes or the enol silyl ethers were formed.

218

Examples:

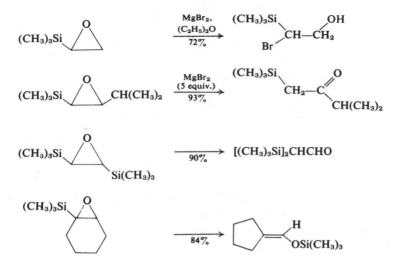

α,β-Epoxysilanes can also be rearranged to isomeric silyl enol ethers by pyrolysis.[3,4]

Examples:

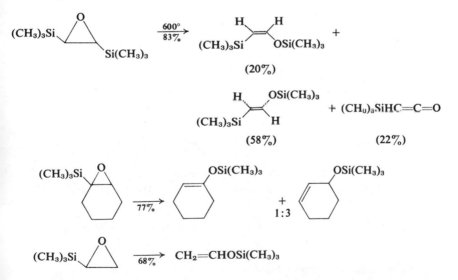

α-Bromo aldehydes and ketones.[5] An attractive route to these substances is the reaction of aldehydes or ketones with an α-chloro sulfone under phase-transfer conditions to form an α,β-epoxy sulfone, which is converted into an α-bromo aldehyde or ketone by treatment with $MgBr_2$[6] (5°, CH_2Cl_2 or ether).

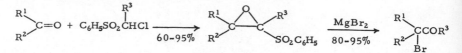

[1] P. F. Hudrlik, R. N. Misra, G. P. Withers, A. M. Hudrlik, R. J. Rona, and J. P. Arcoleo, *Tetrahedron Letters*, 1453 (1976).
[2] For recent references see B. Rickborn and R. M. Gerkin, *Am. Soc.*, **93**, 1693 (1971).
[3] P. F. Hudrlik, C.-N. Wan, and G. P. Withers, *Tetrahedron Letters*, 1449 (1976).
[4] A. R. Bassindale, A. G. Brook, P. Chen, and J. Lennon, *J. Organometal. Chem.*, **94**, C21 (1975).
[5] F. de Reinach-Hirtzbach and T. Durst, *Tetrahedron Letters*, 3677 (1976).
[6] Prepared by the reaction of Mg with $BrCH_2CH_2Br$ in ether.

Magnesium bromide etherate–Hydrogen peroxide.

α-*Bromo carbonyl compounds*. Carbonyl compounds can be brominated at the α-position in fair to high yield by reaction with magnesium bromide etherate (1 equiv.) and a peroxide (1 equiv.) in THF or ether.

Examples:

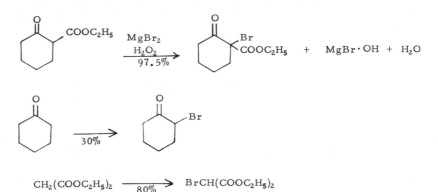

α-Chloro ketones are obtained by use of LiCl as the metal halide; α-iodo ketones are obtained with LiI.[1]

[1] N. Inukai, H. Iwamoto, T. Tamura, I. Yanagisawa, Y. Ishii, and M. Murakami, *Chem. Pharm. Bull. Japan*, **24**, 820 (1976).

Magnesium methoxide, $Mg(OCH_3)_2$.

This base is prepared by dissolving magnesium turnings in methanol.

Triacetic acid methyl ester (2).[1] This ester can be prepared easily by heating dehydroacetic acid (1) with magnesium methoxide. The reaction is slow when sodium or lithium methoxide is used.

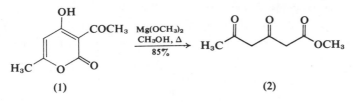

(1)   (2)

[1] J. G. Batelaan, *Syn. Comm.*, **6**, 81 (1976).

Magnesium oxide, 1, 633–637; **2,** 256–257.

Hydrogenation catalyst. Japanese chemists[1] report the unusual properties of a magnesium oxide obtained by evacuating $Mg(OH)_2$ at 1100° for 2 hours. It is an active catalyst for hydrogenation of 1,3-butadiene to *cis*-2-butene. When D_2 is used instead of H_2, deuterium is located at C_1 and C_4 (equation I).

(I) $$CH_2{=}CHCH{=}CH_2 \xrightarrow[\text{MgO}]{D_2} \quad \underset{DH_2C}{\overset{H}{\diagdown}}C{=}C\underset{CH_2D}{\overset{H}{\diagup}}$$

[1] H. Hattori, Y. Tanaka, and K. Tanabe, *Am. Soc.*, **98**, 4652 (1976).

Manganese(III) acetate, 2, 263–264; **4,** 318; **6,** 355–357.

α′-Acetoxylation of α,β-enones. This reaction has been effected by manganic acetate hydrate. Yields are somewhat low, but the reaction may be useful in cases where lead tetraacetate is not satisfactory.[1]

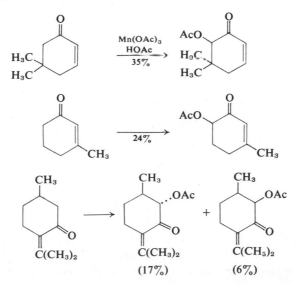

[1] G. J. Williams and N. R. Hunter, *Canad. J. Chem.*, **54**, 3830 (1976).

Manganese(II) chloride, $MnCl_2$. Mol. wt. 125.84. Suppliers: Alfa, ROC/RIC.

Reduction of aryl or vinylic halides. Aryl or vinylic halides are reduced by Grignard reagents in THF in the presence of catalytic amounts of manganese(II) chloride. The choice of solvent is crucial.

Examples:

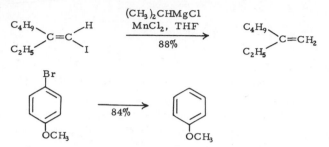

1,3-Dienes.[1] 1-Iodoalkenes couple to form 1,3-dienes when treated with *n*-butyllithium with catalysis by manganese(II) chloride and a trace of lithium

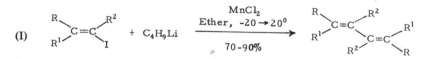

chloride in ether (equation I). Use of THF as solvent results in alkylation. The paper includes a suggested mechanism.[2]

[1] G. Cahiez, D. Bernard, and J. F. Normant, *J. Organometal. Chem.*, **113**, 107 (1976).
[2] *Idem, ibid.*, **113**, 99 (1976).

Manganese(II) iodide, MnI_2. Mol. wt. 308.75. Supplier: ROC/RIC (anhydrous).

Ketone synthesis. Ketones can be prepared by reaction of MnI_2 with Grignard reagents to form an organomanganese species; this reacts with an acid chloride to form a ketone (equation I).[1]

$$\text{(I)} \qquad \text{RMgBr} \xrightarrow[\text{– MgBrI}]{\overset{MnI_2}{\text{Ether}}} \text{[RMnI]} \xrightarrow[60-90\%]{\text{R'COCl}} \text{RCOR'}$$

[1] G. Cahiez, A. Masuda, D. Bernard, and J. F. Normant, *Tetrahedron Letters*, 3155 (1976).

Mercuric acetate, 1, 644–652; **2,** 264–267; **3,** 194–196; **4,** 319–323; **5,** 424–427; **6,** 358–359.

Oxymercuration–acetylation–demercuration of bicyclo[3.1.0]hexanes. Cleavage of the zero bridge is a major pathway in this reaction. This cleavage is promoted markedly by a 3-hydroxy substituent, either *cis* or *trans*, as shown in equations (I) and (II). At present this novel neighboring group effect is not understood; it is not observed in the case of bicyclo[4.1.0] heptanes. The cleavage reaction proceeds with inversion at the carbon atom undergoing electrophilic attack.[1]

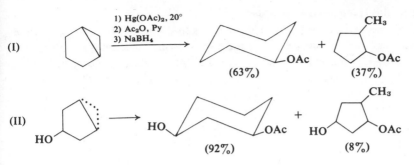

$$(63\%) \qquad (37\%)$$

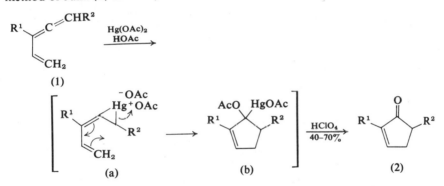

$$(92\%) \qquad (8\%)$$

Acetoxymercuration.[2] Cyclopentenones (2) can be obtained by acetoxy-mercuration of vinylallenes (1). This route is an extension of the cyclization method of Julia (**5**, 426–427).

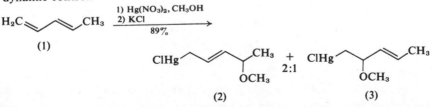

[1] R. G. Salomon and R. D. Gleim, *J. Org.*, **41**, 1529 (1976).
[2] F. Delbecq and J. Goré, *Tetrahedron Letters*, 3459 (1976).

Mercuric nitrate, $Hg(NO_3)_2$, Mol. wt. 280.63. Suppliers: Alfa, ROC/RIC.

Oxymercuration (**2**, 265–267; **3**, 194; **4**, 319–320; **5**, 425–426; **6**, 359). 1,4-Oxymercuration has been observed for the first time in the reaction of 1,3-pentadiene (1) with mercuric nitrate in methanol, which results in (2) and (3), both with (E)-stereochemistry. If mercuric acetate is used, only the 1,2-adduct is formed unless nitric acid is added; thus the 1,4-adduct is the product of thermodynamic control.[1]

$$H_2C \diagdown \diagup \diagdown CH_3 \quad \xrightarrow[\text{89\%}]{\begin{array}{l}1)\ Hg(NO_3)_2,\ CH_3OH\\2)\ KCl\end{array}}$$

(1)

$$ClHg \diagup \diagdown \diagup \underset{\underset{OCH_3}{|}}{CH} CH_3 \quad + \quad ClHg \diagup \diagdown \underset{\underset{OCH_3}{|}}{CH} \diagdown CH_3$$

$$\qquad \qquad \qquad 2:1$$

$$(2) \qquad \qquad \qquad (3)$$

[1] A. J. Bloodworth, M. G. Hutchings, and A. J. Sotowicz, *J.C.S. Chem. Comm.*, 578 (1976).

Mercuric nitrate–Hydrogen peroxide.

Cyclic peroxides. Cyclic peroxides have been prepared[1] by a variant of the peroxymercuration of alkenes.[2] The method is formulated for the synthesis of the dimethyl-1,2-dioxacyclohexane as a mixture of *cis-* and *trans*-isomers.

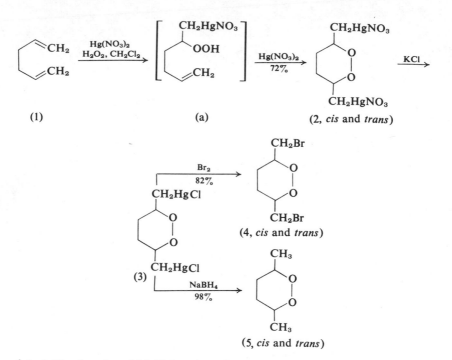

(1)   (a)   (2, *cis* and *trans*)

(5, *cis* and *trans*)

[1] A. J. Bloodworth and M. E. Loveitt, *J.C.S. Chem. Comm.*, 94 (1976).
[2] A. J. Bloodworth and I. M. Griffin, *J.C.S. Perkin I*, 195 (1975).

Mercuric oxide, 1, 655–658; 2, 267–268; 4, 323–324; 5, 428; 6, 360.

Decarboxylation. Della and Patney[1] recommend a two-step method for decarboxylation of bridgehead carboxylic acids. The first step is the Cristol-Firth modification of the Hunsdiecker reaction (1, 657, improved by use of methylene bromide as solvent). The second is photochemical reduction of the bromides by tri-*n*-butyltin hydride (1, 1192–1193) with azobisisobutyronitrile as initiator. Yields are 80–95% in the first step and 80–90% in the second step.

[1] E. W. Della and H. K. Patney, *Synthesis*, 251 (1976).

Methanesulfinyl chloride, 5, 434–435.

Reduction of sulfoxides to sulfides. Sulfoxides are reduced to sulfides by methanesulfinyl chloride at room temperature in 15 minutes. The reduction involves a direct transfer of oxygen.[1]

$$\underset{\uparrow}{\overset{O}{}}\ \ \underset{}{\overset{O}{\uparrow}}$$

$$\text{RSR}' + \text{CH}_3\text{SCl} \xrightarrow[80-90\%]{\text{CCl}_4} \text{RSR}' + \text{CH}_3\text{SCl}$$

[1] T. Numata, K. Ikura, Y. Shimano, and S. Oae, *Org. Prep. Proc. Int.*, **8**, 119 (1976).

Methanesulfonyl chloride, 1, 662–664; **2,** 268–269; **4,** 326–327; **5,** 435–436; **6,** 362–363.

Oxoalkane nitriles. Oximes of the α-hydroxy ketones (1) undergo fragmentation[1] when treated with methanesulfonyl chloride or *p*-toluenesulfonyl chloride and pyridine.[2] The keto nitriles (2) are obtained in about 50–95% yield.

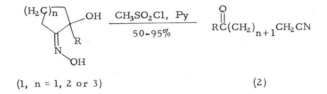

(1, n = 1, 2 or 3)   (2)

Allylic chlorides. Japanese chemists[3] have reported examples of an abnormal conversion of allylic alcohols into chlorides with inversion of configuration by treatment with mesyl chloride or tosyl chloride in pyridine (equation I).

(I)
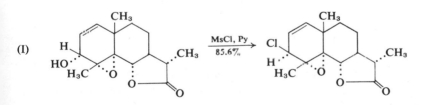

[1] R. K. Hill, *J. Org.*, **27**, 29 (1962); A. Hassner and W. A. Wentworth, *J.C.S. Chem. Comm.*, 44 (1965).
[2] T. Wakamatsu, M. Fukui, and Y. Ban, *Synthesis*, 341 (1976).
[3] Y. Fujimoto, T. Shimizu, and T. Tatsuno, *Chem. Pharm. Bull. Japan*, **24**, 365 (1976).

Methoxyallene, $CH_3OCH{=}C{=}CH_2$. Mol. wt., 70.09, b.p. 52°.
   Preparation[1]:

$$HC{\equiv}CCH_2OCH_3 \xrightarrow[82\%]{\underset{70°}{KOC(CH_3)_3}} H_2C{=}C{=}CHOCH_3$$

Furanes. Brandsma *et al.*[2] have reported a new route to furanes and dihydrofuranes, formulated in scheme (I).

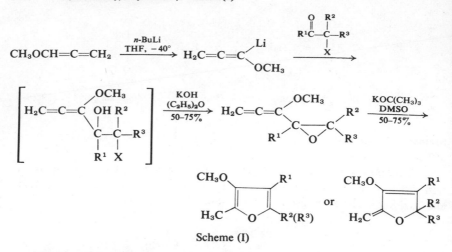

Scheme (I)

Vinylic ethers. Vermeer *et al.*[3] have extended the conversion of methoxyallene to 1-alkynes by reaction with Grignard reagents catalyzed by copper(I) halides **(6, 146)** to a synthesis of vinylic ethers, formulated in equation (II). The ethers

$$\text{(II)}\quad CH_3OCH=C=CH_2 + \begin{cases} R_2CuMgX \\ \text{or} \\ [RCuBr]MgX \end{cases} \xrightarrow[\substack{THF \\ CuBr \\ 80-90\%}]{} RCH_2CH=CHOCH_3$$

are obtained as mixtures of (E)- and (Z)-isomers. The E/Z ratio is strongly dependent on the type of organocopper reagent used. Use of $[(CH_3)_2CHCuBr]$-MgCl results in formation of the (Z)-isomer exclusively; when $[(CH_3)_2CH]_2$-CuMgCl is used, the E/Z ratio is 30:70.

[1] S. Hoff, L. Brandsma, and J. F. Arens, *Rec. trav.*, **87**, 916 (1968).
[2] P. H. M. Schreurs, J. Meijer, P. Vermeer, and L. Brandsma, *Tetrahedron Letters*, 2387 (1976).
[3] H. Klein, H. Eijsinga, H. Westmijze, J. Meijer, and P. Vermeer, *ibid.*, 947 (1976).

2-(2-Methoxy)-allylidene-1,3-dithiane(1). Mol. wt. 188.31, b.p. 92–94°/0.07mm.
    Preparation:

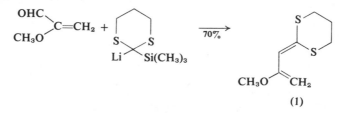

(1)

Michael reactions; Diels-Alder reactions.[1] This diene can form both Michael adducts and [4 + 2] cycloadducts with unsaturated systems. The factors that determine the mode of reaction are not understood as yet, but less reactive electrophiles seem to be more prone to form cycloadducts.

Examples:

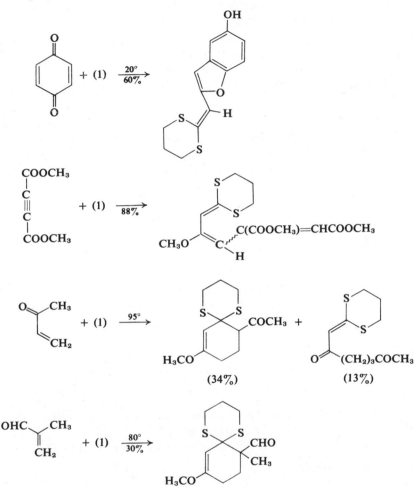

¹ S. Danishefsky, R. McKee, and R. K. Singh, *J. Org.*, **41**, 2934 (1976).

β-Methoxyethoxymethyl chloride, $CH_3OCH_2CH_2OCH_2Cl$. Mol. wt. 124.57, liquid, supplier: Aldrich.

Preparation:

$$CH_3OCH_2CH_2OH + (CH_2O)_3 \xrightarrow[90\%]{HCl} CH_3OCH_2CH_2OCH_2Cl$$

(1)

Protection of hydroxyl groups.[1] Alcohols (primary, secondary, tertiary) can be protected as β-methoxyethoxymethyl (MEM) ethers. These ethers can be prepared by reaction of (1), slight excess, with either the sodio or lithio derivative of the alcohol in THF or DME at 0° (argon). Alternatively, the ethers can be prepared by the reaction of (1) with alcohols in the presence of ethyldiisopropylamine. A third method for etherification is reaction of alcohols with the triethylammonium salt of (1), $CH_3OCH_2CH_2OCH_2N^+(C_2H_5)_3Cl^-$, in CH_3CN at reflux. Yields by the three methods are > 90%.

Deprotection can be effected with zinc bromide (CH_2Cl_2, 25°) or with $TiCl_4$ (0°, 20 minutes).

Selective reaction. Reaction of 1.5 equiv. of $CH_3OCH_2CH_2OCH_2Cl$ and 1.9 equiv. of ethyldiisopropylamine with brefeldin A (1) in CH_2Cl_2 at 25° for 24 hours results in the 7-mono ether (2) in high yield. The same ether (2) is obtained in 85% yield from the bis-4,7-MEM ether on carefully controlled

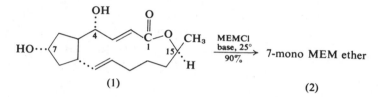

(1)   (2)

cleavage ($ZnBr_2$). The MEM group at C_7 has a notable effect on reduction of a C_4-carbonyl group. Thus reduction of (3) with sodium borohydride at $-78°$

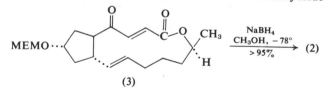

(3)

results in (2) in > 95% yield. The unprotected keto alcohol is reduced under the same conditions mainly to 4-epibrefeldin A.[2]

This steric effect was used to advantage in a total synthesis of ($\pm$)-(1).[3]

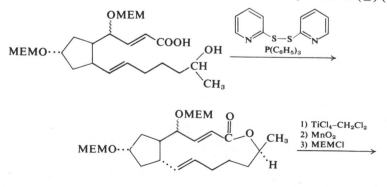

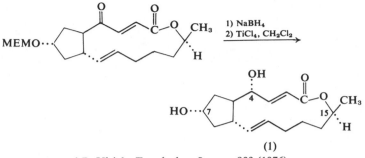

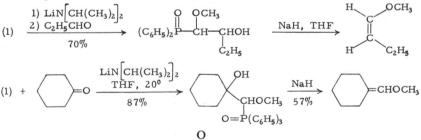

(1)

¹ E. J. Corey, J.-L. Gras, and P. Ulrich, *Tetrahedron Letters*, 809 (1976).
² E. J. Corey and R. H. Wollenberg, *ibid.*, 4701 (1976).
³ *Idem, ibid.*, 4705 (1976).

Methoxymethyldiphenylphosphine oxide, $(C_6H_5)_2\overset{\overset{O}{\|}}{P}CH_2OCH_3$ (1). Mol. wt. 246.24, b.p. 138–139°/0.1 mm.

Preparation:

$$ClCH_2OCH_3 + (C_6H_5)_2POC_2H_5 \xrightarrow[89\%]{} (1)$$

$$90\% \downarrow \begin{array}{l} 1)\ (C_6H_5)_3P \\ 2)\ OH^-,\ H_2O \end{array}$$

$$(1)$$

Vinyl ethers. The ylides $(C_6H_5)_3P{=}CHOR$ are unstable and are not particularly useful for preparation of vinyl ethers. However, the phosphine oxide (1) is stable, and it is useful for Wittig-Horner reactions with both aldehydes and ketones.

Examples:

(1) $\xrightarrow[70\%]{\begin{array}{l}1)\ LiN[CH(CH_3)_2]_2 \\ 2)\ C_2H_5CHO\end{array}}$ $(C_6H_5)_2\overset{\overset{O}{\|}}{P}{-}\overset{\overset{OCH_3}{|}}{C}H{-}\underset{\underset{C_2H_5}{|}}{C}HOH$ $\xrightarrow{NaH,\ THF}$ [vinyl ether product]

(1) + [cyclohexanone] $\xrightarrow[87\%]{\begin{array}{l}LiN[CH(CH_3)_2]_2 \\ THF,\ 20^0\end{array}}$ [product with OH and CHOCH₃, O=P(C₆H₅)₃] $\xrightarrow[57\%]{NaH}$ [=CHOCH₃ product]

The homologous reagent $(C_6H_5)_2\overset{\overset{O}{\|}}{P}{-}CH(CH_3)OCH_3$ can be used in the same way.¹

¹ C. Earnshaw, C. J. Wallis, and S. Warren, *J.C.S. Chem. Comm.*, 314 (1977).

(4S,5S)-(−)-4-Methoxymethyl-2-methyl-5-phenyl-2-oxazoline, m.p. 65–68°; **6**, 386–389.

Asymmetric syntheses of acids and lactones. This readily available oxazoline

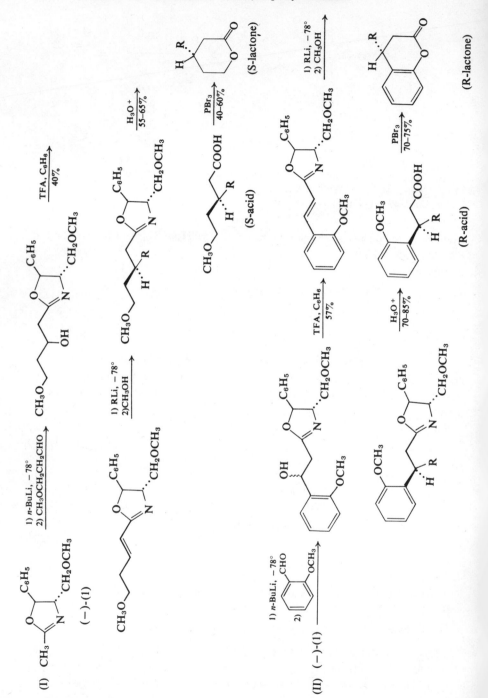

(1) has been used by Meyers and Whitten for synthesis of 5-methoxy-3-substituted acids and related lactones (equations I and II). The acids were obtained with 75–89% enantiomeric excess.

Asymmetric synthesis of dialkylacetic acids (**6**, 387). The complete details have been published.[2]

[1] A. I. Meyers and C. E. Whitten, *Tetrahedron Letters*, 1947 (1976).
[2] A. I. Meyers, G. Knaus, K. Kamata, and M. E. Ford, *Am. Soc.*, **98**, 567 (1976).

4-Methoxy-3-pentene-2-one, $CH_3COCH{=}C$
$\overset{CH_3}{\underset{OCH_3}{\diagup\diagdown}}$ (1). Mol. wt. 114.14.

This enol ether is prepared by treatment of acetylacetone with trimethyl orthoformate and *p*-TsOH (catalyst) in benzene–methanol (20°, 2 days).

Vinylogous aldols and polyenones.[1] The lithium enolate (2) of (1), prepared with LDA, reacts with a ketone (*e.g.*, β-ionone, 3) to form the vinylogous aldol (4). The aldol can be converted into the polyenone (6) by reduction with sodium

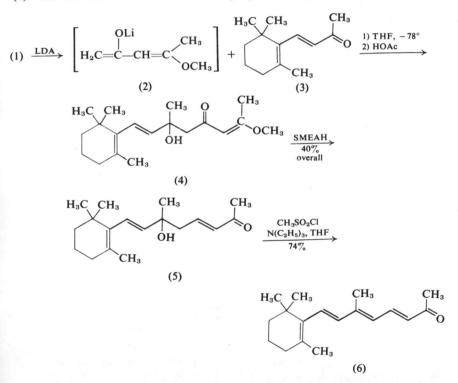

bis-2-methoxyethoxyaluminum hydride (SMEAH) to (5), followed by dehydration with mesyl chloride (2 equiv.) and triethylamine (7 equiv.) in THF at 0°.

Other examples:

$(CH_3)_2C{=}CHCH_2CH_2COCH_3 + (2) \xrightarrow[50\%]{}$

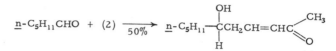

$n\text{-}C_5H_{11}CHO + (2) \xrightarrow[50\%]{}$ n-C₅H₁₁...

[1] G. Stork and G. A. Kraus, *Am. Soc.*, **98**, 2351 (1976).

2-Methoxy-3-phenylthiobuta-1,3-diene, 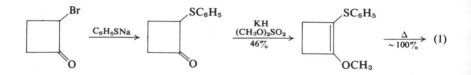 (1). Mol. wt. 192.28, polymerizes readily.

Preparation:

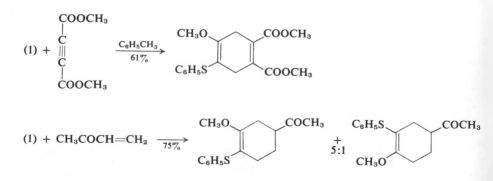

Annelation.[1] This diene undergoes Diels-Alder reactions with regioselectivity
(second and third examples). Rate acceleration with a Lewis acid, however, results
in loss of this selectivity.
Examples:

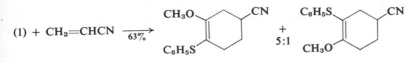

(1) + CH₂=CHCN $\xrightarrow{63\%}$

The products are useful for further transformations, for example, for synthesis of carvone (equation I).

(I)

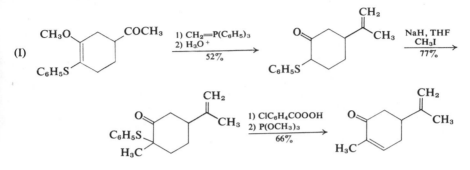

Use of the keto sulfide unit for synthesis of a protected α-diketone is formulated in equation (II).

(II)

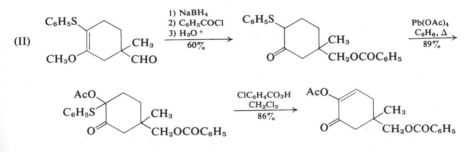

[1] B. M. Trost and A. J. Bridges, *Am. Soc.*, **98**, 5017 (1976).

(E)-1-Methoxy-3-trimethylsilyloxy-1,3-butadiene, 6, 370–371.

Details for the preparation of this useful diene are available. The crude product obtained without chromatography is suitable for most purposes; the yield of purified diene is only 46% when prepared on a larger scale.

[1] S. Danishefsky, T. Kitahara, and P. Schuda, *Org. Syn.*, submitted (1976).

α-Methoxyvinyllithium, 6, 372–373.

Dihydroxyacetone substituents. Baldwin and co-workers[1] have reported two sequences for elaboration of ketones into dihydroxyacetone substituents (equations I and II).

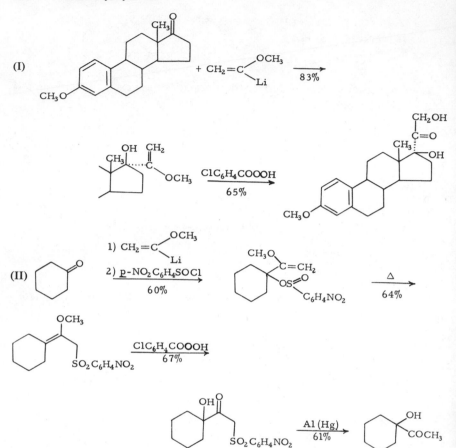

(I)

(II)

Reaction with trialkylboranes; dialkylmethylcarbinols.[2] Dialkylmethyl-carbinols can be obtained in high yield by the reaction formulated in equation (I). If treatment with acid is omitted, methyl ketones, RCOCH₃, are obtained in good yield because the second R group does not migrate from boron to carbon.

(I) $R_3B + Li-C=CH_2 \xrightarrow{-80 \to 20°}$
                 $\quad\quad\quad\quad |$
                 $\quad\quad\quad OCH_3$

$$\left[R_2\bar{B}-\overset{R}{\underset{OCH_3}{\underset{|}{\overset{|}{C}}}}=CH_2Li^+ \xrightarrow{HCl} R-B-\overset{R}{\underset{R}{\underset{|}{\overset{|}{C}}}}-CH_3 \right] \xrightarrow[85-100\%]{\underset{NaOH}{H_2O_2}} R_2\overset{OH}{\underset{}{\overset{|}{C}}}CH_3$$

[1] J. E. Baldwin, O. W. Lever, Jr., and N. R. Tzodikov, *J. Org.*, **41**, 2312 (1976).
[2] A. B. Levy and S. J. Schwartz, *Tetrahedron Letters*, 2201 (1976).

Methyl bis[methylthio]sulfonium hexachloroantimonate, 6, 375.
    Improved synthesis[1]:

$$3CH_3SSCH_3 + 3SbCl_5 \xrightarrow[\;92\%\;]{CH_2Cl_2,\;0°} 2 \quad \begin{matrix}CH_3S \\ CH_3S\end{matrix}\!\!\!\!\overset{+}{S}\!\!-\!CH_3\;SbCl_6^- + SbCl_3$$

m.p. 118–120° dec.

[1] R. Weiss and C. Schlierf, *Synthesis*, 323 (1976).

Methyl α-bromocrotonate, $CH_3CH=C\overset{\displaystyle Br}{\underset{\displaystyle COOCH_3}{\big<}}$ (1).

    This substance, a mixture of (E)- and (Z)-isomers, is obtained by bromination of methyl crotonate in CCl_4 followed by dehydrobromination (quinoline).
    Bisannelation. Tricyclo[3.2.1.0^{2,7}]octanes (3) can be obtained by reaction of enolates of cyclohexenones (2) with (1). The reaction is successful with substituents at C_2, C_3, and C_5 in (2).[1] This synthesis is comparable to one reported

(I)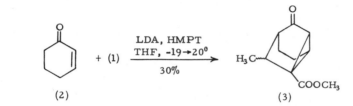

    (2)   (3)

previously that involves the reaction of vinyltriphenylphosphonium bromide with a cross-conjugated dienolate as shown in equation II.[2] Although yields are low, a tricyclic ring system is formed in one step from a monocyclic precursor.

(II)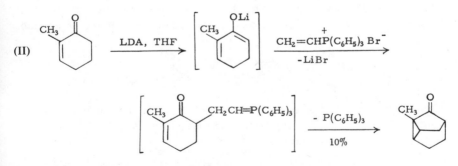

[1] H. Hagiwara, T. Kodama, H. Kosugi, and H. Uda, *J.C.S. Chem. Comm.*, 413 (1976).
[2] R. M. Cory and D. M. T. Chan, *Tetrahedron Letters*, 4441 (1975).

Methyl carbazate, 4, 333–334.
    Ketones → nitriles. The method can be used to convert cyclohexanone into cyclohexylnitrile in about 90% overall yield.[1]

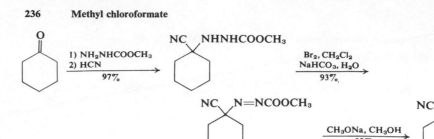

[1] P. A. Wender, M. A. Eissenstat, N. Sapuppo, and F. E. Ziegler, *Org. Syn.*, submitted (1976).

Methyl chloroformate, $ClCOOCH_3$. Mol. wt. 94.5, b.p. 70–72°. Suppliers: Aldrich, Eastman, others.

N-Demethylation. Morphine and codeine have been demethylated by treatment with this chloroformate to form a carbamate which is then cleaved with hydrazine (yields about 70%).[1] The method is a slight variation of that of Rice,[2] who used phenyl chloroformate. In the Rice method phenol rather than methanol is the by-product.

This procedure is also recommended for conversion of apomorphine to N-norapomorphine (87% yield).[3]

[1] G. A. Brine, K. G. Boldt, C. K. Hart, and F. I. Carroll, *Org. Prep. Proc. Int.*, **8**, 103 (1976).
[2] K. E. Rice, *J. Org.*, **40**, 1850 (1975).
[3] J. C. Kim, *Org. Prep. Proc. Int.*, **9**, 1 (1977).

Methylcopper, 4, 334–335; **5,** 148; **6,** 377.

Alkenes; 1,4-dienes. Methylcopper induces the cross-coupling of dialkenyl-chloroboranes (**5,** 465–466) with allyl halides to form 1,4-dienes (equation I). Coupling with an alkyl halide requires the presence of a phosphorus compound,

(I)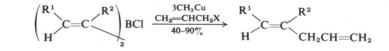

such as $P(OC_2H_5)_3$, or of C_6H_5SLi. In this case alkenes are obtained, but the yield is only moderate (equation II). The dienes obtained in (I) have almost exclusively the (E)-configuration.[1]

(II)

Coupling with (E)-2-iodo-1-alkenyl sulfones (1).[2] Monoalkyl- and monoaryl-copper reagents couple in good to high yield with these compounds with retention of configuration (equation I). Coupling with cuprous phenylacetylide[3] requires temperatures of 20–100°; yields are somewhat lower than with RCu reagents.

(I)

$$RSO_2\!\!\diagdown\!\!\underset{H}{C}\!\!=\!\!\overset{R^1}{\underset{I}{C}} + CH_3Cu \xrightarrow[50-90\%]{THF,\ O^0} RSO_2\!\!\diagdown\!\!\underset{H}{C}\!\!=\!\!\overset{R^1}{\underset{CH_3}{C}}$$

[1] Y. Yamamoto, H. Yatagai, A. Sonoda, and S.-I. Murahashi, *J.C.S. Chem. Comm.*, 452 (1976).
[2] W. E. Truce, A. W. Borel, and P. J. Marek, *J. Org.*, **41**, 401 (1976).
[3] R. D. Stephens and C. E. Castro, *ibid.*, **28**, 3313 (1963).

Methyl cyanodithioformate, $S\!\!=\!\!C\diagup^{\!\!CN}_{\diagdown SCH_3}$ (1). Mol. wt. 117.19, b.p. 50–60° (1.0 mm.), deep purple.

Preparation[1]:

$$CS_2 + NaCN \xrightarrow{DMF} S\!\!=\!\!C\diagup^{\!\!CN}_{\diagdown S^-\ Na^+} \xrightarrow[100\%]{(C_2H_5)_4N^+\ Br^-}$$

$$S\!\!=\!\!C\diagup^{\!\!CN}_{\diagdown S^-}\ ^+N(C_2H_5)_4 \xrightarrow[65\%]{\overset{CH_3I}{\underset{CH_3CN}{}}} \quad (1)$$

Ene reactions; cyanobis(methylthio)methylation.[2] The reagent undergoes ene reactions with alkenes. The product is readily converted into an anion (LDA or *n*-butyllithium), which, in the presence of HMPT, undergoes a [2,3]sigmatropic rearrangement; the resulting thiolate anion can be alkylated. This allylic cyanobis-(methylthio)alkylation is illustrated for the case of β-pinene (2).

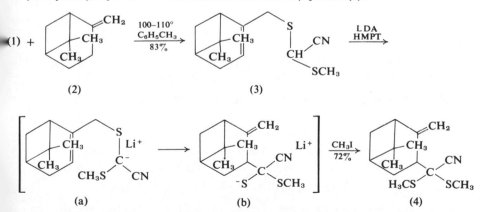

(a)   (b)   (4)

[1] H. E. Simmons, D. C. Blomstrom, and R. D. Vest, *Am. Soc.*, **84**, 4756 (1962).
[2] B. B. Snider, N. J. Hrib, and L. Fuzesi, *ibid.*, **98**, 7115 (1976).

(R)-4-Methylcyclohexylidenemethylcopper, (1).

This reagent is prepared in THF–pentane solution as shown in equation (I).

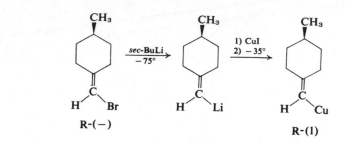

(I)

A *dissymmetric conjugated diene.* When the dark suspension of (1) is coupled oxidatively, the optically active diene (2, α_{Hg} + 37.9°) is formed in 53% yield.

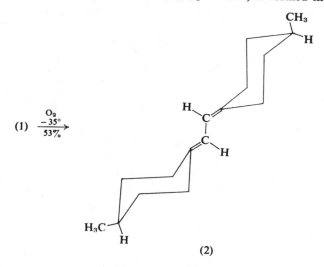

(1) $\xrightarrow[53\%]{\underset{-35°}{O_2}}$

(2)

The product is a new class of dissymmetric conjugated dienes in which the dissymmetry results from combination of two chiral olefinic axes.[1]

[1] R. B. Banks and H. M. Walborsky, *Am. Soc.*, **98**, 3732 (1976).

N-Methyl-N,N′-dicyclohexylcarbodiimidium iodide, 4, 336.

Desoxyiodo sugars. The reagent has been used for iodination of primary OH groups of carbohydrates. A yield of 71% was reported in one instance.[1]

[1] A. R. Gibson, D. M. Vyas, and W. A. Szarek, *Chem. Ind.*, 67 (1976).

L-N-Methyl-N-dodecylephedrinium bromide, L-MDEB. $C_6H_5CH(OH)CH(CH_3)-N^+(CH_3)_2C_{12}H_{14}Br^-$ (1). Mol. wt. 347.57.

This phase-transfer catalyst is prepared from L-ephedrine.[1]

Asymmetric reduction of ketones. This chiral surfactant catalyzes the reduction of ketones by $NaBH_4$ in H_2O–1,2-dichloroethane. More interestingly, the reaction is stereoselective; (+)-alcohols are formed in excess, the optical yield depending on the concentration of the catalyst. L-N-Methyl-N-hexadecylephedrinium bromide was also used, but optical yields of alcohols were lower with this reagent. Presumably the alkylephedrinium tetrahydroborate is formed in the aqueous phase and passes into the organic phase, where reduction occurs with stereoselectivity.[2]

[1] C. A. Bunton, L. Robinson, and M. F. Stam, *Tetrahedron Letters*, 121 (1971).
[2] J. P. Massé and E. R. Paraye, *J.C.S. Chem. Comm.*, 438 (1976).

Methylene chloride, 1, 676–677.

Alkylations. Methylene chloride can be used for alkylation of alcohols and phenols at 20° in the presence of solid, pulverized KOH and a phase-transfer catalyst. Aliquat 336 is more effective than benzyltriethylammonium chloride.[1]

$$2ArOH + CH_2Cl_2 \xrightarrow[80-97\%]{\substack{KOH \\ Cat.}} ArOCH_2OAr$$

$$2ROH + CH_2Cl_2 \xrightarrow[65-80\%]{} ROCH_2OR$$

[1] E. V. Dehmlow and J. Schmidt, *Tetrahedron Letters*, 95 (1976).

Methylene chloride–*n*-Butyllithium, 4, 337–339.

Benzvalene synthesis (**4**, 337–338). Swiss chemists[1] have presented evidence that the formation of benzvalene from cyclopentadienyllithium involves the intermediates depicted in equation (I).

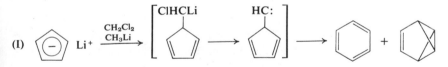

(I)

[1] U. Burger and F. Mazenod, *Tetrahedron Letters*, 2881 (1976).

Methylene chloride–Cesium fluoride.

Methylenation of catechols. This reaction can be carried out in high yield with methylene chloride (or bromide) and CsF (or KF) in DMF (110–120°).

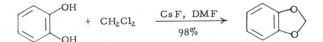

The successful outcome is believed to be a result of formation of a hydrogen bond ($O-H \cdots F^-$) between F^- and the hydroxyl group of the arene.[1]

[1] J. H. Clark, H. L. Holland, and J. M. Miller, *Tetrahedron Letters*, 3361 (1976).

(−)-N-Methylephedrine–Lithium tetra-*n*-butylaluminate (1).

The reaction of (D)-N-methylephedrine and the ate complex at 20° in cyclo-hexane leads to the chiral lithium alkoxytri-*n*-butyl aluminate (1) (equation I).

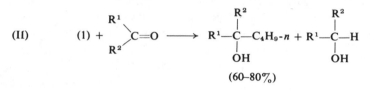

(I) $C_6H_5\cdots\overset{H}{\underset{HO}{C}}-\overset{H}{\underset{N(CH_3)_2}{C}}\cdots CH_3$ + $LiAl(n\text{-}C_4H_9)_4$ $\xrightarrow{C_6H_{12}}$ $LiAl(n\text{-}C_4H_9)_3OR^*$ + C_4H_{10}

  (R*OH)   (1)

Reaction with carbonyl compounds.[1] This chiral ate complex, like lithium tetra-*n*-butylaluminate itself, reacts with carbonyl compounds to form secondary and tertiary alcohols (equation II); optical yields are 8–31% (three examples).

(II)   (1) + $\overset{R^1}{\underset{R^2}{>}}C=O$ $\longrightarrow$ $R^1-\overset{R^2}{\underset{OH}{C}}-C_4H_9\text{-}n$ + $R^1-\overset{R^2}{\underset{OH}{C}}-H$

                          (60–80%)

[1] G. Boireau, D. Abenhaím, J. Bourdais, and E. Henry-Basch, *Tetrahedron Letters*, 4781 (1976).

Methyl fluoride–Antimony(V) fluoride, 3, 201–202.

Peterson[1] and Olah[2] and their co-workers have presented evidence that the actual species formed when methyl fluoride and antimony(V) fluoride are dis-solved in SO_2 is (1), the product of methylation of SO_2.

$$[CH_3O{=}S{=}O]^+ \, SbF_6^-$$

(1)

However, a different species is formed from CH_3F and SbF_5 in SO_2ClF or from CH_3F and $HF–SbF_5$. Olah has suggested the formula (2) for the species formed in SO_2ClF.

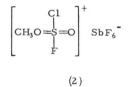

$$\left[CH_3O{=}\overset{Cl}{\underset{F}{S}}{=}O \right]^+ SbF_6^-$$

(2)

[1] P. E. Peterson, R. Brockington, and D. W. Vidrine, *Am. Soc.*, **98**, 2660 (1976).
[2] G. A. Olah, D. J. Donovan, and H. C. Lin, *ibid.*, **98**, 2661 (1976).

Methyl fluorosulfonate, 3, 202; 4, 339–340; 5, 445–446; 6, 381–383.

Caution: Dutch chemists warn that a colleague has died of pulmonary edema after exposure to the vapors of the methylating agent.[1]

Reaction with an oxaziridine. The reagent reacts with the steroidal oxaziridine (1), derived from conamine, to give the quaternary oxaziridinium salt (2). The salt is stable at 20° but decomposes at higher temperatures to (3).

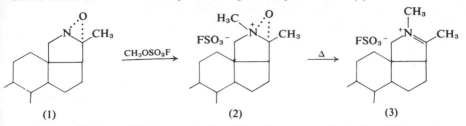

(1)   (2)   (3)

The salt (2) has oxidizing properties. Thus it converts the imine (4) into the nitrone (5). Oxidation of (4) with *p*-nitroperbenzoic acid is different in that it results in formation of (1).[2]

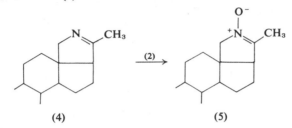

(4)   (5)

[1] D. M. W. van den Ham and D. van der Meer, *Chem. Ind.*, 782 (1976).
[2] P. Milliet, A. Picot, and X. Lusinchi, *Tetrahedron Letters*, 1573, 1577 (1976).

N-Methylhydroxylamine hydrochloride, 1, 478–481.

1,3-Cycloadditions of nitrones.[1] *cis*-N-Methyl-3-oxa-2-azabicyclo[3.3.0]octane can be prepared in 41% yield by the reaction of 5-hexenal and freshly prepared N-methylhydroxylamine (equation I). The reaction involves an intramolecular

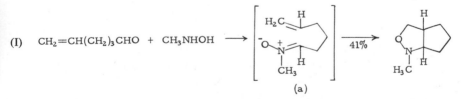

(I) $CH_2=CH(CH_2)_3CHO$ + CH_3NHOH $\longrightarrow$ (a) $\xrightarrow{41\%}$

1,3-cycloaddition of an unsaturated nitrone intermediate (a) and is a general route to fused bicyclic isoxazolidines.[2] The isolation and purification of the hydroxylamine can be avoided by use of the hydrochloride and sodium methoxide as base. It is usually neither necessary nor desirable to preform the intermediate nitrone.[3]

[1] D. S. C. Black, R. F. Crozier, and V. C. Davis, *Synthesis*, 205 (1975).
[2] N. A. Le Bel, M. E. Post, and J. J. Whang, *Am. Soc.*, 86, 3759 (1964).
[3] N. A. Le Bel and D. Hwang, *Org. Syn.*, submitted (1976).

Methyl iodide–Potassium bicarbonate.

Quaternization of amines. Amines (primary, secondary, and tertiary) are efficiently quaternized by methyl iodide and potassium bicarbonate in methanol at 20°. The reagent does not attack hydroxyl or phenolic groups, amides, or amino groups protected as the BOC or CBZ derivatives.[1]

[1] F. C. M. Chen and N. L. Benoiton, *Canad. J. Chem.*, **54**, 3310 (1976).

Methyliodine(III) difluoride, CH_3IF_2. Mol. wt. 179.94.

This only known stable alkyliodine(III) difluoride is prepared[1] by the reaction of xenon difluoride with excess methyl iodide and hydrogen fluoride at 20°.

Iodofluorination.[2] The reagent reacts with phenyl-substituted alkenes (1) to form the corresponding *vic*-iodofluorides (Markownikoff addition). Note that aryliodine(III) difluorides react with rearrangement of the phenyl group.

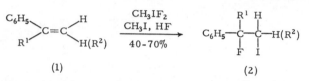

(1)   (2)

Iodofluorination of 1-phenyl-1-alkynes.[3] The reagent, under catalysis with anhydrous hydrogen fluoride, converts 1-phenyl-1-alkynes into a mixture of the (Z)- and (E)-isomers, (1) and (2). Actually (1) is the initial product and is then isomerized to (2) as the reaction proceeds to give the thermodynamically controlled equilibrium mixture.

$$C_6H_5C{\equiv}CR + CH_3IF_2\ (+ CH_3I) \xrightarrow{HF}$$

(1, Z)   +   (2, E)

[1] J. A. Gibson and A. F. Janzen, *J.C.S. Chem. Comm.*, 739 (1973).
[2] M. Zupan and A. Pollak, *J. Org.*, **41**, 2179 (1976).
[3] M. Zupan, *Synthesis*, 473 (1976).

Methyllithium, 1, 686–689; **2,** 274–278; **3,** 202–204; **5,** 448–454; **6,** 384–385.

Allenes. The first known optically active allenone (4) has been prepared by Crabbé and co-workers[1] by the route indicated. The optical purity is only 8%; the ketone loses activity rapidly at 20°.

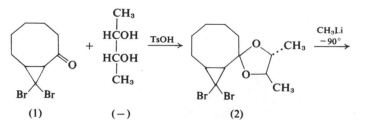

(1)   (−)   (2)

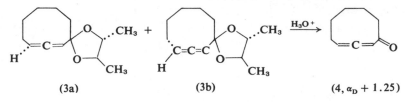

(3a)   (3b) $(4, \alpha_D + 1.25)$

Trisubstituted alkenes. Methyllithium reacts with the β-ketosilane (1) to give mainly one addition product (a), which on treatment with base yields the (E)-alkene (2) with 91% stereoselectivity. If (a) is subjected to acid treatment, the

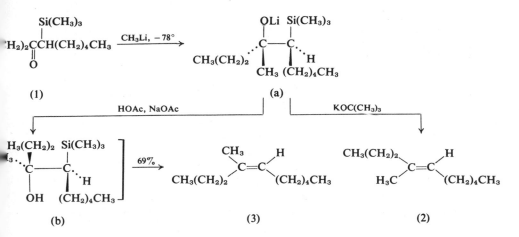

(Z)-alkene (3) is obtained with 88% stereoselectivity. Yields are lower if methyl-lithium is replaced by methylmagnesium bromide. Other alkyllithiums were also used.[2]

[1] J.-C. Damiano, J.-L. Luche, and P. Crabbé, *Tetrahedron Letters*, 779 (1976).
[2] K. Utimoto, M. Obayashi, and H. Nozaki, *J. Org.*, **41**, 2940 (1976).

Methyl methanethiosulfonate, $CH_3SO_2SCH_3$ (1), **5**, 454–455. Mol. wt. 126.20, b.p. 96–97°/4.5 mm. The reagent can also be obtained by reaction of methyl-sulfenyl chloride, CH_3SCl, with water.[1]

γ-Hydroxylation of α,β-unsaturated esters. Dienolate anions of α,β-unsaturated acids react with the usual electrophilic oxygenating reagents almost exclusively at the α-position. Ortiz de Montellano and Hsu[2] have reported an indirect method of accomplishing hydroxylation at the γ-position via rearrangement of allylic sulfoxides to isomeric allylic sulfenates (**6**, 30). The method is outlined in equation (I). The reagent (1) cannot be replaced by dimethyl disulfide.

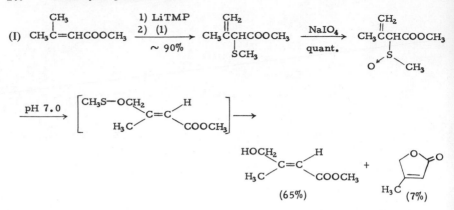

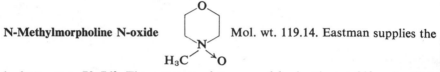

[1] W. A. Slusarchyk, H. E. Applegate, P. Funke, W. Koster, M. S. Puar, M. Young, and J. E. Dolfini, *J. Org.*, **38**, 943 (1973).

[2] P. R. Ortiz de Montellano and C. K. Hsu, *Tetrahedron Letters*, 4215 (1976).

N-Methylmorpholine N-oxide   Mol. wt. 119.14. Eastman supplies the hydrate, m.p. 73–76°. The water can be removed by heating at 90° under high vacuum.

Ruthenium catalyzed oxidation of alcohols. Primary and secondary alcohols are oxidized by this amine oxide in the presence of various ruthenium compounds. This particular amine oxide is superior to a number of others. Acetone, DMF, and HMPT are the most satisfactory solvents. Only ruthenium compounds can function as catalysts. $RuCl_3$ is the cheapest catalyst, but is poorly soluble in acetone. $RuCl_2[P(C_6H_5)_3]$ and $Ru_3(CO)_{12}$ do not suffer from this limitation. Saturated primary and secondary alcohols are oxidized in high yield; allylic alcohols are generally oxidized in satisfactory yields, but homoallylic alcohols are inactive and tend to give low conversions. Cholesterol is not oxidized. There is little discrimination between axial and equatorial hydroxyl groups.[1]

[1] K. B. Sharpless, K. Akashi, and K. Oshima, *Tetrahedron Letters*, 2503 (1976).

3-Methyl-1-phenyl-2-phospholene, 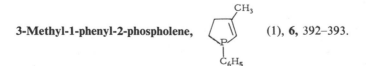 (1), **6**, 392–393.

Aromatic 1-deuterioaldehydes. The reaction of aromatic acid chlorides with (1) to form aldehydes has been extended to deuterioaldehydes.[1]

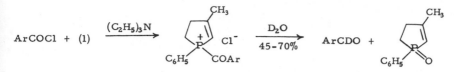

$$\text{ArCOCl} + (1) \xrightarrow{(C_2H_5)_3N} \quad \xrightarrow[45-70\%]{D_2O} \quad \text{ArCDO} + $$

[1] C. A. Scott, D. G. Smith, and D. J. H. Smith, *Syn. Comm.*, **6**, 135 (1976).

3-Methyl-2-selenoxobenzothiazole, 113°, red crystals. =Se (1). Mol. wt. 228.03, m.p.

The reagent is prepared by the reaction of 3-methylbenzothiazolium iodide with selenium (87% yield).

Deoxygenation of epoxides; desulfurization of episulfides.[1] In the presence of trifluoroacetic acid, this heterocycle converts epoxides into olefins within minutes even at 0°; yields are high (90–100%). Episulfides are also converted into olefins, but more slowly. In both cases, the configuration about the C—C bond is retained; e.g., *trans*-stilbene oxide → *trans*-stilbene. 3-Methyl-2-thioxobenzothiazole has also been used for deoxygenation of epoxides.[2]

[1] V. Calò, L. Lopez, A. Mincuzzi, and G. Pesce, *Synthesis*, 200 (1976).
[2] V. Calò, L. Lopez, L. Marchese, and G. Pesce, *J.C.S. Chem. Comm.*, 621 (1975).

S-Methyl p-toluenethiosulfate, 5, 460; **6,** 400–403.

The complete paper on the synthesis of samandarine-type alkaloids (**6**, 401) by Beckmann fragmentation has been published.[1]

[1] Y. Shimizu, *J. Org.*, **41**, 1930 (1976).

Methyl N-tosylmethylthiobenzimidate, $TsCH_2N{=}C{<}^{C_6H_5}_{SCH_3}$ Mol. wt. 319.46, m.p. 95–97°. N-Tosylmethylimino compounds are fairly stable crystalline solids. They should be stored at −20° under N_2.

Preparation:

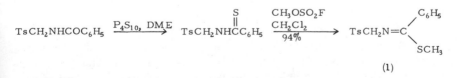

$$TsCH_2NHCOC_6H_5 \xrightarrow{P_4S_{10}, \ DME} TsCH_2NH\overset{S}{\overset{\|}{C}}C_6H_5 \xrightarrow[94\%]{\substack{CH_3OSO_2F \\ CH_2Cl_2}} TsCH_2N{=}C{<}^{C_6H_5}_{SCH_3}$$

(1)

Synthesis of heterocycles. Typical syntheses of oxazoles (equation I), imidazoles (equation II), and pyrroles (equation III) are formulated.[1]

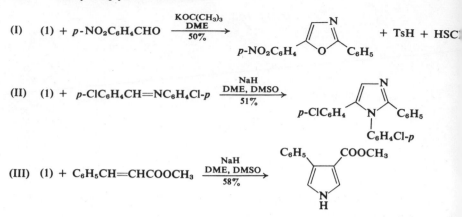

(I)   (1) + p-NO₂C₆H₄CHO $\xrightarrow[50\%]{\underset{DME}{KOC(CH_3)_3}}$ + TsH + HSC

(II)   (1) + p-ClC₆H₄CH=NC₆H₄Cl-p $\xrightarrow[51\%]{\underset{DME, DMSO}{NaH}}$

(III)   (1) + C₆H₅CH=CHCOOCH₃ $\xrightarrow[58\%]{\underset{DME, DMSO}{NaH}}$

[1] H. A. Houwing, J. Wildeman, and A. M. van Leusen, *Tetrahedron Letters*, 143 (1976).

Methyltricaprylammonium chloride (Aliquot 336), **4**, 28, 30; **5**, 460; **6**, 404–406.

Alkyl azides.[1] This quaternary ammonium salt is a suitable catalyst for the phase-transfer preparation of alkyl azides from alkyl bromides (iodides) and sodium azide in an aqueous medium; yields are 74–93%.

N-Alkylindoles. This salt can function as a catalyst for alkylation of indoles in an aqueous alkaline medium; an organic phase is not necessary and indeed is detrimental. This method fails in the case of tertiary alkyl halides and unactivated aryl halides.[2]

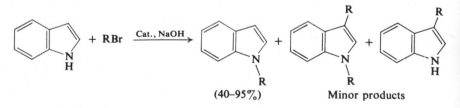

+ RBr $\xrightarrow{Cat., NaOH}$

(40–95%)   Minor products

[1] W. P. Reeves and M. L. Bahr, *Synthesis*, 823 (1976).
[2] V. Bocchi, G. Casnati, A. Dossena, and F. Villani, *Synthesis*, 414 (1976).

Methyl trifluoroacetate, CF₃COOCH₃. Mol. wt. 128.05, b.p. 43–43.5°.

The ester is prepared by esterification of the acid with methanol (slight excess) in the presence of sulfuric acid.[1]

Trifluoroacetylation of amino groups.[2] Trifluoroacetic anhydride is the traditional reagent for trifluoroacetylation of amino acids, but this method is wasteful since only one of the two trifluoroacetic units is utilized.[3] Methyl trifluoroacetate can be used for trifluoroacetylation of amino acid esters and peptide esters. The ester can also be used for trifluoroacetylation of amino acids themselves if tetramethylguanidine (1.5 molar equiv., **1**, 1145; **4**, 489–490)[4] is present.

[1] E. Gryszkiewicz-Trochimowski, A. Sporzynski, and J. Wnuk, *Rec. trav.*, **66**, 413 (1947).

[2] W. Steglich and S. Hinze, *Synthesis*, 399 (1976).

[3] F. Weygand and R. Geiger, *Ber.*, **92**, 2099 (1959).

[4] This base has been used in the preparation of BOC-amino acids: A. Ali, F. Fahrenholz, and B. Weinstein, *Angew. Chem., Int. Ed.*, **11**, 289 (1972).

Methyl vinyl ketone, 1, 697–703; **2,** 283–285; **5,** 464; **6,** 407–409.

Spirocyclohexadienones. The first step in a recent synthesis of compounds of this type involves Michael addition of an enamine to methyl vinyl ketone; an enone is obtained on hydrolysis from *in situ* intramolecular aldol condensation. The final step involves dehydrogenation with DDQ in dioxane.[1]

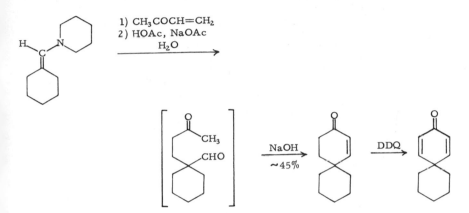

[1] V. V. Kane, *Syn. Comm.*, **6**, 237 (1976).

Molybdenum hexacarbonyl (Hexacarbonylmolybdenum), 2, 287; **3,** 206–297; **4,** 346.

Reaction with activated halides. Molybdenum hexacarbonyl reacts with α-halo ketones (equimolar amounts) in refluxing DME as shown in equation (I). The

$$(I) \quad RCOCH_2X + Mo(CO)_6 \xrightarrow[\text{2) H}_2\text{O}]{\text{1) DME}} RCOCH_3 + RCOCH=C\begin{smallmatrix}R\\ \\CH_3\end{smallmatrix}$$

$$X = Cl, Br \qquad\qquad (25\text{–}75\%) \qquad (5\text{–}25\%)$$

reaction differs from the reaction of iron pentacarbonyl with the same substrates in that monoketones, $RCOCH_3$, are the major products.

This metal carbonyl also effects coupling of some activated halides (equations II and III).[1]

$$(II) \quad (C_6H_5)_2CCl_2 \xrightarrow[65\%]{Mo(CO)_6} (C_6H_5)_2C=C(C_6H_5)_2$$

(III)

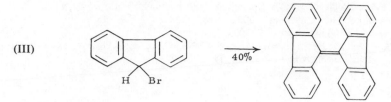

$\xrightarrow{40\%}$

Rearrangement of epoxides. This relatively inexpensive metal carbonyl can serve as a homogeneous catalyst for rearrangement of epoxides to aldehydes and for isomerization of β,γ- to α,β-unsaturated aldehydes. The olefin corresponding to the epoxide is a by-product.[2]

Examples:

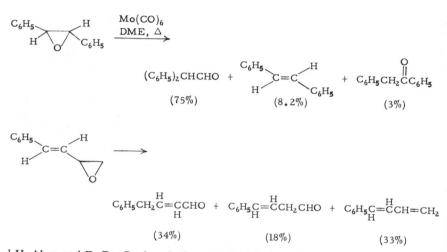

$(C_6H_5)_2CHCHO$ + [olefin] + $C_6H_5CH_2CC_6H_5$

(75%)   (8.2%)   (3%)

$C_6H_5CH_2C=CCHO$ + $C_6H_5C=CCH_2CHO$ + $C_6H_5C=CCH=CH_2$

(34%)   (18%)   (33%)

[1] H. Alper and D. Des Roches, *J. Org.*, **41**, 806 (1976).
[2] H. Alper, D. Des Roches, T. Durst, and R. Legault, *ibid.*, **41**, 3611 (1976).

Molybdenum(V) trichloride oxide (Molybdenyl chloride), $MoOCl_3$. Mol. wt. 218.3, m.p. 100° (subl.). Suppliers: Alfa, ROC/RIC.

Oxidative cleavage of hydrazones. Dimethylhydrazones and tosylhydrazones are reductively cleaved to carbonyl compounds by either $MoOCl_3$ or MoF_6. Yields are generally in the range of 70–90%.[1] *Cf.* **Tungsten hexafluoride,** this volume.

Reductive cleavage of oximes. The combination of $MoOCl_3$ and zinc in H_2O–THF cleaves oximes to carbonyl compounds (75–85% yield).[2]

Deoxygenation of sulfoxides. $MoOCl_3$–zinc or VCl_3 reduces sulfoxides to sulfides in 75–90% yield.[2]

[1] G. A. Olah, J. Welch, G. K. Surya Prakash, and T.-L. Ho, *Synthesis*, 808 (1976).
[2] G. A. Olah, G. K. Surya Prakash, and T.-L. Ho, *ibid.*, 810 (1976).

Monochloroborane diethyl etherate, 4, 346–347; **5,** 465–467.

Hydroboration of α,ω-*dienes.* Cyclic hydroboration of α,ω-dienes can be effected with this reagent. The initial product is often polymeric, but six- to eight-membered cyclic B-chloroboracycloalkanes can be obtained after de-polymerization by distillation in good yields. The B-chloroboracycloalkanes (2) and (3) from 1,4-pentadiene (1) were transformed into the cyclic ketones (6) and (7) by reaction with α,α′-dichloromethyl methyl ether (5, 200–203) followed by oxidation.[1]

Example:

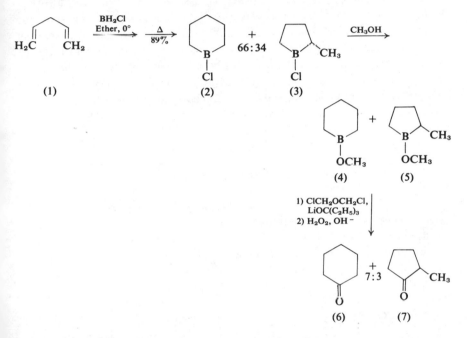

Hydroboration. The definitive paper on hydroboration of alkenes and alkynes with this reagent has been published.[2]

[1] H. C. Brown and M. Zaidlewicz, *Am. Soc.,* **98,** 4917 (1976).
[2] H. C. Brown and N. Ravindran, *ibid.,* **98,** 1785 (1976).

N

Nickel(II) acetylacetonate [Bis(2,4-pentanedionato)nickel(II)], 5, 471; **6,** 417.

1,8-Diarylnaphthalenes. Two nonequivalent aromatic nuclei can be joined by the coupling of aryl halides with aryl Grignard reagents, catalyzed by organo-metallic compounds, particularly nickel(II) acetylacetonate and dichloro[1,2-bis(diphenylphosphino)ethane]nickel(II). The latter is less useful since it is not available commercially. The Ni(acac)$_2$ catalyst does cause some homogeneous Grignard coupling, since excess Grignard reagent is necessary for satisfactory results. Roberts *et al.*[1] used this coupling reaction particularly for synthesis of the sterically crowded 1,8-diarylnaphthalenes and obtained, surprisingly, fairly high yields. For example, the yield of 1,8-diphenylnaphthalene from 1,8-diiodo-naphthalene and (excess) phenyl magnesium iodide was 70% and reproducible. The order of reactivity of the halides is I > Br ≫ Cl.[1]

[1] R. L. Clough, P. Mison, and J. D. Roberts, *J. Org.*, **41**, 2252 (1976).

Nickel carbonyl (Tetracarbonylnickel), 1, 720–723; **2,** 290–293; **3,** 210–212; **4,** 353–355; **5,** 472–474; **6,** 417–419.

Reaction of π-allylnickel bromide complexes with quinones (**4,** 354–355). Complete details are available for this method of synthesis of isoprenoid quinones.[1] Yields in general are moderate to high. The reaction is sensitive to both substrate and conditions, and each case requires individual attention. In some cases, unusual enediones are obtained (example I).

(I)   CH$_2$=CCH$_2$Br   ⟶

π-complex +   ⟶ (47%)

[1] L. S. Hegedus, B. R. Evans, D. E. Korte, E. L. Waterman, and K. Sjöberg, *Am. Soc.*, **98**, 3901 (1976).

Nickel peroxide, 1, 731–732; **5,** 474.

Oxidative cyclization of some hydrazones. Japanese chemists[1] have reported the high-yield reactions formulated in equations I and II effected with nickel peroxide in benzene at 20°.

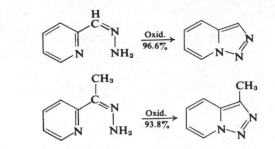

(I)

(II)

[1] S. Mineo, S. Kawamura, and K. Nakagawa, *Syn. Comm.*, **6**, 69 (1976).

Nitrobenzene, $C_6H_5NO_2$, **2**, 295.

1,3-Cycloaddition. Nitrobenzene can undergo cycloaddition to a strained alkene such as (Z,E)-1,5-cyclooctadiene (1) to form a 1,3,2-dioxazolidine (2). The reaction proceeds in THF at room temperature over a period of 5 days. The adducts are formed more rapidly and in higher yield when the benzene ring

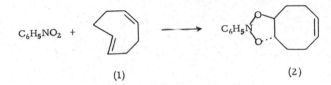

(1)   (2)

carries additional electron-withdrawing groups.[1] This cycloaddition has been observed previously only as a photochemical reaction.

[1] J. Leitich, *Angew. Chem., Int. Ed.*, **15**, 372 (1976).

p-Nitrobenzenesulfonyl peroxide, $\left(\underset{\substack{\parallel\\O}}{\overset{\substack{O\\\parallel}}{p\text{-}NO_2C_6H_4S}}O\right)_2$. Mol. wt. 202.17, m.p. 127° (explosive decomposition).

The peroxide is prepared[1] by oxidation of p-nitrobenzenesulfonyl chloride with 30% H_2O_2 (C_2H_5OH, H_2O, K_2CO_3); yield, 40%.

Oxidation of amines.[2] Primary and secondary amines are oxidized, often in high yield, to imines, usually isolated as the corresponding carbonyl compound. The reaction is carried out in ethyl acetate at $-78°$ under N_2 (equation I).

(I) $R^1CH_2NHR^2 + (ArSO_2O)_2 \longrightarrow \left[R^1CH_2\overset{\displaystyle OSO_2Ar}{\underset{\displaystyle |}{N}}R^2 \right] + ArSO_3H$

$$R^1CHO \longleftarrow R^1CH{=}NR^2 + ArSO_3H$$

Examples:

$$(C_6H_5CH_2)_2NH \xrightarrow[96\%]{} C_6H_5CHO$$

$$\underset{\underset{NH_2}{|}}{C_6H_5CHCH_3} \xrightarrow[66\%]{} C_6H_5COCH_3$$

[1] R. L. Dannley, J. E. Gagen, and O. J. Stewart, *J. Org.*, **35**, 3076 (1970).
[2] R. V. Hoffman, *Am. Soc.*, **98**, 6702 (1976).

Nitrogen dioxide, NO_2. Mol. wt. 46.01, b.p. 21.2°.

Chlorination of anthracene. Nitrogen dioxide (6 mole %) catalyzes the chlorination of anthracene by aluminum chloride to form 9-chloroanthracene (equation I). 9,10-Dichloroanthracene is obtained (52% yield) if 2 equiv. of

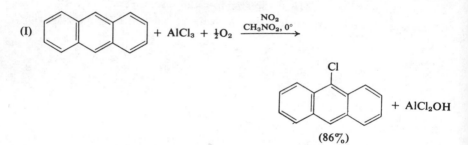

(I) ... $+ \ AlCl_3 + \tfrac{1}{2}O_2 \xrightarrow[CH_3NO_2, \ 0°]{NO_2}$ ... $+ \ AlCl_2OH$

(86%)

$AlCl_3$ are used. $TiCl_4$, $SnCl_4$, $SiCl_4$, and $FeCl_3$ can replace $AlCl_3$. $CuCl_2$ is ineffective.[1]

[1] T. Sugiyama, K. Tanioka, A. Ohno, and S. Oka, *Chem. Letters*, 307 (1976).

Nitrogen trioxide, N_2O_3; Nitrogen tetroxide, N_2O_4, 1, 737–738.

Nitrosation. Nitric oxide (NO) reacts very slowly with primary or secondary amines unless oxygen is present to convert NO into N_2O_3 and N_2O_4. These two oxides of nitrogen convert primary and secondary amines into N-nitrosamines rapidly in either neutral or alkaline aqueous solutions. This reaction is insensitive to the basicity of the amine. Nitrogen oxides are common pollutants and some N-nitrosamines (*e.g.*, N-nitrosodimethylamine) are known to be carcinogens.[1]

[1] B. C. Challis and S. A. Kyrtopoulos, *J.C.S. Chem. Comm.*, 877 (1976).

***o*-Nitrophenyl selenocyanate, 6, 420–421.**

Alkyl o-nitrophenyl selenides. These selenides can be prepared directly in high yield from primary alcohols and *o*-nitrophenyl selenocyanate in THF or pyridine at 20° in the presence of tri-*n*-butylphosphine (equation I). Yields are somewhat lower for secondary alcohols.

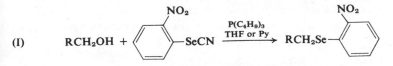

(I) RCH_2OH + [o-nitrophenyl-SeCN] $\xrightarrow[\text{THF or Py}]{P(C_4H_9)_3}$ RCH_2Se-[o-nitrophenyl]

Phenyl selenocyanate can also be used in this reaction, but not phenylselenenyl chloride, di-o-nitrophenyl diselenide, or diphenyl diselenide.[1]

[1] P. S. Grieco, S. Gilman, and M. Nishizawa, *J. Org.*, **41**, 1485 (1976).

2-Nitropropene, $CH_3C{=}CH_2$, $\underset{NO_2}{|}$. Mol. wt. 87.08, b.p. 58° (90 mm.).

Nitroethane and formalin react to form 2-nitro-1-propanol (84% yield),[1] which is dehydrated with phthalic anhydride[2] (72% yield).

1,4-Diketones.[2] Nitroolefins and silyl enol ethers undergo Michael addition under acidic conditions ($SnCl_4$ or $TiCl_4$). The reaction is carried out in CH_2Cl_2 first at −78° and finally at 0°. On addition of water, the adduct is hydrolyzed to a 1,4-diketone. The diketone can undergo aldol condensation (KOH, C_2H_5OH) to a cyclopentenone.

Examples:

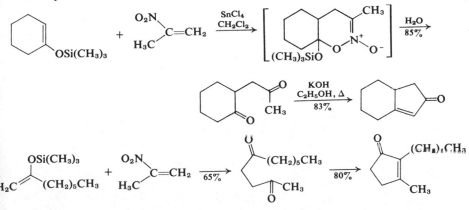

[1] G. D. Buckley and C. W. Scaife, *J. Chem. Soc.*, 1471 (1947).
[2] M. Miyashita, T. Yanami, and A. Yoshikoshi, *Am. Soc.*, **98**, 4679 (1976).

Nitrosonium tetrafluoroborate, 1, 747–748.

Oxidation.[1] Benzylic alcohols are oxidized by this reagent in methylene chloride to aldehydes or ketones in 60–80% yield. Simple alcohols cannot be oxidized in this way because they are converted into nitrite esters. However, trimethylsilyl ethers or tri-n-butylstannyl ethers are oxidized in satisfactory yields (70–85%).

Deoximation.[2] Oximes are oxidatively cleaved to carbonyl compounds by either nitrosonium tetrafluoroborate ($NO^+BF_4{}^-$) or nitronium tetrafluoroborate

$(NO_2{}^+BF_4{}^-)$ in CH_2Cl_2 and pyridine. Yields are generally higher with the former salt (55–85%). N,N-Dimethylhydrazones are also cleaved oxidatively (60–85% yield), but in this case a base is not necessary.[2]

[1] G. A. Olah and T.-L. Ho, *Synthesis*, 609 (1976).
[2] *Idem, ibid.*, 610 (1976).

Nitrosyl chloride, 1, 748–755; **2,** 298–299; **3,** 214.

Nitrosamines. Nitrosamines are usually prepared by reaction of *sec*-amines with nitrous acid in water. They can also be prepared by conversion of the amine into the sodium amide (NaH, THF) followed by reaction with nitrosyl chloride. *sec*-Amines also react directly with nitrosyl chloride in THF if pyridine is added as base. Yields by either method are fair to good.[1]

[1] R. E. Lyle, J. E. Saavedra, and G. G. Lyle, *Synthesis*, 462 (1976).

O

Orthophosphoric acid, H_3PO_4. Mol. wt. 98.00. Suppliers: Alfa and others.

Debenzylation. Two classic reagents for debenzylation are sodium in liquid NH_3 and conc. HBr. In a synthesis of α-dehydrobiotin (1, an antagonist of biotin), both reagents gave unsatisfactory results. The problem was resolved by use of orthophosphoric acid, (2) → (1).[1]

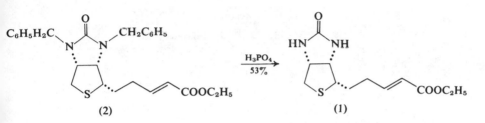

$$(2) \qquad\qquad (1)$$

[1] G. F. Field, W. J. Zally, L. H. Sternbach, and J. F. Blount, *J. Org.*, **41**, 3853 (1976).

Osmium tetroxide–*t*-Butyl hydroperoxide.

Vicinal hydroxylation. Osmium tetroxide has been the reagent of choice for effecting this reaction. However, because of the toxicity and great expense of OsO_4, procedures have been developed that use only catalytic amounts of this reagent, such as the $KClO_3$–OsO_4 reagent of K. A. Hofmann (**1**, 759) and the H_2O_2–OsO_4 reagent of Milas (**1**, 475–476). However, both of these reagents give products of overoxidation, as well as diols. Sharpless and Akashi[1] have reported yet another hydroxylation of alkenes in which a catalytic amount (0.2%) of OsO_4 is used. The reagent is *t*-butyl hydroperoxide; a base [$(C_2H_5)_4NOH$,[2] soluble in organic media] and OsO_4 in *t*-butanol are catalysts. Yields of *cis,vic*-diols with this combination are in the range of 50–75%. The yields are comparable to those obtained by Milas or Hofmann hydroxylation of mono- or disubstituted olefins, but are definitely higher for hydroxylation of tri- and tetrasubstituted olefins, which show low reactivity in the Milas and Hofmann methods.

The paper notes that RuO_4 is not a satisfactory substitute for OsO_4 in this reaction because of low yields.

[1] K. A. Sharpless and K. Akashi, *Am. Soc.*, **98**, 1987 (1976).
[2] A base is essential for this new reagent, probably for hydrolysis of intermediate osmate esters.

Osmium tetroxide–Chloramine-T, 4, 75, 445–446; **5,** 104.

Vicinal oxyamination. In the presence of catalytic amounts (1%) of osmium tetroxide the trihydrate of chloramine-T reacts with alkanes to form vicinal hydroxy *p*-toluenesulfonamides (equation I). The effective reagent is considered to be (1). In some instances addition of silver nitrate was found to be advantageous.[1]

(1)

(I) $TsNClNa \cdot 3H_2O$ +
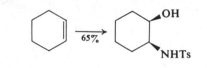
+ NaCl

Examples:

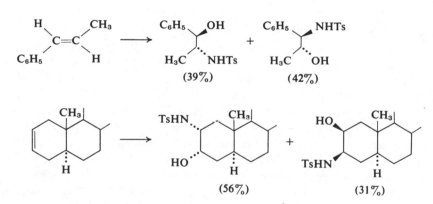

[1] K. B. Sharpless, A. O. Chong, and K. Oshima, *J. Org.,* **41,** 177 (1976).

Osmium tetroxide–N-Methylmorpholine N-oxide.

cis-1,2-Diols. Upjohn chemists[1] report that *cis*-hydroxylation of alkenes can be effected with only catalytic amounts of osmium tetroxide if a tertiary amine oxide is present to regenerate OsO_4. N-Methylmorpholine N-oxide was used routinely because it generally increases the reaction rate and is easily prepared.[2] The oxidation is carried out in aqueous acetone or *t*-butanol.

Examples:

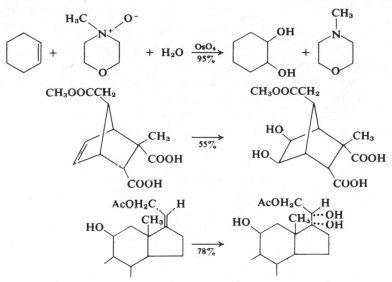

Detailed directions for the preparation of *cis*-1,2-cyclohexanediol (first example) are available.[3] In general, yields are higher than those obtained with OsO_4–$NaClO_3$ or by Schneider's procedure (**1**, 690).

[1] V. Van Rheenen, R. C. Kelly, and D. Y. Cha, *Tetrahedron Letters*, 1973 (1976).
[2] Prepared by reaction of N-methylmorpholine with 50% H_2O_2 at 50–75°; available from Eastman, Fluka.
[3] V. Van Rheenen, D. Y. Cha, and W. M. Hartley, *Org. Syn.*, submitted (1977).

Oxalyl chloride, 1, 767–772; **2**, 301–302; **3**, 216–217; **4**, 361; **5**, 481–482; **6**, 424.

β-Chloro-α,β-unsaturated ketones (**5**, 481–482). Details for the preparation of these substances from β-diketones are now available.[1]

Salicylaldehydes. Formylation of phenols generally gives mixtures of *o*- and

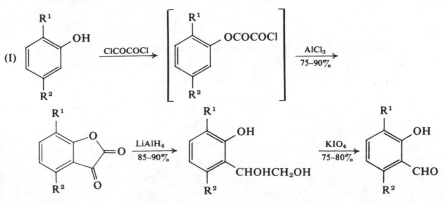

p-formylated products and is subject to steric hindrance. A new method is formulated in equation (I) and is even satisfactory when $R^1 = R^2 = C(CH_3)_3$.[2]

[1] R. D. Clark and C. H. Heathcock, *J. Org.*, **41**, 636 (1976).
[2] D. J. Zwanenburg and W. A. P. Reynen, *Synthesis*, 624 (1976).

Oxygen, 4, 362; **5,** 482–486; **6,** 246–430.

Oxygenation of 4-aryl-2,6-di-t-butylphenols. Oxygenation of these phenols in the presence of potassium *t*-butoxide in *t*-butanol results in ring contraction to 3-aryl-2,5-di-*t*-butylcyclopentadienones.[1] Evidence has now been presented that the oxygenation of (1) to (5) proceeds through (2), (3), and (4).[2]

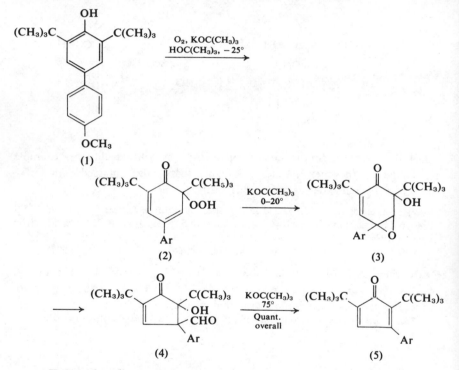

α-Keto carboxylic esters. Oxygenation, under irradiation and in the presence of tetra-*n*-butylammonium iodide, of 1-(1-ethoxycarbonylalkyl)pyridinium iodides gives α-keto carboxylic esters (equation I). A similar reaction is shown in equation (II).[3]

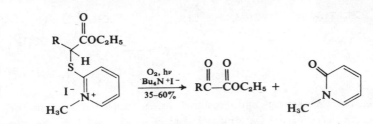

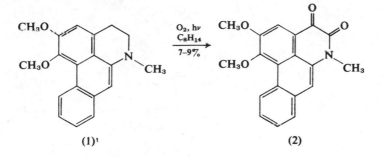

(II)

Cepharadione B (2). This aporphine alkaloid has been obtained, albeit in low yield, by photooxygenation of dehydromuciferine (1) in hexane. The solvent polarity is critical in this reaction.[4]

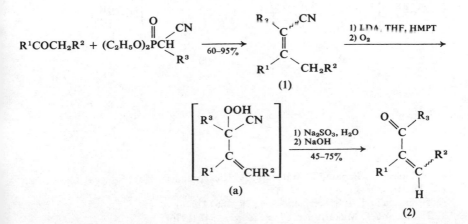

(1)[1]   (2)

α,β-Unsaturated ketones. Wroble and Watt[5] have described a new route to α,β-unsaturated ketones (2) starting with α,β-unsaturated nitriles (1), prepared from carbonyl compounds by a Wittig-Horner reaction. This oxidative decyanation is not suitable for synthesis of α,β-unsaturated aldehydes or for synthesis of

α,β-unsaturated ketones that do not possess at least one β-hydrogen. It is useful for preparation of an unsymmetrical ketone such as (5) from (3).

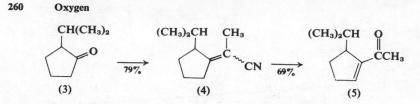

(3)   (4)   (5)

β-Lactams. Wasserman and Lipshutz[6] have reported a new method for synthesis of β-lactams from azetidinecarboxylic acids (1), which can be prepared as formulated in equation (I). The azetidine (1) is converted by LDA into the

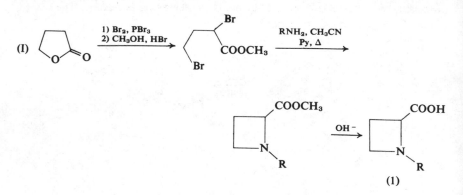

dianion (a), which is then oxygenated to give a dilithium salt of a hydroperoxy acid (b); this loses CO_2 and H_2O on addition of p-TsOH to give the β-lactam (2).

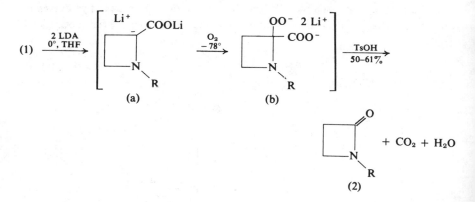

[1] A. Nishinaga, T. Itahara, T. Matsuura, A. Rieker, and D. Koch, *Angew. Chem., Int. Ed.*, **15**, 160 (1976).
[2] A. Nishinaga and A. Rieker, *Am. Soc.*, **98**, 4667 (1976).
[3] T. Takeda and T. Mukaiyama, *Chem. Letters*, 347 (1976).
[4] J. M. Saá, M. J. Mitchell, and M. P. Cava, *Tetrahedron Letters*, 601 (1976).
[5] R. R. Wroble and D. S. Watt, *J. Org.*, **41**, 2939 (1976).
[6] H. H. Wasserman and B. H. Lipshutz, *Tetrahedron Letters*, 4613 (1976).

Oxygen, singlet, 4, 362–363; 5, 486–491; 6, 431–436.

Cycloaddition to a 1,3-cyclohexadiene. Racemic cybullol (4), a metabolite of bird's nest fungi (*cyathus bulleri*), has been synthesized by the route shown in equation (I).[1]

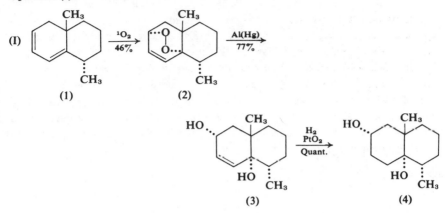

(I)

(1)   (2)

H₂
PtO₂
Quant.

(3)   (4)

Cleavage of indoles. Singlet oxygen cleaves both N-unsubstituted[2] and N-substituted[3] indoles.

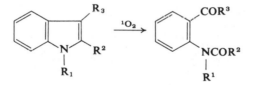

Tryptophan–kynurenine conversion. Photosensitized oxygenation of (1) followed by work-up under carefully controlled conditions gives (2) in 41% yield, originally believed to be formed from a dioxetane but now considered to arise from 3-hydroperoxyindolenine (a).

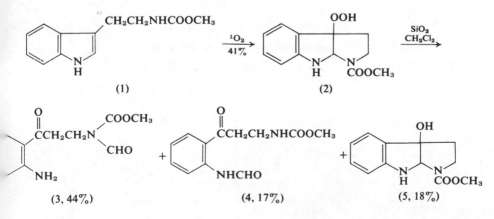

(1)   (2)

(3, 44%)   (4, 17%)   (5, 18%)

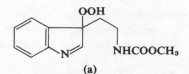

(a)

The oxygenation product (2) is convertible under mild conditions into the kynurenine derivatives (3) and (4).[4] This work may have a bearing on the known enzymic conversion of tryptophan into kynurenine.[5]

Reaction with indenes.[6] The reaction of singlet oxygen with indenes (1) at room temperature is slow and leads to complex products. However, at $-78°$ in acetone the reaction is rapid and leads to an unusual oxygenated product (2) in yields of 40–93%. Products similar to (2) are also obtained from the photo-oxygenation of 1,2-dihydronaphthalenes (3), but in this case, an ene product (6) is also formed, sometimes as the only product.[7]

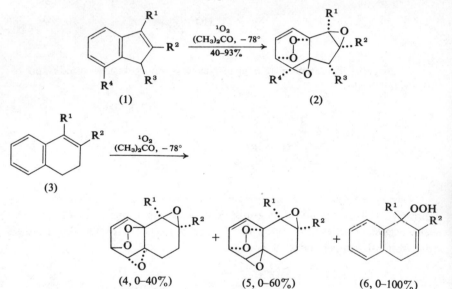

Oxygenation of pyrazines and pyrimidines. endo-Peroxides have been obtained on photosensitized oxygenation of these heterocycles.[8]

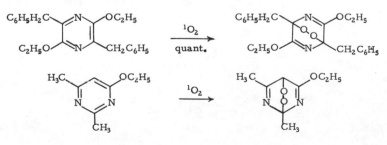

Photosensitized oxygenation of a chalcone. Photooxygenation of the chalcone (1) with Methylene Blue as sensitizer gives the flavanol (2).[9] Rose Bengal is a less efficient sensitizer. The reaction may be a model for the biological oxidation of chalcones to flavanoids.

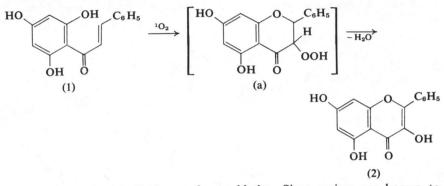

Allylic hydroperoxidation; endoperoxidation. Since amines are known to quench singlet oxygen, Paquette *et al.*[10] have examined the effect on singlet oxygen reactions of a nitrogen-containing group within compounds related to 3-norcarane. Photooxidation of (1) results exclusively, after reduction, in the allylic alcohol (2) owing to attack from the less hindered face of the double bond. Photooxidation of (3), however, is much slower and leads almost entirely to the isomeric *syn*-allylic alcohol. The hydrazide group therefore has directed attack from the more hindered side.

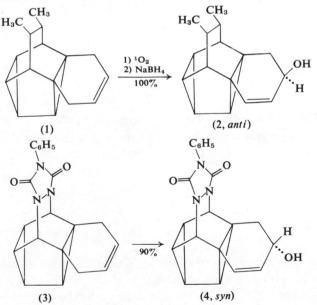

The same effect of a hydrazide substituent has also been observed in 1,4-endo-peroxidation of conjugated dienes related to norcaradienes.[11] The results have been discussed in terms of frontier molecular orbital theory.

Oxidation of 1-naphthols. These substrates (1) are photooxidized (Methylene Blue) to 1,4-naphthoquinones (2); no 1,2-naphthoquinones are detected.

(I)

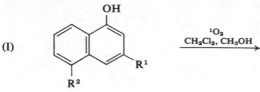

$(1, R^1, R^2 = H, OH, OCH_3)$

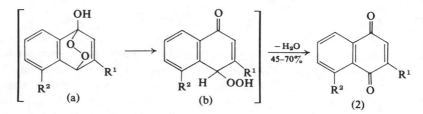

(a)   (b)   (2)

1-Methoxynaphthalene (3), however, gives, as the only isolable product, 4-methoxy-1,2-naphthoquinone (equation II). A dioxetane intermediate (c) is postulated.[12]

(II)

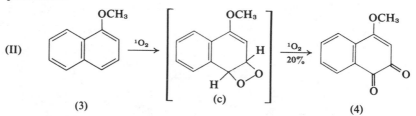

(3)   (c)   (4)

2-Naphthol is not amenable to this photoxidation.

Oxidation of selenides. Selenides are converted into selenoxides in high yield by oxidation with singlet oxygen (Rose Bengal sensitizer, CH_3OH). In the case of β-hydroxy selenides, elimination of RSeOH occurs to some extent to form allylic alcohols.[13]

$$R^1—Se—R^2 \xrightarrow{{}^1O_2} R^1—\overset{\overset{O}{\uparrow}}{Se}—R^2$$

Oxidative decarboxylation. Reaction of α-keto acids, for example, α-keto-valeric acid (1), with singlet oxygen leads to evolution of CO_2 and formation of the corresponding carboxylic acid (2). The same reaction can be carried out, but

(I) $CH_3(CH_2)_2COCOOH \xrightarrow{{}^1O_2} CH_3(CH_2)_2COOH + CO_2$

(1)   (2, 40%)   (35%)

more efficiently, by reaction of (1) with perbenzoic acid (equation II). The reaction in equation (I) therefore involves intermediate formation of the peracid, $CH_3(CH_2)_2CO_3H$, which then reacts with the starting (1) to form the product (2) and CO_2.[14]

(II)   (1) + $C_6H_5CO_3H \longrightarrow$ (2)   + CO_2 + C_6H_5COOH

(Quant.)

This work is of interest because some monooxygenases require α-ketoglutaric acid as cofactor.[15]

Photochemical ene reactions. Photooxygenation of 1,3-cholestadiene (1) does not lead to the expected endoperoxide, but to $\Delta^{1,4}$-cholestadienone-3 as the major product (equation I). Similarly, photooxygenation of 6,11-βH-eudesma-1,3-diene-6,13-olide (1) gives the hydroperoxide (2) and hyposantonin (3), which undoubtedly is formed from (2) (equation II). In both examples, the ene reaction is

(I)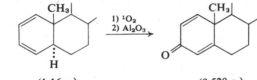

(1.16 g.)   (0.528 g.)

(II)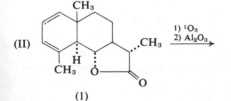

(1)

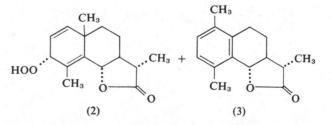

(2)   (3)

observed rather than formation of endoperoxides.[16] The paper rationalizes these results in terms of steric effects.

Naphthalene 1,4-endoperoxide. This endoperoxide (3) cannot be prepared directly by oxygenation of naphthalene, but has been prepared indirectly from 1,6-imino[10]annulene (1) as shown in equation (I). The endoperoxide (3) loses

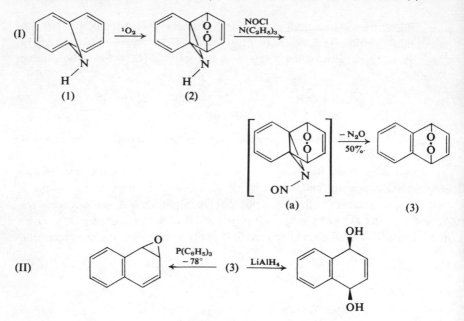

singlet oxygen with formation of naphthalene when heated and may be useful as a source of $^1O_2$.[17] Other reactions are formulated in equation (II). These reactions have been used in a synthesis of *syn*-naphthalene-1,2;3,4-dioxide.[18] Arene oxides have recently gained interest because of involvement in metabolism of carcinogenic arenes and of some antibiotics.

   Furanoterpenes. Perillene, (3), one of the simplest furanoterpenes, has been synthesized as shown in scheme (1) from β-myrcene (1) by way of an endo-peroxide (2).[19]

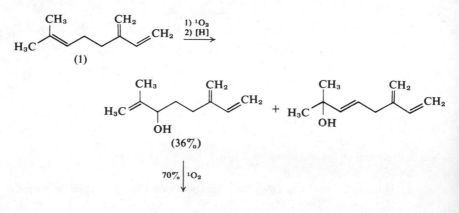

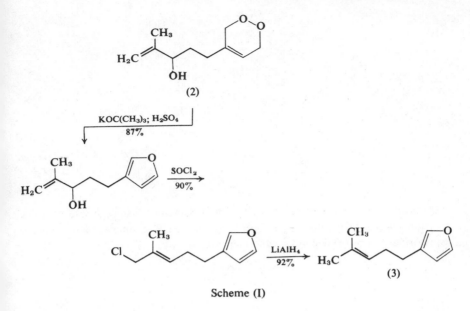

(2)

KOC(CH₃)₃; H₂SO₄
87%

SOCl₂
90%

LiAlH₄
92%

(3)

Scheme (I)

Photooxygenation of biadamantylidene (correction to **5**, 487–488). The earlier report that this alkene is photooxygenated to a dioxetane and an epoxide, the latter being formed by involvement of the solvent (pinacolone), has recently been clarified.[20] The solvent actually plays no significant role, but the photosensitizer controls the nature of the products. Use of *meso*-tetraphenylporphin results in formation of the dioxetane as major or even exclusive product. With Rose Bengal as sensitizer the epoxide is formed almost exclusively. Jefford considers that the dioxetane is the normal product.

Photooxygenation of 2-phenylcycloalkenes. The reaction of singlet oxygen with 2-phenylcyclohexene, 2-phenylcyclopentene, and 2-phenylcyclobutene has been reported (equation I–III). The last reaction (III) is solvent dependent; the

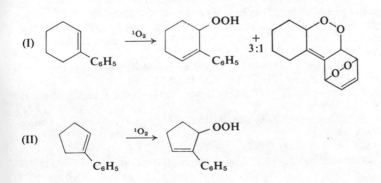

(I)

$^1O_2$

+
3:1

(II)

$^1O_2$

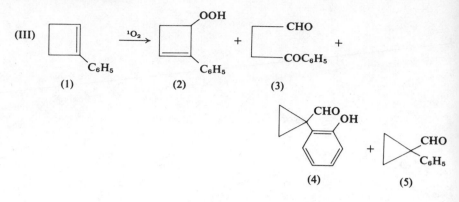

(1)   (2)   (3)

(4)   (5)

main product is usually (2) or (3). Jefford and Rimbault[21] suggest that all three reactions probably involve a perepoxide (a) as the first intermediate. Further transformations of (a) depend on the ring size.

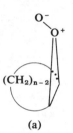

(a)

[1] W. A. Ayer, L. M. Browne, and S. Fung, *Canad. J. Chem.*, **54**, 3276 (1976).
[2] M. Nakagawa, T. Kaneko, K. Yoshikawa, and T. Hino, *Am. Soc.*, **96**, 624 (1974).
[3] I. Saito, M. Imuta, S. Matsugo, H. Yamamoto, and T. Matsuura, *Synthesis*, 255 (1976).
[4] M. Nakagawa, H. Okajima, and T. Hino, *Am. Soc.*, **98**, 635 (1976); M. Nakagawa, H. Okajima, and T. Hino, *ibid.*, **99**, 4424 (1977).
[5] A. Ek, H. Kissman, J. B. Patrick, and B. Witkop, *Experientia*, **8**, 36 (1952).
[6] P. A. Burns, C. S. Foote, and S. Mazur, *J. Org.*, **41**, 899 (1976).
[7] P. A. Burns and C. S. Foote, *ibid.*, **41**, 908 (1976).
[8] J. L. Markham and P. G. Sammes, *J.C.S. Chem. Comm.*, 417 (1976).
[9] H. M. Chawla and S. S. Chibber, *Tetrahedron Letters*, 2171 (1976).
[10] L. A. Paquette, C. C. Liao, D. C. Liotta, and W. E. Fristad, *Am. Soc.*, **98**, 6412 (1976).
[11] L. A. Paquette, D. C. Liotta, C. C. Liao, T. G. Wallis, N. Eickman, J. Clardy, and R. Gleiter, *ibid.*, **98**, 6413 (1976).
[12] J. Griffiths, K.-Y. Chu, and C. Hawkins, *J.C.S. Chem. Comm.*, 676 (1976).
[13] L. Hevesi and A. Krief, *Angew. Chem., Int. Ed.*, **15**, 381 (1976).
[14] C. W. Jefford, A. F. Boschung, T. A. B. M. Bolsman, R. M. Moriarty, and B. Melnick, *Am. Soc.*, **98**, 1017 (1976).
[15] M. T. Abott and S. Udenfriedin, *Molecular Mechanism of Oxygen Activation*, Academic Press, New York, 1974.
[16] J. W. Huffman, *J. Org.*, **41**, 3847 (1976).
[17] M. Schäfer-Ridder, U. Brocker, and E. Vogel, *Angew. Chem., Int. Ed.*, **15**, 228 (1976).

[18] E. Vogel, H.-H. Klug, and M. Schäfer-Ridder, *ibid.*, **15**, 229 (1976).
[19] K. Kondo and M. Matsumoto, *Tetrahedron Letters*, 391 (1976).
[20] C. W. Jefford and A. F. Boschung, *ibid.*, 4771 (1976).
[21] C. W. Jefford and C. Q. Rimbault, *ibid.*, 2479 (1976).

Ozone, 1, 773–777; **4,** 363–364; **5,** 491–495; **6,** 436–441.

Oxidation of secondary alcohols. There have been scattered reports of oxidation of *sec*-alcohols to ketones by ozonization. Waters and co-workers[1] have used this method for oxidation of 21 acyclic and cyclic *sec*-alcohols in methylene chloride at 0°. The method has some limitations: the oxidation of $\diagdown$CHOH is much slower than the reaction of O_3 with C=C bonds and hence is not applicable to unsaturated substrates or to substrates containing other groups reactive to O_3 (CHO, NH_2). However, ozonization compares favorably to other methods with respect to yield and expense.[1]

Cleavage of silyloxyalkenes. In the course of studies on the synthesis of vernolepin prototypes, Clark and Heathcock[2] encountered a novel ozonization of a silyloxyalkene (equation I). When the amount of O_3 is controlled, no attack

(I)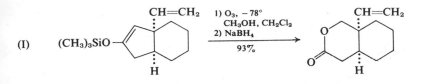

on the vinyl group is apparent. The scope of this high-yield ozonization was further explored; some examples are formulated.

Examples:

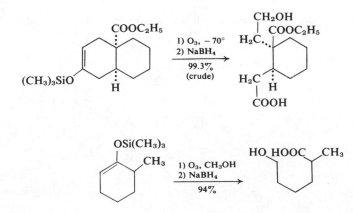

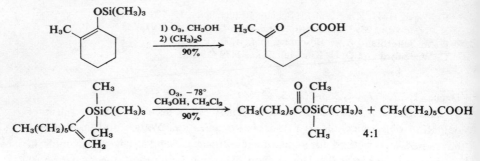

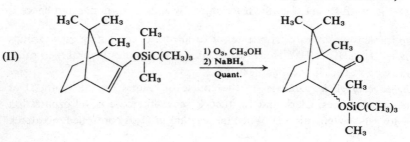

However anomalous results were also observed (equations II and III).

(II)

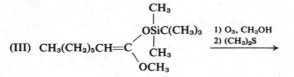

(III)

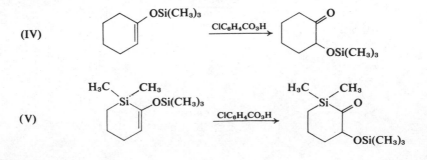

Similar oxidations of silyloxyalkenes with peracids have been reported (equations IV[3] and V[4]).

Singlet oxygen. Murray and co-workers[5] have presented evidence that some organic substrates, for example, benzaldehyde, are converted into hydrotrioxides such as (1) on ozonization. They also showed that (1) decomposes at ~10° to

(I)

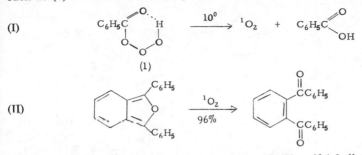

(II)

give benzoic acid and singlet oxygen (equation I). Thus if 1,3-diphenylisobenzofurane is added to the ozonization reaction and the mixture is allowed to come to room temperature, *o*-dibenzoylbenzene is formed in 96% yield. This product is known to result from oxidation of the furane with singlet oxygen.

Ozonization of alkanes in Magic Acid. Olah *et al.*[6] have studied the ozonization of alkanes in Magic Acid (**5**, 312) and SO_2ClF and suggest that the reaction involves the insertion of O_3H^+ into a C–H bond (equation I).

(I) $\overset{-}{O}-\overset{\overset{+}{O}}{\underset{}{O}}$ + H^+ $\longrightarrow$ $HO-O-O^+$

$$R-\underset{R}{\overset{R}{C}}-H \quad + \quad HO-O-O^+ \longrightarrow \left[R-\overset{R}{\underset{R}{C}}\begin{matrix} O-O-OH \\ \cdots \\ H \end{matrix} \right]^+ \xrightarrow{-H^+} R-\underset{R}{\overset{R}{C}}-O-O-OH$$

$$\xrightarrow{H^+} \left[R-\underset{R}{\overset{R}{C}}-O-\overset{+}{\underset{H}{O}}-OH \right] \xrightarrow{-H_2O_2} \underset{R}{\overset{R}{>}}C=\overset{+}{O}-R$$

[1] W. L. Waters, A. J. Rollin, C. M. Bardwell, J. A. Schneider, and T. U. Aanerud, *J. Org.*, **41**, 889 (1976).
[2] R. D. Clark and C. H. Heathcock, *ibid.*, **41**, 1396 (1976).
[3] A. G. Brook and D. M. MacRae, *J. Organometal. Chem.*, **77**, C19 (1974).
[4] G. M. Rubottom, M. A. Vazquez, and D. R. Pelegrina, *Tetrahedron Letters*, 4319 (1974).
[5] F. E. Stary, D. E. Emge, and R. W. Murray, *Am. Soc.*, **98**, 1880 (1976).
[6] G. A. Olah, N. Yoneda, and D. G. Parker, *ibid.*, **98**, 5261 (1976).

Ozone–Silica gel, 6, 440–441.

Tertiary alcohols (**6**, 440). The detailed preparation of 1-adamantanol by hydroxylation of adamantane with ozone adsorbed on silica gel has been submitted to *Organic Syntheses*. In addition to 1-adamantanol (93% yield), adamantane-1,3-diol and adamantanone are formed in minor amounts.[1]

Hydroxylation of alkenes. Tertiary carbons of alkenes can be hydroxylated by prior protection of the double bond by addition of hydrogen bromide or bromine. Bromoalkanes are somewhat less reactive than the hydrocarbons.[2]

Examples:

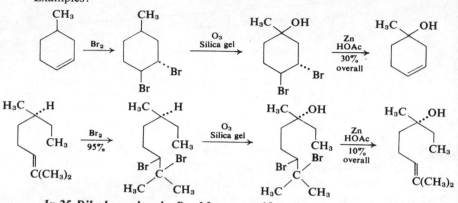

1α,25-Dihydroxyvitamin D_3. Mazur *et al.*[3] have used the dry ozonation-hydroxylation reaction in a key step in a synthesis of this calcium-regulating steroid (3). Thus the dibromide (1) is hydroxylated selectively at C_{25} by ozone absorbed on silica gel to give (2). The product was dehydrobrominated (6, 574) to the $\Delta^{5,7}$-diene, which had been converted previously[4] into the desired dihydroxyvitamin (3).

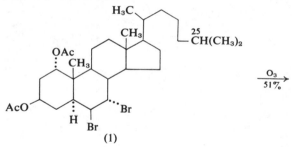

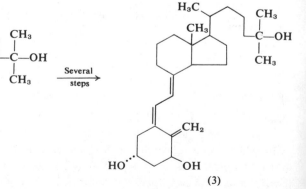

(2)   (3)

[1] Z. Cohen, H. Varkony, E. Keinan, and Y. Mazur, *Org. Syn.*, submitted (1976).

[2] E. Keinan and Y. Mazur, *Synthesis*, 523 (1976).

[3] Z. Cohen, E. Keinan, Y. Mazur, and A. Ulman, *J. Org.*, **41**, 2651 (1976).

[4] E. J. Semmler, M. F. Holick, H. K. Schnoes, and H. F. DeLuca, *Tetrahedron Letters*, 4147 (1972).

P

Palladium(II) acetate, 1, 778; **2,** 303; **4,** 365; **5,** 496–497; **6,** 442–443.

Acetoxylation of arenes. Arenes are acetoxylated by acetic acid (sodium acetate can be added) with potassium persulfate as oxidant and palladium(II) acetate as catalyst. The reaction is unusual in that *meta*-acetoxylation predominates; this selectivity can be enhanced by addition of a complexing amine such as 2,2′-bipyridine. Side-chain acetoxylation can be effected with some arenes. Thus mesitylene and durene are acetoxylated mainly in the α-position of the substituents.[1]

Arylation of allylic alcohols with aryl halides.[2,3] This reaction can be realized with palladium(II) acetate as catalyst, in some cases complexed with triphenylphosphine, and a base (triethylamine).

Example:

$$CH_2{=}CHCH_2OH + C_6H_5I \xrightarrow[71\%]{\substack{Pd(OAc)_2 \\ N(C_2H_5)_3}}$$

$$\underset{(84\%)}{C_6H_5CH_2CH_2CHO} + \underset{(16\%)}{C_6H_5\overset{\overset{\textstyle CH_3}{|}}{C}HCHO} + H\overset{+}{N}(C_2H_5)_3I^-$$

[1] L. Eberson and L. Jönsson, *Acta Chem. Scand.,* **30B,** 361 (1976).
[2] J. B. Melpolder and R. F. Heck, *J. Org.,* **41,** 265 (1976).
[3] A. J. Chalk and S. A. Magennis, *ibid.,* **41,** 273 (1976).

Palladium black, 1, 778–782; **2,** 203; **4,** 365–366; **5,** 498–499; **6,** 443–445.

Dehydrogenation. Recent syntheses of eight isomeric hydroxybenzo[*a*]pyrenes involved dehydrogenation of ketones to phenols (for example, equation I) as the final step. Pd black (Engelhard) was found to be generally superior to sulfur for this reaction, which was carried out at near reflux in high-quality 1-methylnaphthalene under argon. The choice of solvent is critical.[1]

(I)

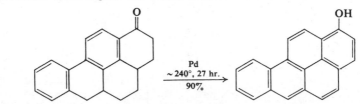

[1] H. Yagi, G. M. Holder, P. M. Dansette, O. Hernandez, H. J. C. Yeh, R. A. LeMahieu, and D. M. Jerina, *J. Org.,* **41,** 977 (1976).

Palladium catalysts, 1, 778–782; **2,** 203; **4,** 368–369; **5,** 499; **6,** 445–446.

Hydrogenation of 1,1-disubstituted cyclopropanes. Gröger and Musso[1] have observed that the hydrogenation of these compounds with 10% Pd on charcoal (or PtO_2, HOAc) is controlled by electronic factors; that is, the weakest bond is severed first when hydrogenation is conducted at the lowest possible temperature.

Examples:

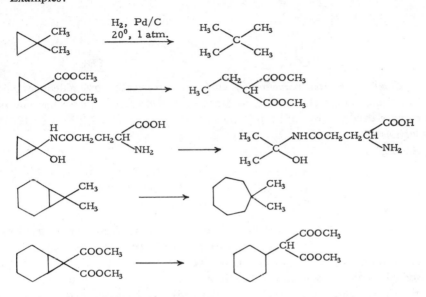

Cleavage of allyl ethers (**1,** 300; **2,** 158; **3,** 246–247; **5,** 736). Allyl ethers are cleaved when heated at 60–80° with Pd on activated charcoal in the presence of an acid (*p*-TsOH) to effect isomerization to an enol ether.[2]

Examples:

$$C_6H_5OCH_2CH{=}CH_2 \xrightarrow[\text{H}^+]{\text{Pd–C}} [C_6H_5OCH{=}CHCH_3] \xrightarrow[>95\%]{} C_6H_5OH$$

$$\begin{array}{l} CH_2OCH_2CH{=}CH_2 \\ | \\ CHOCH_2C_6H_5 \\ | \\ CH_2OCH_2C_6H_5 \end{array} \xrightarrow[>95\%]{} \begin{array}{l} CH_2OH \\ | \\ CHOCH_2C_6H_5 \\ | \\ CH_2OCH_2C_6H_5 \end{array}$$

Rosenmund reduction (**4,** 367, 368).[3] Burgstahler and co-workers[4] recommend use of 2,6-dimethylpyridine as the hydrogen chloride acceptor and THF as solvent. If the solution is dilute (< 0.25 *M*), Pd–C or Pd–$BaSO_4$ can be used as such. Aryl acyl chlorides require a quinoline-S poisoned catalyst (benzene as solvent). Nitro groups are also reduced under these conditions.

Examples:

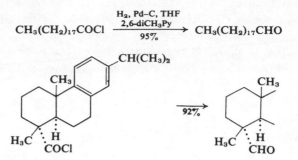

CH₃(CH₂)₁₇COCl $\xrightarrow[95\%]{\substack{\text{H}_2,\text{ Pd–C, THF} \\ \text{2,6-diCH}_3\text{Py}}}$ CH₃(CH₂)₁₇CHO

Catalytic transfer reduction of carbonyl compounds. Aryl aldehydes and ketones are completely reduced by transfer of hydrogen from cyclohexene or limonene catalyzed by 10% Pd–C and a Lewis acid (FeCl₃, AlCl₃, even H₂O). Alcohols are intermediates.[5]

Examples:

C₆H₅CHO $\longrightarrow$ C₆H₅CH₃(80%)
C₆H₅COCH₃ $\longrightarrow$ C₆H₅CH₂CH₃(100%)
(C₆H₅)₂CO $\longrightarrow$ (C₆H₅)₂CH₂(100%)

cis-Enynes. Raphael *et al.*[6] have developed a stereoselective synthesis of the terminal *cis*-enyne (5). The acetylenes (1) and (2) were coupled by the Hay[7] modification of the Glaser procedure (**1**, 168) and the product was hydrogenated over Pd on BaSO₄ poisoned with quinoline (**1**, 566). The *cis*-enyne (4) was the only product observed [some (3) was recovered]. The final step involved removal of the protecting group with tetra-*n*-butylammonium fluoride in THF (**5**, 645).

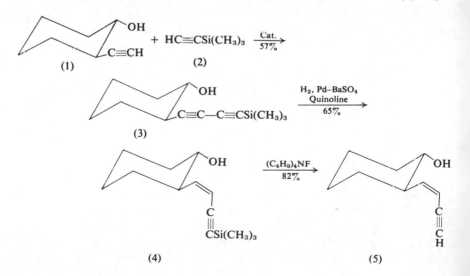

[1] C. Gröger and H. Musso, *Angew. Chem., Int. Ed.*, **15**, 373 (1976).
[2] R. Boss and R. Scheffold, *ibid.*, **15**, 558 (1976).
[3] E. Mosettig and R. Mozingo, *Org. React.*, **4**, 362 (1948).
[4] A. W. Burgstahler, L. O. Weigel, and C. G. Shaefer, *Synthesis*, 767 (1976).
[5] G. Brieger and T.-H. Fu, *J.C.S. Chem. Comm.*, 757 (1976).
[6] A. B. Holmes, R. A. Raphael, and N. K. Wellard, *Tetrahedron Letters*, 1539 (1976).
[7] A. S. Hay, *J. Org.*, **27**, 3320 (1962).

Palladium(II) chloride, 1, 782; **3,** 303–305; **4,** 367–370; **5,** 500–503; **6,** 447–450.

π-Allylpalladium dimers (**4,** 369; **5,** 500–501; **6,** 45–47, 447–449). 3-Oxo-Δ⁴-steroids can be converted conveniently into 6β-derivatives by conversion to the π-allyl complexes followed by reaction with a nucleophile (equation I).[1]

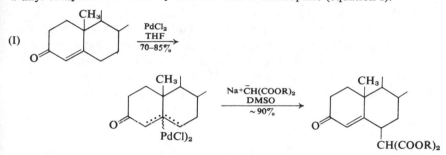

Oxidation of primary amines. Primary amines with at least one α-hydrogen atom are oxidized to carbonyl compounds by $PdCl_2$ in water. Yields range from 5 to 70%, depending on the substrate and reaction pH. Chloroauric acid, $HAuCl_4$, can also be used. Oxidation of secondary amines in this way is not satisfactory. However, indoline can be oxidized to indole by $PdCl_2$ in yields as high as 83%.[2]

Polymer-supported reagent (**6,** 450). This form can be prepared from polymeric diphenylbenzylphosphine and $PdCl_2$. Alkynes and allyl halides can be co-dimerized with this catalyst (equation I).[3] The ratio of the two possible products

(I) $C_6H_5C≡CY + CH_2=CHCH_2X \xrightarrow{Cat.}$
$\begin{cases} CH_2=CHCH_2CY=C(X)C_6H_5 \\ + \\ CH_2=CHCH_2C(C_6H_5)=CXY \end{cases}$

depends on the solvent and the mode of preparation of the polymeric catalyst. The rate of codimerization is somewhat slower and the yields are lower than with the homogeneous catalyst $PdCl_2(C_6H_5CN)_2$.[4]

[1] D. J. Collins, W. R. Jackson, and R. N. Timms, *Tetrahedron Letters*, 495 (1976).
[2] M. E. Kuehne and T. C. Hall, *J. Org.*, **41**, 2742 (1976).
[3] K. Kaneda, T. Uchiyama, M. Terasawa, T. Imanaka, and S. Teranishi, *Chem. Letters*, 449 (1976).
[4] K. Kaneda, F. Kawamoto, Y. Fujiwara, T. Imanaka, and S. Teranishi, *Tetrahedron Letters*, 1067 (1974).

Palladium(II) chloride–Chloranil.

Dehydrogenation. Cyclopentanones can be dehydrogenated to cyclopentenones by reaction with $PdCl_2$ (1 equiv.) and chloranil (1 equiv.) in conc. HCl (110°, 2–3 days).[1]

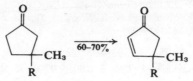

[1] S. Wolff and W. C. Agosta, *Synthesis*, 240 (1976).

Palladium(II) chloride–Copper(I) chloride.

1,4- and 1,5-Diketones. Palladium(II) chloride has been used industrially as a catalyst for oxygenation of alkenes to carbonyl compounds, but this reaction has been of little use in the research laboratory. Japanese chemists have now used this reaction for preparation of 1,4-diketones from α-alkenyl ketones. Cuprous chloride is used as reoxidant (equation I). The method is also suitable for preparation of 1,5-diketones (equation II).[1]

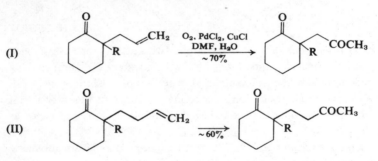

[1] J. Tsuji, I. Shimizu, and K. Yamamoto, *Tetrahedron Letters*, 2975 (1976).

Palladium chloride(II)–Thiourea, 6, 450–451.

α-Methylene lactones, Heathcock *et al.*[1] used Norton's cyclocarbonylation reaction (6, 451) to prepare (2) from (1). The product was obtained in only 21% yield (isolated).[1]

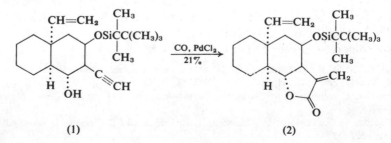

(1)   (2)

[1] C. G. Chavdarian, S. L. Woo, R. D. Clark, and C. H. Heathcock, *Tetrahedron Letters*, 1769 (1976).

Peracetic acid–Sodium acetate.

Allene oxides (**2**, 308–309; **3**, 49–50). Chan et al.[1] have shown that this system can be generated via vinylsilanes, as formulated for 1-*t*-butylallene oxide in equation (I). This allene oxide is stable at 25° for 1 hour; it polymerizes but does

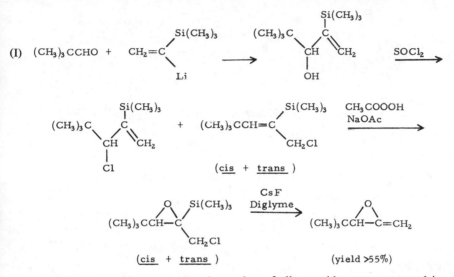

not isomerize to a cyclopropanone. A number of allene oxides were prepared in this way and the stability was shown to depend on the substituent at C_1. When the group is aryl the oxide rearranges very readily to a cyclopropanone.[2]

[1] T. H. Chan, B. S. Ong, and W. Mychajlowskij, *Tetrahedron Letters*, 3253 (1976).
[2] B. S. Ong and T. H. Chan, *ibid.*, 3257 (1976).

Perbenzoic acid, 1, 791–796; **3,** 219; **6,** 453.

Polymer-supported reagent. A polymer-supported reagent has been prepared from carboxy-substituted polystyrene resins. Its properties are similar to those of perbenzoic acid itself; it reacts with di- and trisubstituted olefins in THF at 40° to give epoxides in fairly good yields, but forms epoxides from monosubstituted olefins in low yield. The resins lose activity slowly at 0°.[1] The polymer-supported reagent was used for oxidation of penicillins and desacetoxycephalosporins to (S)-oxides in yields of 70–100%.[2]

[1] C. R. Harrison and P. Hodge, *J.C.S. Perkin I*, 605 (1976).
[2] *Idem, ibid.*, 2252 (1976).

Perchloric acid, 1, 796–802; **2,** 309–310; **3,** 220; **5,** 506–507; **6,** 453–454.

2-Alkyltetronic acids. Schlessinger et al.[1] have developed a route to these mold metabolites (3). Although several steps are involved, the overall yields are surprisingly good (65–85%). The key intermediate is an α-hydroxy thiomethyl ester (1), prepared as shown. This is then allowed to react with an ester enolate together with LDA to form a hydroxy keto ester (2). The last step is elimination of *t*-butanol, best accomplished with 70% perchloric acid.

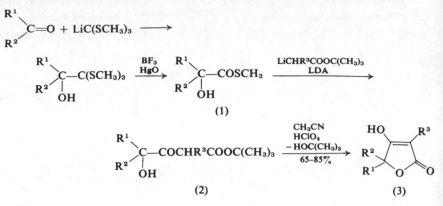

(1)

(2)   (3)

[1] R. E. Damon, T. Luo, and R. H. Schlessinger, *Tetrahedron Letters*, 2749 (1976).

Perchloryl fluoride, 1, 802–803; **2,** 310–311; **6,** 454.

Fluorination of an ester. Grieco *et al.*[1] have successfully fluorinated the enolate of the ester (1) with perchloryl fluoride. *Caution*: A violent explosion occurred in one experiment conducted below −40°. Two products, (2) and (3), were

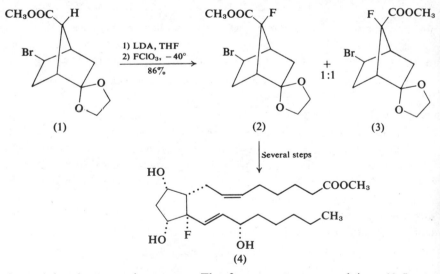

(1)   (2)   (3)

Several steps

(4)

formed in about equal amounts. The former was converted into 12-fluoro-prostaglandin $F_{2\alpha}$ methyl ester (4).

[1] C.-L. J. Wang, P. A. Grieco, and F. J. Okuniewicz, *J.C.S. Chem. Comm.*, 468 (1976).

Peroxybenzimidic acid (Payne's reagent), 1, 469–470; **4,** 375; **5,** 511; **6,** 455–456.

Oxidation of allenic alcohols. French chemists[1] have reported an interesting preparation of 3-oxacyclanones (equations I and II) by oxidation of allenic alcohols with this peracid. Yields are low when the alcohols are primary, around 60% when the alcohols are secondary, and almost quantitative when the alcohols

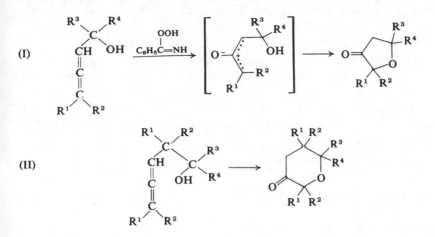

are tertiary. These reactions are believed to involve intermediate zwitterions as postulated for several other reactions by Hoffmann.[2]

Oxidation of β-allenic alcohols of the type (1) with Payne's reagent results in formation of γ-lactones (2) probably via an allenic oxide (a).[3]

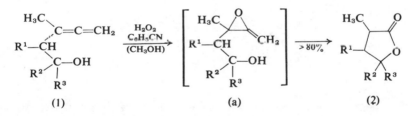

[1] M. Bertrand, J. P. Dulcere, G. Gil, J. Grimaldi, and P. Sylvestre-Panthet, *Tetrahedron Letters*, 1507 (1976).

[2] H. M. R. Hoffmann, *Angew. Chem., Int. Ed.*, **12**, 819 (1973).

[3] M. Bertrand, J. P. Dulcere, G. Gil, J. Grimaldi, and P. Sylvestre-Panthet, *Tetrahedron Letters*, 3305 (1976).

Pertrifluoracetic acid, 1, 821–827; **2,** 316; **3,** 221.

Hydroxylation of alkanes. The reaction of cyclohexane with this peracid at either 25 or 83° results in formation of cyclohexyl trifluoroacetate (73% yield) together with minor amounts of cyclohexanol and two other products (not cyclohexanone or *e*-caprolactone). The position of hydroxylation of alkanes is controlled by electronegative substituents. Thus octanol-1 is hydroxylated mainly at C_7. Hydroxylation of alkanes containing *t*-hydrogen atoms is complex.[1]

[1] N. C. Deno and L. A. Messer, *J.C.S. Chem. Comm.*, 1051 (1976).

p-Phenylbenzoyl chloride (4-Biphenylcarbonyl chloride), —COCl.

Mol. wt. 216.67, m.p. 111–113°. Suppliers: Eastman, Fluka.

Protection of primary and secondary amines. Scribner[1] recommends the *p*-phenylbenzoyl group for protection of these amines. Derivatives of even liquid amines are crystalline solids. Deprotection can be achieved with 3% sodium amalgam in methanol at 25°; known methods for hydrolysis of amides are ineffective or require vigorous conditions. Most amides [*e.g.*, $C_6H_5CON(CH_3)_2$] react only slowly with Na(Hg).

[1] R. M. Scribner, *Tetrahedron Letters*, 3853 (1976).

Phenyl(bromodichloromethyl)mercury, 1, 851–854; 2, 326–328; 5, 225; 4, 385; 5, 514.

α-Chloro-α,β-unsaturated esters. These substances can be prepared as shown in equation (I).[1]

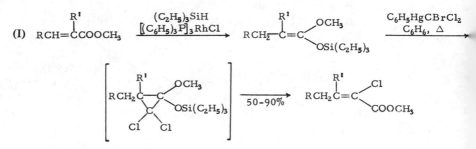

[1] E. Yoshii, T. Koizumi, and T. Kawazoe, *Chem. Pharm. Bull. Japan*, **24**, 1957 (1976).

Phenylcopper, 5, 504; 6, 234.

α-Arylation of α,β-unsaturated ketones. The reaction can be accomplished as illustrated for α-arylation of isophorone oxide (1).[1] The scheme utilizes the

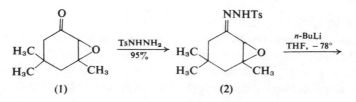

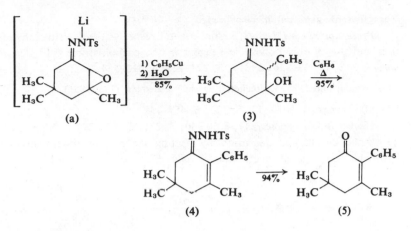

(a)   (3)   (4)   (5)

fragmentation reaction of Eschenmoser (**2**, 419–422; **6**, 232–233). The phenyl-copper is solubilized as the complex with diisopropyl sulfide. This method of Fuchs is also applicable to acyclic enones.[1]

This reaction (as well as α-alkylation) has been carried out in a somewhat similar way by Stork and Ponares,[2] whose method has the advantage that the readily available Grignard reagents are employed. This procedure is also illustrated for the case of isophorone oxide (1); however, it is not useful in the case of secondary Grignard reagents or when the epoxy ketone bears no hydrogen.

$$(1) + H_2NN(CH_3)_2 \xrightarrow[95\%]{\begin{array}{c} CH_3COOH \\ CH_3COOC_2H_5 \end{array}}$$

(6)   (7)

It has been used in addition for preparation of β-hydroxy ketones (equation I).

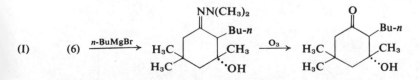

(I)   (6)

[1] P. L. Fuchs, *J. Org.*, **41**, 2935 (1976).
[2] G. Stork and A. A. Ponares, *ibid.*, **41**, 2939 (1976).

Phenylboronic acid, polymer-supported, (P)—B(OH)₂.

Protection of cis-1,3-diols, Seymour and Fréchet[1] have described the preparation and use of this polymeric reagent in the carbohydrate field. *Cf. Phenylboronic acid,* **1**, 833–834; **2**, 317; **3**, 221–222; **5**, 513–514.

[1] E. Seymour and J. M. J. Fréchet, *Tetrahedron Letters*, 1143 (1976).

Phenylhydrazine, 1, 838–842; **2,** 322; **6,** 457–458.

Fischer indole synthesis (**1**, 839–840; **3**, 231; **5**, 396–397). This reaction can
be carried out in one operation by heating the ketone and phenylhydrazine
hydrochloride in glacial acetic acid at reflux.[1]

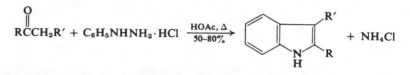

[1] V. Dave, *Org. Prep. Proc. Int.*, **8**, 41 (1976).

Phenyl isocyanate, 1, 842–843; **2,** 322–323; **4,** 378.

Nitriles. α-Oximino acids react with phenyl isocyanate in benzene solution
at ambient temperatures to form nitriles (equation I).[1]

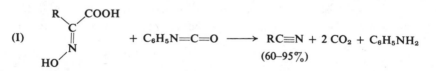

Isoalloxazines. This heterocyclic system (*e.g.,* 4) is of interest because it is
the basic structure of riboflavin and lumiflavins. A new synthetic approach has
been developed by reaction of isocyanates with 1-alkyl-2-aminopyrazines (1).
An example is formulated in equation (I).[2]

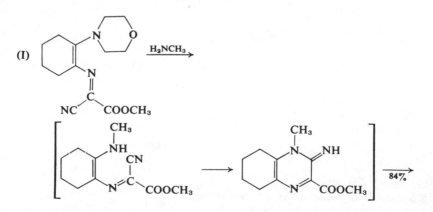

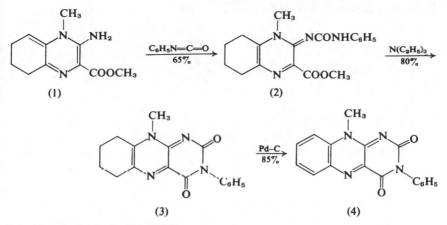

(1)   (2)

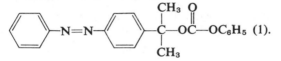

(3)   (4)

[1] A. Ahmad, *Synthesis*, 418 (1976).
[2] D. Clerin, A. Lacroix, and J.-P. Fleury, *Tetrahedron Letters*, 2899 (1976).

Phenyl-[2-(*p*-phenylazophenyl)isopropyl]carbonate (AZOC—OPh),

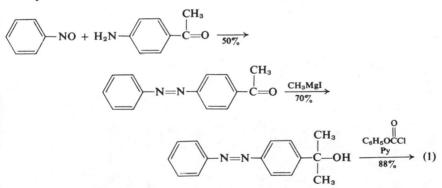

Mol. wt. 360.39, m.p. 100–103°, stable, λ_{max} 320–330 nm.
Preparation:

Protection of amino groups.[1] This reagent has been introduced for protection of amino groups in peptide synthesis. The derivatives are formed preferably with the tetramethylguanidine salts in DMSO (yields are generally 50–75%). The color of the derivatives facilitates isolation and purification and also removal of the protective group, which is carried out with mild acid treatment. AZOC—N_3 (m.p. 49–50°, stable at 0°) can be used in place of (1).

[1] A. Tun-Kyi and R. Schwyzer, *Helv.*, **59**, 1642 (1976).

Phenyl N-phenylphosphoroamidochloridate (1). Mol. wt. 267.7, m.p. 129–133°.
Preparation:

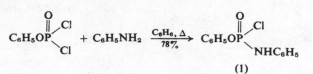

$$\text{(1)}$$

Phosphorylation.[1] The compound reacts with an alcohol or a phenol at 0° in
pyridine to form O-alkyl-O-phenyl N-phenylphosphoromidates (2) in 70–86%
yield. The aniline group in (2) is removed by treatment with isoamyl nitrite (excess).[2]

$$\text{(1)} + \text{ROH} \xrightarrow[70-86\%]{\text{Py}} C_6H_5OP\underset{\text{NHC}_6H_5}{\overset{\text{OR}}{}} \xrightarrow[85-95\%]{\substack{1) \ i\text{-}C_5H_{11}ONO \\ 2) \ \text{NaOH}}} C_6H_5OP\underset{\text{ONa}}{\overset{\text{OR}}{}}$$

$$\text{(2)} \qquad\qquad\qquad\qquad \text{(3)}$$

[1] W. S. Zieliński and Z. Leśnikowski, *Synthesis*, 185 (1976).
[2] M. Ikehara, S. Uesugi, and T. Fukui, *Chem. Pharm. Bull. Japan*, **15**, 440 (1967).

Phenylselenenyl bromide and chloride, 5, 518–522; **6,** 459–460.
    Detailed instructions for preparation of the latter selenium reagent (m.p.
62–64°) from bromobenzene are available.[1] Commercial material is expensive.

$$2 \ C_6H_5Br \xrightarrow[\text{2) Se}]{\text{1) Mg}} \left[2 \ C_6H_5SeMgBr \right] \xrightarrow[64-70\%]{\text{Br}_2}$$

$$C_6H_5SeSeC_6H_5 \xrightarrow[88-93\%]{\text{Cl}_2} 2 \ C_6H_5SeCl$$

β-*Dicarbonyl enones.* Detailed instructions are available for preparation of
2-acetyl-2-hexene-1-one (equation I). The publication includes directions for

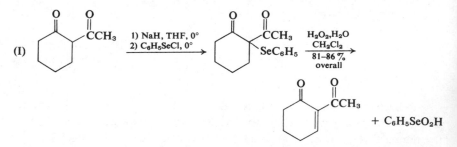

controlling the exothermic oxidation reaction.[2] Other methods for synthesis of
these enones result in enolized material.
    (±)-*Muscone.* An efficient, recent synthesis of (±)-muscone (3) from exaltone
(1, Aldrich) is formulated in equation I.[3]

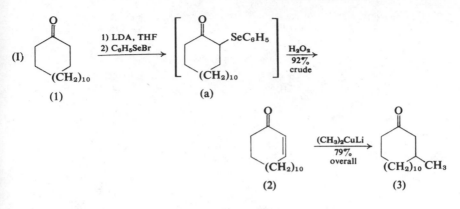

(I)

(1)   (a)   (2)   (3)

[1] H. J. Reich, M. L. Cohen, and P. D. Clark, *Org. Syn.*, submitted (1976).
[2] J. M. Renga and H. J. Reich, *ibid.*, submitted (1976).
[3] S. H. Korzeniowski, D. P. Vanderbilt, and L. B. Hendry, *Org. Prep. Proc. Int.*, **8**, 81 (1976).

N-Phenyl-1,2,4-triazoline-3,5-dione, 1, 849–850; **2,** 324–326; **3,** 223–224; **4,** 381–383; **5,** 528–530; **6,** 467.

Oxidation of alcohols. This triazoline-3,5-dione has been used to oxidize alcohols to carbonyl compounds with formation of 4-phenyl-1,2,4-triazolidine-3,5-dione.[1]

Further studies[2] indicate that this method of oxidation is efficient only with easily oxidized alcohols, especially arylalkylmethanols. Reasonable yields can be obtained in the case of propane-2-ol (68%), cyclopentanol (60%), and benzyl alcohol (80%). Simple primary aliphatic alcohols are converted mainly into N-phenylcarbamates formed from two molecules of the dione and one of the alcohol (equation I). Pyridine catalyzes this reaction and suppresses oxidations of the alcohol.

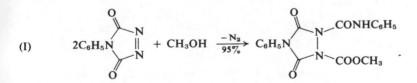

(I)

Reaction with α,β-enones. This enophile reacts with (+)-pulegone (1) at room temperature to give a mixture of two epimeric adducts (2). It also reacts with mesityl oxide (3) to give the adduct (4). However, isophorone (3,5,5-tri-methylcyclohexene-2-one-1) does not react with PTAD under the same conditions. Thus the reaction seems to be limited to α,β-enones which can assume an s-*cis* conformation, and may not be a usual ene reaction.[3]

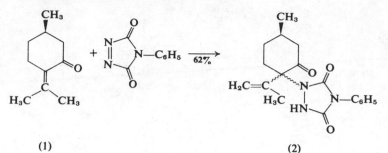

(1)   (2)

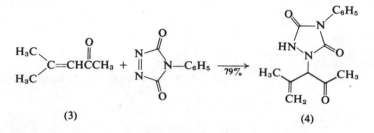

(3)   (4)

[1] R. C. Cookson, I. D. R. Stevens, and C. T. Watts, *J.C.S. Chem. Comm.*, 744 (1966).
[2] L. H. Dao and D. Mackay, *ibid.*, 326 (1976).
[3] J. D. Schiloff and N. R. Hunter, *Tetrahedron Letters*, 3773 (1976).

Phenyltungsten trichloride–Aluminum chloride, $C_6H_5WCl_3$–$AlCl_3$ (1:1).

The complex is prepared from the two components in deoxygenated chlorobenzene under argon. Unstable to oxygen.

Diene and cyclobutane interconversions. Gassman and Johnson[1] have found that treatment of the diene (1) with catalytic amounts of the complex for 5

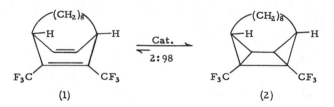

(1)   (2)

minutes results in formation of the cyclobutane derivative (2). Although the catalyst does not effect conversion of (4) or (6) into (3) and (5), it does effect the opposite metathesis of (3) and (5) into (4) and (6), respectively.

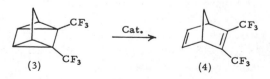

(3)   (4)

(5)   (6)

These experiments are of interest because the mechanism of metathesis of olefins is not well understood.[2]

[1] P. G. Gassman and T. H. Johnson, *Am. Soc.*, **98**, 861 (1976).
[2] R. J. Haines and G. J. Leigh, *Chem. Soc. Rev.*, **4**, 155 (1975); N. Calderon, *Accts. Chem. Res.*, **5**, 127 (1972).

Phosgene (Carbonyl chloride), 1, 856–859; **2,** 328–329; **3,** 225; **5,** 532–534.

Aryl thiocarboxylic acid chlorides. These compounds can be obtained from reaction of aryl carbodithioic acids with phosgene at 0–35° followed by thermolysis of the S-chlorocarbonyl derivatives below 1 torr. The products are stable for several months in the dark at 0°.[1]

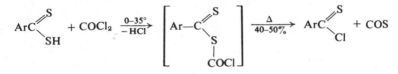

[1] H. Viola and R. Mayer, *Z. Chem.*, **15**, 348 (1975).

Phosphoric acid–Boron trifluoride complex, $H_3PO_4 \cdot BF_3$. Supplier: Fluka.

Cyclization. The bicyclic precursors (1) and (2) are cyclized by this system to the *cis*-fused phenanthridine (3) without any detectable trace of the *trans*-isomer. The formyl derivatives corresponding to (1) and (2) are cyclized to both

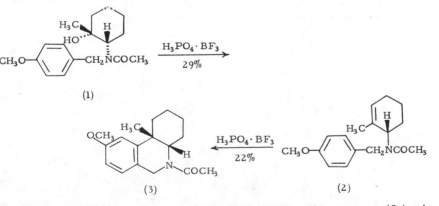

(1)

(3)   (2)

cis- and *trans*-phenanthridines corresponding to (3). The stereospecificity is believed to be controlled by conformational factors.[1]

[1] A. Fischli, H. Hoffmann, and P. M. Müller, *Helv.*, **59**, 1661 (1976).

Phosphorus pentachloride, 1, 866–870; **4,** 388–389; **5,** 534.

Tertiary alkyl chlorides. Tertiary alcohols are converted into alkyl chlorides by reaction with PCl_5 at 0° for about 3 minutes in an inert solvent (usually $CHCl_3$) containing $CaCO_3$ (0.5–1.0 molar equiv.). Yields are usually high; the reaction proceeds with retention of configuration.[1]

[1] R. M. Carman and I. M. Shaw, *Aust. J. Chem.,* **29,** 133 (1976).

Phosphorus pentasulfide, 1, 870–871; **3,** 226–228; **4,** 389; **5,** 534–535; **6,** 470–471.

Deoxygenation. Penicillin and cephalosporin sulfoxides, (1) and (3), are deoxygenated in high yield by P_2S_5 (0.5 mole equiv.) and pyridine at 20° (16 hours).[1]

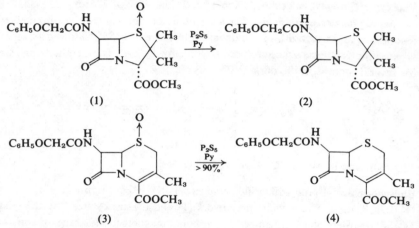

(1)   (2)

(3)   (4)

Deoxygenation of some aldoguloses. The carbonyl group of some sugars has been reduced to methylene by reaction with P_2S_5 to give a sulfur-containing dimer, which is then reduced with Raney nickel.[2]

Example:

[1] R. G. Micetich, *Tetrahedron Letters,* 971 (1976).
[2] P. Köll, R.-W. Rennecke, and K. Heyns, *Ber.,* **109,** 2537 (1976).

Phosphorus pentoxide, 1, 871–872.

Dehydration. British chemists[1] have developed a total synthesis of un-symmetrical pulvinic acid pigments of lichens and fungi. The method is illustrated for the synthesis of the permethylated derivative of gomphidic acid (4), the

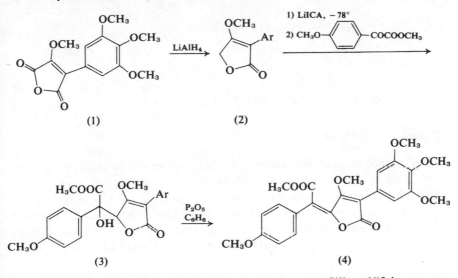

(1)   (2)

(3)   (4)

pigment of *Gomphidius glutinosur.* The dehydration step [(3) → (4)] is accomplished only with extreme difficulty.

[1] D. W. Knight and G. Pattenden, *J.C.S. Chem. Comm.,* 660 (1976).

Phosphorus pentoxide–Methansulfonic acid, 5, 535.

α-*Carboalkoxy-γ-alkylidene-Δ*^{α,β}-*butenolides.*[1] The reaction of diethyl keto-malonate (1) with the morpholine enamine (2) followed by acidic work-up gives (3) in 80% yield. Treatment of (3) with P_2O_5 and methanesulfonic acid (1:10 by

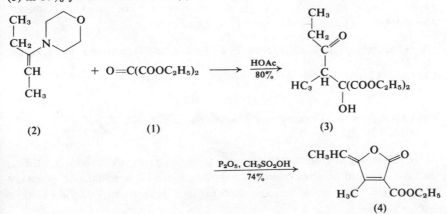

(2)   (1)   (3)

(4)

weight) effects enol lactonization–dehydration to give (4). Actually reaction of 3-pentanone itself with (1) in P_2O_5–CH_3SO_2OH gives (4) directly, but only in 19–23% yield. A somewhat higher yield was obtained in the direct synthesis of (5).

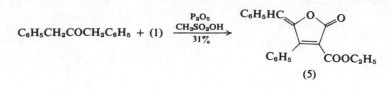

$$C_6H_5CH_2COCH_2C_6H_5 \; + \; (1) \quad \xrightarrow[31\%]{\substack{P_2O_5 \\ CH_3SO_2OH}}$$

(5)

[1] A. G. Schultz and Y. K. Yee, *J. Org.*, **41**, 561 (1976).

Phosphorus tribromide, 1, 873–874; **2,** 330.

Optically active sec-alkyl bromides. These halides can be prepared in ~60–75% yield by reaction of chiral alcohols and excess pyridine in ether at −25° with PBr_3 over a 1-hour period, followed by reaction for 1–2 days at 4° for cleavage of the intermediate phosphite ester. The pyridine is converted by liberated HBr into pyridinium hydrobromide necessary for the cleavage reaction. The optical purity of the bromides is generally in the range of 80–95%.[1]

[1] R. O. Hutchins, D. Masilamani, and C. A. Maryanoff, *J. Org.*, **41**, 1071 (1976).

Phosphorus trichloride–Dimethylformamide

Alkyl chlorides (6, 220). Anderson *et al.*[1] have prepared a Vilsmeier-like reagent by addition of PCl_3 (0.058 moles) to DMF (100 ml.). The thick colorless solid is allowed to stand for 40 minutes before use.[2] This material converts unhindered, primary alcohols into alkyl chlorides in 70–87% yield. Yields are 39–84% for secondary alcohols. Tertiary alcohols are converted into alkenes. The exact structure of the chlorinating reagent is uncertain.

[1] A. G. Anderson, Jr., N. E. T. Owen, F. J. Freenor, and D. Erickson, *Synthesis*, 398 (1976).
[2] The ageing is essential for maximum yields.

Phosphoryl chloride (Phosphorus oxychloride), 1, 876–882; **2,** 330–331; **3,** 228; **4,** 390; **5,** 535–537.

Iminium salts. When α-tertiary amino acids are heated briefly in $POCl_3$, the resulting acid chloride is decarbonylated to give an iminium salt, generally in high yield.[1] The second example demonstrates the usefulness of these salts in synthesis.

Examples:

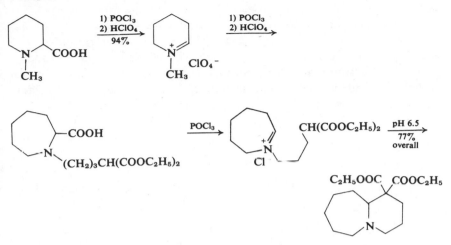

¹ R. T. Dean, H. C. Padgett, and H. Rapoport, *Am. Soc.*, **98**, 7448 (1976).

Picolinic acid, 5, 538.

Chromic acid oxidation. Complete details for use of this acid as a catalyst for chromic acid oxidation have been published.¹

¹ T.-Y. Peng and J. Roček, *Am. Soc.*, **98**, 1026 (1976).

Piperidine, 1, 886–890; **2**, 332; **4**, 393; **6**, 472.

Cycloenone synthesis. McCurry and Singh ¹ have reported a one-flask synthesis of 3-methyl-5-alkyl-4-carboethoxycyclohexene-2-ones-1 from ethyl acetoaceate and an aldehyde using piperidine as catalyst. The reaction involves Knoevenagel

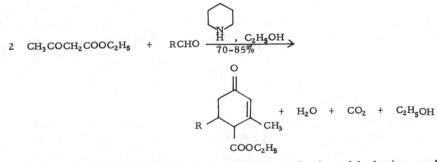

condensation, Michael addition, intramolecular cyclization, dehydration, and selective decarboethoxylation. These can be carried out in sequence by keeping the temperature at 0° for 3 days, then at 20° for 1 day, and finally at 100° for 6 hours.

¹ P. M. McCurry, Jr., and R. K. Singh, *Syn. Comm.*, **6**, 75 (1976).

Polyhydrogen fluoride–Pyridine (Pyridinium polyhydrogen fluoride), **5**, 538–539; **6**, 473–474.

Deprotection in peptide synthesis (2, 216).[1] This reagent, like hydrogen fluoride itself, can be used for cleavage of most protecting groups currently employed in peptide synthesis, and, of course, it is easier and safer to handle.

5-Fluoro steroids. 5-Hydroxycholestanes react with this complex or with hydrogen fluoride in methylene chloride to form 5-fluoro steroids. The former reagent is more regioselective. Generally the configuration is retained.[2]

Examples:

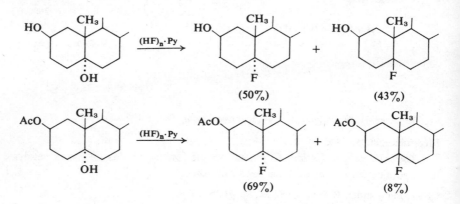

(50%)   (43%)

(69%)   (8%)

[1] S. Matsuura, C.-H. Niu, and J. S. Cohen, *J.C.S. Chem. Comm.*, 451 (1976).
[2] A. Ambles and R. Jacquesy, *Tetrahedron Letters*, 1083 (1976).

Polyphosphoric acid (PPA), **1**, 894–895; **2**, 334–336; **3**, 231–233; **4**, 395–397; **5**, 590–592; **6**, 474–475.

trans- *to* **cis-***Isomerization.* *trans*-Cinnamic acids are isomerized to the *cis*-isomers in high yield when heated in PPA at 80–95° for 90 minutes (equation I).[1] This reaction may well be involved in the cyclization of *trans*-cinnamic acids to

(I)

flavones.[2] The isomerization has been extended to the diene *trans,trans*-5-phenylpenta-2,4-dienoic acid (equation II) and appears to require both an aryl

(II)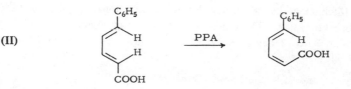

group and a carbonyl function. It was first encountered with azlactones (equation III).[3]

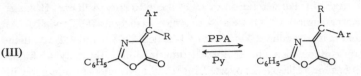

(III)

Cyclization of diarylmethylmalonic acids. Two examples of double cyclization of acids of this type with commercial PPA have been reported by British chemists (equations I and II).[4] The reaction formulated in equation (II) involves dehydrogenation as well as double cyclization. This reaction had been observed by Clar *et al.*,[5] who used phosphoric oxide dissolved in nitrobenzene (130°, 10 minutes).

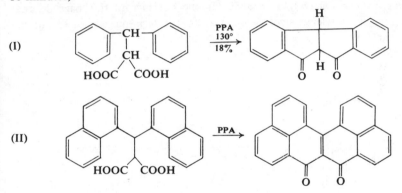

(I)

(II)

Uracil. Uracil (3) is obtained in 61% yield by the reaction of propiolic acid (1) and urea (2) in PPA at 85°. The yield is slightly less (59%) if PPA is replaced by H_2SO_4.[6]

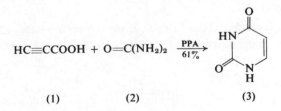

$$HC{\equiv}CCOOH + O{=}C(NH_2)_2 \xrightarrow[61\%]{PPA}$$

(1)   (2)   (3)

[1] Y. S. Rao and R. Filler, *J.C.S. Chem. Comm.*, 471 (1976).

[2] H. O. House, *Modern Synthetic Reactions*, 2nd ed., Benjamin, Menlo Park, CA, 1972, p. 795.

[3] Y. S. Rao, *J. Org.*, **41**, 722 (1976).

[4] J. M. Allen, K. M. Johnston, and R. Z. Shotter, *Chem. Ind.*, 108 (1976).

[5] E. Clar, W. Kemp, and D. G. Steward, *Tetrahedron*, **3**, 325 (1958).

[6] K. Harada and S. Suzuki, *Tetrahedron Letters*, 2321 (1976).

Potassium in graphite, 4, 397.

$C_{24}K$ is a moderately active catalyst for hydrogenation of alkenes. It is a very efficient catalyst for isomerization of *cis*-stilbene to *trans*-stilbene. It is also effective for isomerization of alkynes: 2-octyne → 1-octyne (86% yield). An allene is a probable intermediate: 2-decyne → 1-decyne (20%) and 1,2-decadiene (9%). Potassium alone also effects these isomerizations, but only at higher temperatures. Benzene is slowly converted into biphenyl when in contact with $C_{24}K$ in refluxing cyclohexane.[1] This reaction has also been reported for C_8K in DMF at 20° (61% yield).[2]

[1] J. M. Lalancette and R. Roussel, *Canad. J. Chem.*, **54**, 2110 (1976).
[2] F. Béguin and R. Setton, *J.C.S. Chem. Comm.*, 611 (1976).

Potassium 3-aminopropylamide, 6, 476.

Isomerization of acetylenic alcohols. In the presence of this base internal triple bonds of acetylenic alcohols migrate to the terminus remote from the —OH group.[1]

Examples:

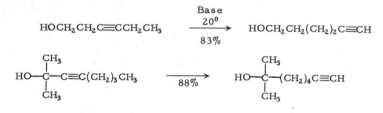

[1] C. A. Brown and A. Yamashita, *J.C.S. Chem. Comm.*, 959 (1976); J. C. Lindhoudt, G. L. van Mourik, and H. J. J. Pabon, *Tetrahedron Letters*, 2565 (1976).

Potassium *t*-butoxide, 1, 911–916; **2,** 338–339; **3,** 233–234; **4,** 399–405; **5,** 544– 553; **6,** 477–479.

Dehydrohalogenation. The 1-cyclopropene-1-carboxylate (2) has been obtained by dehydrochlorination of (1) with potassium *t*-butoxide in THF. This

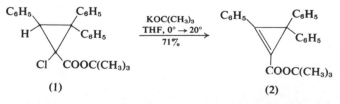

reaction is only successful when the ring bears bulky groups, because 1-cyclo-propene-1-carboxylates are potent Michael acceptors. Thus dehydrochlorination of (3) gives mainly polymer and 25–30% of the adduct (4).[1]

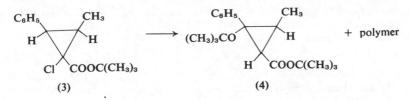

Cyclofragmentation of epoxy sulfones. Swiss chemists[2] have developed this new method of ring expansion for a synthesis of muscone (7). The starting material was cyclododecanone (1), which was converted by known reactions into the hydroxy sulfone (2). Dehydration of (2) results in (3) as the major product. This is epoxidized to give (4), together with the isomeric epoxy sulfone.

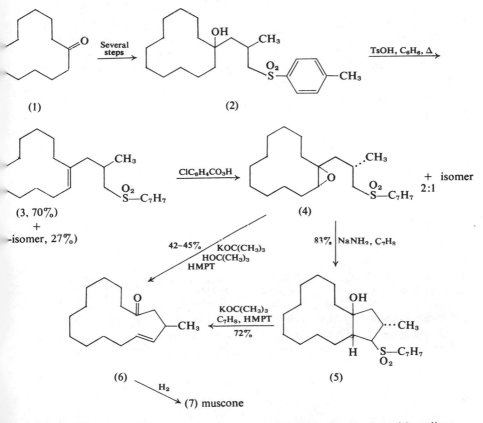

Both isomers are converted into (E)-4-muscenone (6) by treatment with sodium amide to give a hydroxy sulfone (5) and then with potassium *t*-butoxide. The conversion of (4) into (6) can be accomplished in one step using potassium *t*-butoxide in *t*-butanol and HMPT (42–45% yield).

[1] V. Sander and P. Weyerstahl, *Angew. Chem., Int. Ed.*, **15**, 244 (1976); see also K.-O. Henseling and P. Weyerstahl, *Ber.*, **108**, 2803 (1975).
[2] A. Fischli, Q. Branca, and J. Daly, *Helv.*, **59**, 2443 (1976).

Potassium *t*-butoxide–Dimethyl sulfoxide, 6, 479.

2-Chlorobenzocyclopropene. Treatment of (1), 3,4,7,7-tetrachlorobicyclo-[4.1.0]heptane, with potassium *t*-butoxide in DMSO (1 hour, 20°) gives 2-chlorobenzocyclopropene (2) in about 50% yield. The original paper should be consulted for a possible mechanism.[1]

(1)   (2)

Cycloalkynes. Medium-ring cycloalkynes can be prepared by treatment of 2-bromo-3-methoxy-*trans*-cycloalkenes with potassium *t*-butoxide in anhydrous DMSO (*anti*-elimination).[2] For the preparation of the precursors *see* **4**, 432–433.
Examples:

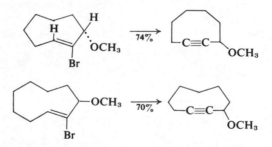

[1] A. Kumar, S. R. Tayal, and D. Devaprabhakara, *Tetrahedron Letters*, 863 (1976).
[2] C. B. Reese and A. Shaw, *J.C.S. Perkin I*, 890 (1976).

Potassium chromate, K_2CrO_4. Mol. wt. 194.20. Suppliers: Alfa, Baker, ROC/RIC.

Aldehydes. Allylic and benzyl halides are oxidized to aldehydes in about 80% yield by potassium chromate in anhydrous HMPT in the presence of

$$(I) \qquad RCH_2X \; + \; KO-\overset{\overset{O}{\|}}{\underset{\underset{O}{\|}}{Cr}}-OK \; \longrightarrow \; R\overset{\overset{H}{|}}{\underset{\underset{H}{|}}{C}}-O-\overset{\overset{O}{\|}}{\underset{\underset{O}{\|}}{Cr}}-OK \; \longrightarrow \; RC\overset{\diagup O}{\diagdown_H}$$

dicyclohexyl-18-crown-6 (equation I). Yields are low with alkyl halides (20% in the case of *n*-octyl bromide).[1]

[1] G. Cardillo, M. Orena, and S. Sandri, *J.C.S. Chem. Comm.*, 190 (1976).

Potassium cyanide, 5, 553.

Modified benzoin condensation. Benzoin-type condensation[1] of terephthal-aldehyde (1) apparently involves only one of the two aldehyde groups. However,

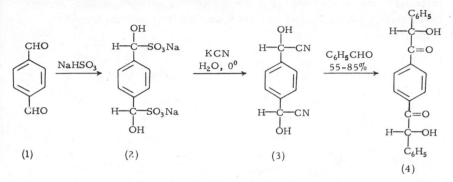

(1)   (2)   (3)   (4)

benzoin condensation with benzaldehyde proceeds as expected if the aldehyde (1) is first converted into biscyanohydrin (3) by way of the bisbisulfite adduct (2).[2]

5'-Deoxy-5'-cyanonucleosides. 2',3'-O-Isopropylidene-5'-O-tosylates of nucleosides (1) react with potassium cyanide complexed with 18-crown-6 to form (2). There is no reaction in the absence of a crown ether. The protective group can be removed by formic acid to give (3).[3] The product can be converted into the amide or the carboxylic acid by usual methods.

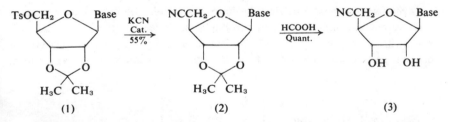

(1)   (2)   (3)

Photocyanation of arenes. This reaction occurs easily with potassium cyanide solubilized in acetonitrile by means of 18-crown-6. Yields are about twice as high as those obtained with KCN dissolved in H_2O.[4]

[1] W. S. Ide and J. S. Buck, *Org. React.*, **4**, 269 (1948).
[2] R. H. Kratzer, K. L. Paciorek, and D. W. Karle, *J. Org.*, **41**, 2230 (1976).
[3] W. Meyer, E. Böhnke, and H. Follmann, *Angew. Chem., Int. Ed.*, **15**, 499 (1976).
[4] R. Beugelmans, M.-T. LeGoff, J. Pusset, and G. Roussi, *J.C.S. Chem. Comm.*, 377 (1976); *idem, Tetrahedron Letters*, 2305 (1976).

Potassium(sodium) decacarbonylhydridochromate, $K(Na)HCr_2(CO)_{10}$.

The potassium complex can be prepared by the reaction of $Cr(CO)_6$ with potassium-graphite (C_8K) in THF followed by addition of water. The sodium complex is obtained by the reaction of $Cr(CO)_6$ with sodium amalgam in THF followed by addition of water.

Conjugate reduction of enones. The alkali metal carbonylchromates reduce α,β-unsaturated carbonyl compounds to the corresponding saturated carbonyl compounds in 40–80% yield. They are comparable to potassium hydridotetracarbonylferrate (**6**, 483–486), but are simpler to prepare because chromium hexacarbonyl is a stable solid and less toxic than iron pentacarbonyl.[1]

Examples:

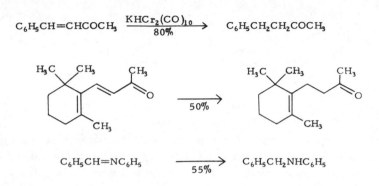

$$C_6H_5CH=CHCOCH_3 \xrightarrow[80\%]{KHCr_2(CO)_{10}} C_6H_5CH_2CH_2COCH_3$$

$$C_6H_5CH=NC_6H_5 \xrightarrow[55\%]{} C_6H_5CH_2NHC_6H_5$$

[1] G. P. Boldrini, A. Umani-Ronchi, and M. Panunzio, *Synthesis*, 596 (1976).

Potassium ferricyanide, 1, 929–933; **2**, 345; **4**, 406–407; **5**, 554–555; **6**, 480–482.

Acetylene synthesis. 3,4-Dialkyl-4-halo-2-pyrazoline-5-ones (1) are known to be converted into α,β-unsaturated carboxylic acids (equation I),[1] with the (E)-isomer (2) often predominating.[2] Kocienski *et al.*[3] reasoned that this stereochemical preference could result from a vinyl radical intermediate (a). Indeed,

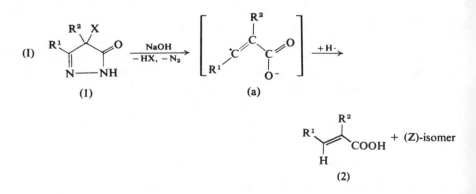

addition of an oxidizing agent [$K_3Fe(CN)_6$] results also in loss of CO_2 and formation of a disubstituted alkyne. This new reaction was used in a new synthesis of junipal (3, equation II).[4]

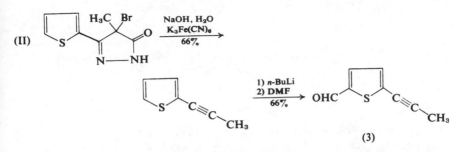

(II)

(3)

Stereospecific phenol coupling. Oxidation of (S)-(+)-7-hydroxy-1,5,6-tri-methyl-1,2,3,4-tetrahydronaphthalene (1) in ether with $K_3Fe(CN)_6$ in 0.2 N NaOH at 20° yields the optically active dimer shown to be the (SS)-*trans*-enantiomer (2). The *dl*-form of the biaryl is also the major product of oxidation of (RS)-(1) together with smaller amounts of two diastereomeric forms. It

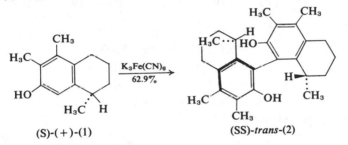

(S)-(+)-(1)   (SS)-*trans*-(2)

appears that one enantiomer influences the stereochemical course of the dimerization of the other enantiomer.[4]

[1] L. A. Carpino and E. G. S. Rundberg, Jr., *J. Org.*, **34**, 1717 (1969).
[2] P. J. Kocienski, J. M. Ansell, and R. W. Ostrow, *ibid.*, **41**, 3625 (1976).
[3] P. J. Kocienski, J. M. Ansell, and B. E. Norcross, *ibid.*, **41**, 3650 (1976).
[4] B. Feringa and H. Wynberg, *Am. Soc.*, **98**, 3372 (1976).

Potassium hydride, 1, 935; **2,** 346; **4,** 409; **5,** 557; **6,** 482–483.

Alkylation of enamino ketones and esters. The potassium enolates of enamino ketones and esters and related systems (generated with KH in THF at 0°) are alkylated almost exclusively at the γ-position (80–90% yield), as shown in the two examples.[1] When the lithium enolates of (1) and (2) are used, the yields of

Examples:

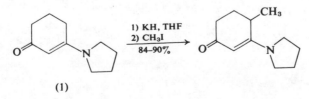

(1)

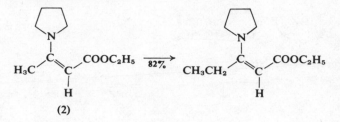

(2)

products are 40–60% and 52%, respectively. The lithium enolates were generated with LDA in THF–HMPT; α'-Alkylation products also obtain in this case.

Metallation of weakly acidic hydrocarbons. Potassium hydride can metallate triphenylmethane (pK_a 31.4), but only in the presence of a crown ether (equation I). Diphenylmethane (pK_a 33.1) and di-*p*-tolylmethane (pK_a 35.1) are also

$$(I) \qquad (C_6H_5)_3CH + KH \xrightarrow[\text{THF}]{\text{18-Crown-6}} (C_6H_5)_3CK + H_2$$

metallated in this way, but di-2,4-xylylmethane (pK_a 36.3) is only partially metallated. Still weaker hydrocarbons are not metallated.[2]

[1] R. B. Gammill and T. A. Bryson, *Synthesis*, 401 (1976).
[2] E. Buncel and B. Menon, *J.C.S. Chem. Comm.*, 648 (1976).

Potassium hydride–Hexamethylphosphorus triamide.

Sigmatropic shifts. The oxy-Cope system (1) undergoes a [3,3] sigmatropic shift within minutes at 65° to give (2) when treated with potassium hydride in HMPT or THF. The reaction of the sodium salt of (1) is markedly slower; the lithium alkoxide shows no evidence of rearrangement after 24 hours.[1]

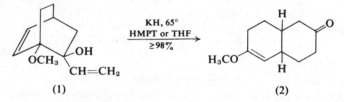

More recently, [1,3] shifts at room temperature have been reported under these conditions with medium-sized cyclic vinylic carbinols. For example, (3) rearranges predominantly to the ring enlarged ketones (4) and (5) (equation I).

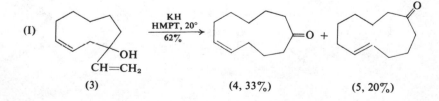

These products, together with (7), are also obtained by thermolysis of the trimethylsilyloxy derivative at about 300° (equation II). In this particular case

(II) 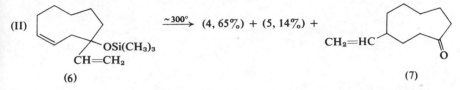 $\xrightarrow{\sim 300°}$ (4, 65%) + (5, 14%) +

(6)   (7)

yields are higher by the thermal route, but rearrangement of (8) to (9) proceeds in much higher yield by the KH–HMPT method. This rearrangement may well be the method of choice for syntheses of benzo-substituted ring systems.[2]

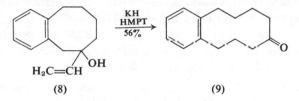

(8)   (9)

[1] D. A. Evans and A. M. Golob, *Am. Soc.*, **97**, 4765 (1975).
[2] R. W. Thies and E. P. Seitz, *J.C.S. Chem. Comm.*, 846 (1976).

Potassium hydroxide, 5, 557–560; **6,** 486.

α,β-Unsaturated nitriles. Acetonitrile undergoes condensation with aldehydes or ketones under catalysis with KOH in the presence or absence of 18-crown-6 to yield α,β-unsaturated nitriles.[1]

Examples:

$$C_6H_5CHO + CH_3CN \xrightarrow[82\%]{KOH} \begin{matrix} C_6H_5 \\ H \end{matrix} C=CHCN$$

$$C_6H_5COC_6H_5 + CH_3CN \xrightarrow[84\%]{} (C_6H_5)_2C=CHCN$$

Deblocking of the benzylidene group. A benzylidene group adjacent to C=O can be cleaved by a retroaldol reaction catalyzed by 4-aminobutyric acid. Only DMSO or HMPT can be used as solvent (equation I).[2] The yield is improved by use of a crown ether to solvate the KOH.[3] The reaction was used in a total synthesis of (±)-α- and β-pinene.

(I)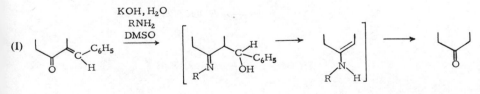

Transannular 1,4-hydride shift. When (1) is heated with KOH in *t*-butanol–water (35:1), it rearranges to (2). Evidence was obtained that a transannular

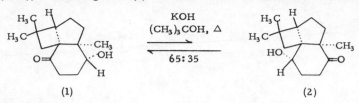

(1)   (2)

1,4-hydride shift is involved. Such shifts are known in several systems, but had never been observed previously in cyclohexyl systems. However, (1) and (2) can attain the boat conformation necessary for such a shift.[4]

Conversion of nitriles to amides (1, 469–470). This reaction has been carried out traditionally with NaOH and H_2O_2 in $C_2H_5OH–H_2O$. Actually H_2O_2 is not necessary; the reaction can be conducted in good yield with finely powdered KOH or NaOH in *t*-butanol. Hydrolysis with KOH in methanol (homogeneous solution) results in further hydrolysis to the acid.[5]

[1] G. W. Gokel, S. A. DiBiase, and B. A. Lipisko, *Tetrahedron Letters*, 3495 (1976).

[2] M. T. Thomas and A. G. Fallis, *Am. Soc.*, **98**, 1227 (1976).

[3] M. T. Thomas, E. G. Breitholle, and A. G. Fallis, *Syn. Comm.*, **6**, 113 (1976).

[4] E. W. Warnhoff, *J.C.S. Chem. Comm.*, 517 (1976).

[5] J. H. Hall and M. Gisler, *J. Org.*, **41**, 3769 (1976).

Potassium superoxide, 6, 488–490.

Cleavage of esters. Esters, but not amides, are cleaved in high yield by potassium superoxide in the presence of 18-crown-6. This reaction is relevant to certain biological oxidations that apparently involve O_2^-.

Oxidative cleavage. α-Keto, α-hydroxy, and α-halo ketones, esters, and carboxylic acids are cleaved in fair to high yield by excess potassium superoxide solubilized in benzene with 18-crown-6 (25°, 12 hours). The reaction resembles the behavior of some dioxygenases in several aspects.

Examples:

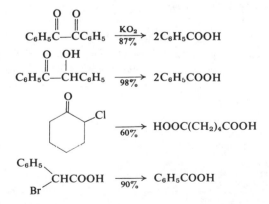

Oxidation of enones. Tetracyclone (1) reacts with KO_2 (crown ether complex) to yield the 3-furanone (2).[3] This reaction has been extended to chalcones and has been found to lead mainly to carboxylic acids (equation I).[4] These reactions

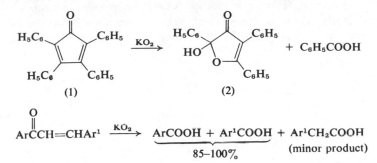

(1)

(2)

(I) $\underset{\displaystyle \text{ArCCH=CHAr}^1}{\overset{O}{\parallel}}$ $\xrightarrow{\text{KO}_2}$ $\underbrace{\text{ArCOOH} + \text{Ar}^1\text{COOH}}_{85-100\%} + \underset{\text{(minor product)}}{\text{Ar}^1\text{CH}_2\text{COOH}}$

appear to involve formation of radical anions, which then react with atmospheric oxygen.

Oxidation of hydroquinones and quinones. Oxidation of these substrates with solubilized KO_2 produces strongly colored semiquinones identified by the ESR spectra. The radical anions of *o*-dihydroxyarenes are oxidized further to dicarboxylic acids.[5]

Example:

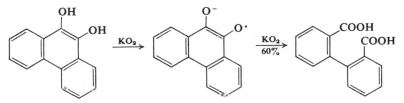

Diacyl peroxides. These substances can be prepared conveniently by reaction of acyl chlorides with potassium superoxide solubilized in benzene with a crown ether (equation I).[6]

(I) $2 \text{ RCOCl} + 2 \text{ KO}_2 \xrightarrow[50-75\%]{}$ $\underset{}{\overset{O \qquad\quad O}{\overset{\parallel \qquad\quad \parallel}{\text{RC}-\text{O}-\text{O}-\text{CR}}}} + 2 \text{ KCl} + \text{O}_2$

Reaction with halonitrobenzenes. Halonitrobenzenes are converted into nitrophenols by KO_2 complexed with a crown ether. The order of halogen displacement is $F > Br \sim I > Cl$. The phenolic oxygen is derived from atmospheric oxygen, which reacts with an intermediate anion radical. A nitro group

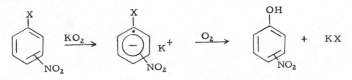

can also be displaced by KO_2. Thus 1,4-dinitrobenzene is converted into 4-nitrophenol (90% yield).[7]

Oxidation of catechols.[8] Oxidation of catechol itself by KO_2 is complicated by extensive concomitant polymerization. On the other hand, oxidation of 3,5-di-*t*-butylcatechol (1) is relatively simple. The first intermediate is the *o*-benzo-semiquinone as evidenced by the blue color that develops rapidly. An early

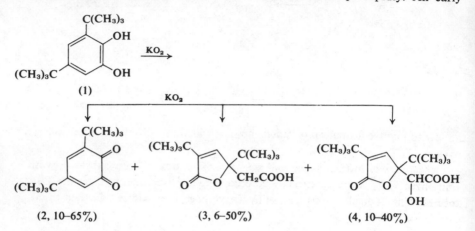

(2, 10–65%)   (3, 6–50%)   (4, 10–40%)

isolated product is the quinone (2). Products of more extensive oxidation are the lactones (3) and (4), derived from the dicarboxylic acids (5) and (6), formed by

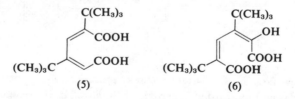

(5)   (6)

cleavage of the C—C bond bearing the hydroxyl groups. This reaction is observed in the oxidation of 9,10-dihydroxyphenanthrene by KO_2 to diphenic acid (equation I) and in the *in vivo* oxidation of catechol to *cis,cis*-muconic acid by the enzyme pyrocatecholase.

(I)

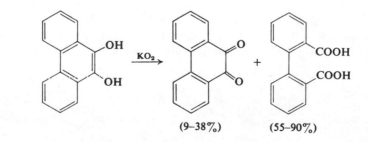

(9–38%)   (55–90%)

[1] J. San Filippo, Jr., L. J. Romano, C.-I. Chern, and J. S. Valentine, *J. Org.*, **41**, 588 (1976).

[2] J. San Filippo, Jr., C.-I. Chern, and J. S. Valentine, *ibid.*, **41**, 1077 (1976).

[3] I. Rosenthal and A. Frimer, *Tetrahedron Letters*, 3731 (1975).

[4] *Idem, ibid.*, 2805 (1976).

[5] E. Lee-Ruff and A. B. P. Lever, *Canad. J. Chem.*, **54**, 1837 (1976).

[6] R. A. Johnson, *Tetrahedron Letters*, 331 (1976).

[7] A. Frimer and I. Rosenthal, *ibid.*, 2809 (1976).

[8] Y. Moro-oka and C. S. Foote, *Am. Soc.*, **98**, 1510 (1976).

Potassium(lithium) tri-*sec*-butylborohydride, 6, 490–492.

Reduction of α,β-unsaturated carbonyl compounds (**6,** 491–492). The final paper has now been published. In general, β-substituted cyclohexenones undergo exclusive 1,4-reduction with either Selectride. Acyclic enones generally undergo 1,2-reduction to allylic alcohols. The Selectrides are particularly useful for 1,4-reduction of enoates.[1] Super-Hydride (lithium triethylborohydride) is less useful for this purpose. Unfortunately L-Selectride reduces α,β-acetylenic esters only to propargylic alcohols.

[1] J. M. Fortunato and B. Ganem, *J. Org.*, **41**, 2194 (1976).

Proline, 6, 492 493.

Asymmetric cyclization (**6,** 410–411). The use of optically active amines as catalysts for asymmetric cyclizations, as well as an application for synthesis of optically active estrone, has been mentioned previously. This cyclization has now been used in construction of the asymmetric aldehyde (4), a known precursor to 12-methylprostaglandin (5). Thus cyclization of the trione (1) in DMF with D-proline as catalyst gave the aldol (2), which was dehydrated to the enedione (3), obtained with 96% optical purity.[1]

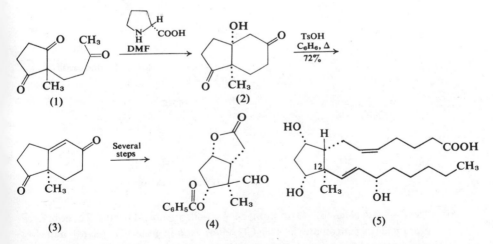

[1] P. A. Grieco, N. Fukamiya, and M. Miyashita, *J.C.S. Chem. Comm.*, 573 (1976).

Pyridinium chloride, 1, 964–966; **2,** 352–353; **3,** 239–240; **4,** 415–418; **5,** 566–567; **6,** 497–498.

Cleavage of cyclopropane rings. Treatment at reflux of (1), or of the corresponding oxime, with pyridinium chloride in pyridine (20 hours) results in formation of (2), or of the corresponding oxime. This cleavage reaction requires

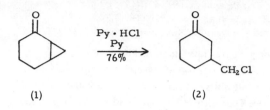

(1)   (2)

two factors: conjugation of the cyclopropane ring with C=O or C=NOH and a strained ring system. Thus cyclopropyl methyl ketone is cleaved very slowly. Hydrochloric acid itself (12 *M*) does not react with (1).[1]

[1] L. Pellacani, P. A. Tardella, and M. A. Loreto, *J. Org.,* **41,** 1282 (1976).

Pyridinium chlorochromate, 6, 498–499.

Oxidation–rearrangement. Tertiary vinylcarbinols are converted by reaction with this reagent into α,β-unsaturated aldehydes in yields of 80–90%.[1]
Examples:

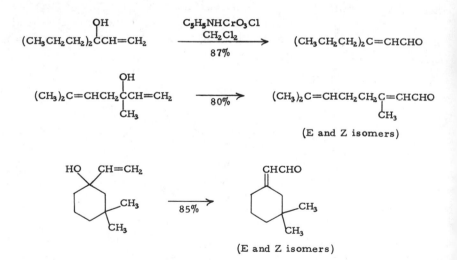

(E and Z isomers)

(*S*)-(−)-*Pulegone* (3). Oxidation of (−)-citronellol (1) with 2.5 equiv. of pyridinium chlorochromate in CH_2Cl_2 gives isopulegone (2) via (a) and (b). The product is isomerized to (3) by base.[2]

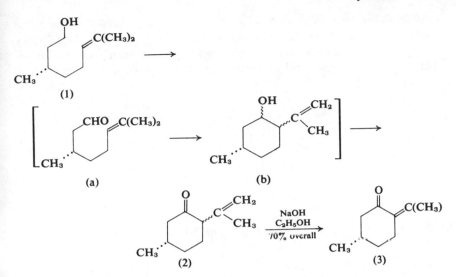

(1)

(a)   (b)

$$\xrightarrow[\text{70\% overall}]{\begin{array}{c}\text{NaOH}\\ \text{C}_2\text{H}_5\text{OH}\end{array}}$$

(2)   (3)

[1] J. H. Babler and M. J. Coghlan, *Syn. Comm.*, **6**, 469 (1976).
[2] E. J. Corey, H. E. Ensley, and J. W. Suggs, *J. Org.*, **41**, 380 (1976).

Pyrrolidine, 1, 972–974; **2,** 354–355; **3,** 240.

Spirocyclization. 6-Formylcyclohexenones of type (1), (3), or (5) undergo spirocyclization when refluxed with pyrrolidine in acetic acid. The choice of base is critical in this cyclization.[1]

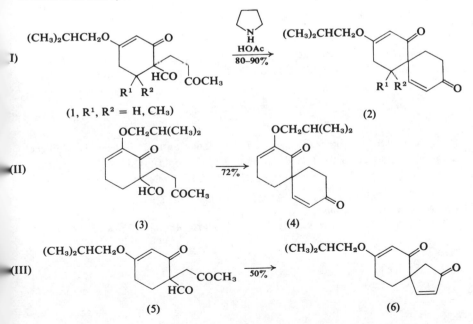

I)

$(1, R^1, R^2 = H, CH_3)$

(2)

(II)

(3)   (4)

(III)

(5)   (6)

Pyrrolidinoenamines. Review.[2] Pyrrolidinoenamines of cyclohexanone and cyclopentanone can be prepared in about 82% yield by reaction of the ketone and pyrrolidine neat or in ether at room temperature. Under these conditions 1-benzyl-3-pyrrolidinone was converted into the enamine (1) in 68% yield.

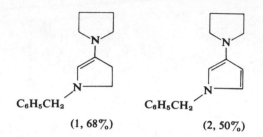

(1, 68%)   (2, 50%)

Reaction in refluxing benzene resulted in dehydrogenation to give (2) in 50% yield.[3]

[1] Ae. de Groot and B. J. M. Jansen, *Tetrahedron Letters*, 2709 (1976).
[2] A. G. Cook, *Enamines: Synthesis, Structure, and Reactions*, Marcel Dekker, New York, 1969.
[3] P. A. Zoretic, F. Barcelos, and B. Branchaud, *Org. Prep. Proc. Int.*, **8**, 211 (1976).

Pyruvyl chloride, $CH_3COCOCl$. Mol. wt. 106.51, b.p. 53°/126 torr.

Preparation[1]:

$$\underset{O}{\overset{\overset{\displaystyle O}{\|}}{CH_3CCOOH}} + Cl_2CHOCH_3 \longrightarrow \underset{(54\%)}{\overset{\overset{\displaystyle O}{\|}}{CH_3CCOCl}} + HCOOCH_3 + HCl$$

Oxidation of alcohols.[2] Alcohols can be oxidized by conversion to the pyruvate esters followed by irradiation (equation I). This method has some attractive

$$(I)\quad \underset{R^2}{\overset{R^1}{>}}CHOH \longrightarrow \underset{R^2}{\overset{R^1}{>}}CHO\overset{O}{\overset{\|}{C}}-\overset{O}{\overset{\|}{C}}CH_3 \xrightarrow[75-100\%]{\overset{h\nu}{C_6H_6}} \underset{R^2}{\overset{R^1}{>}}C{=}O + CO + CH_3CHO$$

features. Allylic alcohols are oxidized normally, and cholesterol is oxidized without isomerization to Δ^5-cholestenone-3 (89% yield). This method is based on an earlier report of Hammond[3] on the photolysis of a few pyruvates.

[1] H. C. J. Ottenheijm and J. H. M. de Man, *Synthesis*, 163 (1975).
[2] R. W. Binkley, *Syn. Comm.*, **6**, 281 (1976).
[3] G. S. Hammond, P. A. Leermakers, and N. J. Turro, *Am. Soc.*, **83**, 2395 (1961).

Q

Quinine, 6, 501.

Optically active epoxides. Wynberg *et al.*[1] have achieved asymmetric induction in epoxidations under phase-transfer conditions using the quaternary ammonium salt (1) derived from quinine. The oxidation was carried out with 30% aqueous H_2O_2 (*t*-butyl hydroperoxide was used in one instance) with

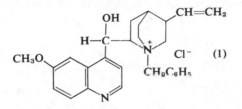

toluene as the organic phase. Optically active epoxides were obtained from chalcones, a quinone, and certain alkenes. The enantiomeric excess determined in one case was 25%. Use of the quaternary salt from quinidine and $C_6H_5CH_2Cl$ gave oxides enantiomeric with those obtained with (1). Asymmetric induction was low with quinine itself.

Examples:

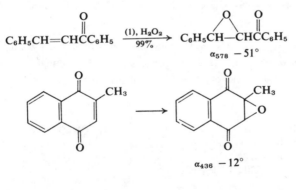

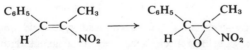

[1] R. Helder, J. C. Hummelen, R. W. P. M. Laane, J. S. Wiering, and H. Wynberg, *Tetrahedron Letters*, 1831 (1976).

311

R

Raney nickel, 1, 723–731; **2,** 293–294; **5,** 570–571; **6,** 502.

Reductive denitrosation. Secondary amines can be regenerated from their nitroso derivatives by hydrogenation with Raney nickel catalysts. Olefinic double bonds and β-ketonic groups are also reduced, but other functional groups are inert.[1]

Examples:

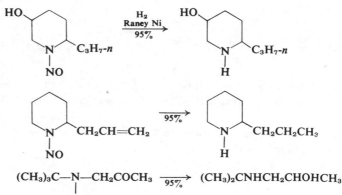

Selective reduction of —NO₂. The nitro group of the pyranone (1) can be selectively reduced, without hydrogenolysis of the benzyl ether linkage, by hydrogen and Raney nickel. The blocking group is cleaved by hydrogen and Pd–C.[2]

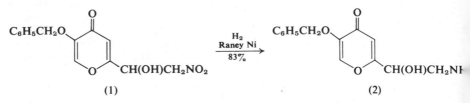

(1)   (2)

Selective hydrogenation of α,β-*unsaturated carbonyl compounds.*[3] Several substrates of this type have been selectively hydrogenated to α,β-unsaturated alcohols with catalysis by a chromium-promoted Raney nickel catalyst.[4]

[1] D. Enders, T. Hassel, R. Pieter, B. Renger, and D. Seebach, *Synthesis*, 548 (1976).
[2] H. W. R. Williams, *Canad. J. Chem.*, **54**, 3377 (1976).
[3] P. S. Gradeff and G. Formica, *Tetrahedron Letters*, 4681 (1976).
[4] A typical catalyst is Raney No. 24 (W. R. Grace and Co.).

Rhodium(II) carboxylates, Rh(OCOR)₂.

Rhodium(II) methoxyacetate (R = CH₂OCH₃), rhodium(II) *n*-butanoate (R = *n*-C₃H₇), and rhodium(II) pivalate [R = C(CH₃)₂] can be prepared in the same way as rhodium(II) acetate.[1]

Cyclopropanation with alkyl diazoacetates.[2] Rhodium(II) acetate is an efficient catalyst for the insertion of carboethoxycarbene into activated C–H-bonds (**5**, 571–572), but it is less effective than rhodium(II) *n*-butanoate or pivalate for catalysis of cyclopropanation of alkenes with alkyl diazoacetates, possibly because the latter carboxylates are more soluble in organic solvents. However, rhodium(II) trifluoroacetate is readily soluble, but it is a poor catalyst for cyclopropanation. The alkyl group of the diazoacetate strongly influences the yields, which are highest with *n*-butyl diazoacetate.

Examples:

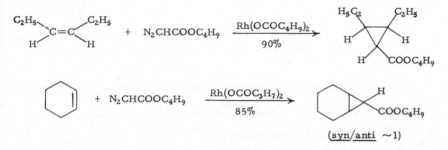

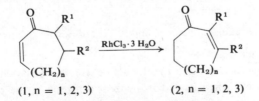

(syn/anti ~1)

[1] P. Legzdins, R. W. Mitchell, G. L. Rempel, J. D. Ruddick, and G. Wilkinson, *J. Chem. Soc. (A)*, 3322 (1970).
[2] A. J. Hubert, A. F. Noels, A. J. Anciaux, and P. Teyssié, *Synthesis*, 600 (1976).

Rhodium(III) chloride, 3, 242–243.

Enone transposition. Alkenones of type (1) are isomerized to the more stable isomers (2) when heated at 100° (sealed tube) in ethanol with RhCl₃·3 H₂O.[1]

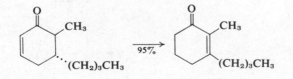

Examples:

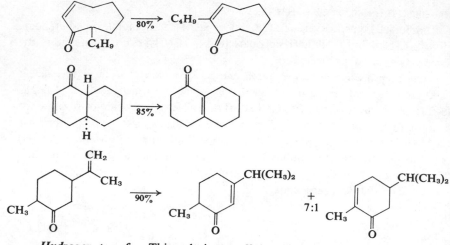

Hydrogen transfer. This salt is an efficient catalyst for the reduction of aromatic nitro compounds to amines by indoline. $RhCl[P(C_6H_5)_3]_3$ is only weakly active.[2]

[1] P. A. Grieco, M. Nishizawa, N. Marinovic, and W. J. Ehmann, *Am. Soc.*, **98**, 7102 (1976).

[2] H. Imai, T. Nishiguchi, and K. Fukuzumi, *Chem. Letters*, 655 (1976).

Rhodium(III) chloride–Triphenylphosphine.

Dehydration of 1,3-diols. 1,3-Diols are converted into monoketones with various rhodium catalysts; highest activity was shown by $RhCl_3–P(C_6H_5)_3$ (1:1). Yields of ketones decrease as the steric bulk of R^1 and R^2 is increased.

$$R^1\!-\!\underset{HO}{\underset{|}{C}}\!-\!\underset{H}{\underset{|}{C}}\!-\!\underset{OH}{\underset{|}{C}}\!-\!H \xrightarrow[\substack{-H_2O}]{\substack{Cat.\\80^\circ,\ 48\ hr.}} R^1\underset{O}{\underset{\|}{C}}\!-\!CH\!-\!CH_2R^3$$

Ethers are obtained under these conditions from 1,3-, 1,4-, and 1,5-diols; 1,2- and 1,6-diols are unreactive.[1]

[1] K. Kaneda, M. Wayaku, T. Imanaka, and S. Teranishi, *Chem. Letters*, 231 (1976).

Ruthenium on alumina.

Catalytic oxygenation. Dibenzothiophene (1) is converted into the 5,5-dioxide in high yield when heated in benzene with air in the presence of 5% ruthenium on alumina (100° and 70 atm.). A suitable catalyst can also be obtained

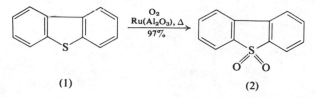

(1)   (2)

by treatment of alumina with $RuCl_3$, $Ru_3(CO)_{12}$, or $Ru(acac)_3$. This oxidation is not general for dialkyl sulfides.[1]

[1] M. A. Ledlie and I. V. Howell, *Tetrahedron Letters*, 785 (1976).

Ruthenium tetroxide, 1, 986–989; **2,** 357–359; **3,** 243–244; **4,** 420–421; **6,** 504–506.

Oxidation of tert-*amines.* Ruthenium tetroxide oxidizes N-benzyl derivatives of piperidines (1, n = 2) and pyrrolidines (1, n = 1) to imides (2). The products can be correlated with known dicarboxylic acids (3) by hydrolysis.[1]

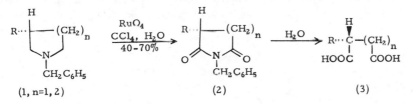

This oxidation provides an alternative to von Braun degradation (**1,** 174–175).

[1] G. Bettoni, C. Franchini, F. Morlacchi, N. Tangari, and V. Tortorella, *J. Org.*, **41,** 2780 (1976).

S

Salcomine, 2, 360; **3,** 245.

Oxidation of 2,6-di-t-butylphenol (**3,** 245). 4-Hydroxysalcomine (**1**), orange solid, slightly soluble in water, readily soluble in 1 N Na_2CO_3, is a more effective

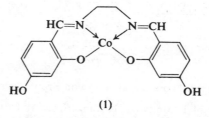

(1)

catalyst for oxidation of 2,6-di-t-butylphenol (equation I) in $CH_3OH–H_2O$ (9:1) than salcomine itself, probably because of increased solubility in water.[1]

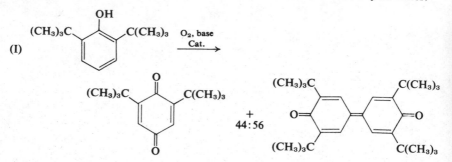

(I)

+
44:56

[1] T. J. Fullerton and S. P. Ahern, *Tetrahedron Letters,* 139 (1976).

Selenium–Chloramine-T.

Allylic amination.[1] The reaction of selenium metal with 2 equiv. of anhydrous chloramine-T (TsNClNa)[2] in CH_2Cl_2 gives a solution of a reagent formulated as (**1**). A similar, but less reactive, reagent is formed by reaction of selenium tetrachloride with p-toluenesulfonamide. The reagent reacts with alkenes at 20°

(1)

or below to form allylic amination products with almost complete absence of products of allylic rearrangement. The position selectivity resembles that of allylic oxygenation by selenium dioxide (**1,** 994).

Examples:

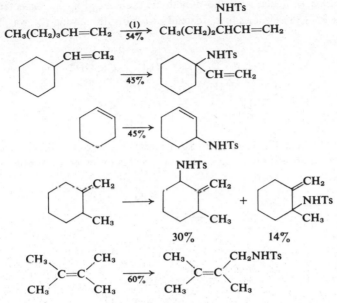

$$CH_3(CH_2)_3CH=CH_2 \xrightarrow[54\%]{(1)} CH_3(CH_2)_2\overset{\overset{\displaystyle NHTs}{|}}{C}HCH=CH_2$$

The corresponding sulfur compound (2) has also been prepared and has been shown to effect allylic amination of alkenes and alkynes.[3]

$$TsN \diagdown \underset{\underset{\displaystyle \ddot{S}:}{\|}}{} \diagup NTs$$

(2)

Examples:

Diamination of 1,3-dienes. Sharpless and Singer[4] allowed this selenium diimide species to react with 2,3-dimethylbutadiene with the expectation of obtaining a [4 + 2] cycloadduct. Instead, a *vic*-disulfonamide (2) was formed

(I)

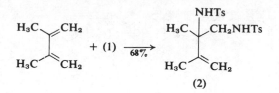

(2)

(equation I). The reaction was found to be general. The two examples indicate that the sulfonamide groups are *cis* to each other; yields in general are poor to fair.

Examples:

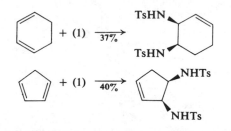

[1] K. B. Sharpless, T. Hori, L. K. Truesdale, and C. O. Dietrich, *Am. Soc.*, **98**, 269 (1976).
[2] On one occasion an explosion occurred on drying the trihydrate in a vacuum oven. A drying pistol with the temperature carefully controlled to 90° is recommended.
[3] K. B. Sharpless and T. Hori, *J. Org.*, **41**, 176 (1976).
[4] K. B. Sharpless and S. P. Singer, *ibid.*, **41**, 2504 (1976).

Selenium(I) chloride, Se_2Cl_2. Mol. wt. 228.83, b.p. 130° dec. Suppliers: Alfa, ROC/RIC.

Addition to alkynes. German chemists[1] have reported the reactions of Se_2Cl_2 with alkynes formulated in equations (I) and (II).

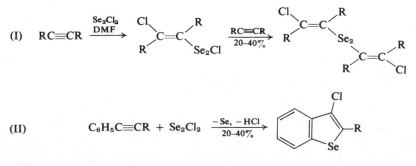

(I) $RC{\equiv}CR$

(II) $C_6H_5C{\equiv}CR + Se_2Cl_2$ $\xrightarrow[20-40\%]{-Se, -HCl}$

[1] W. Ried and G. Sell, *Synthesis*, 447 (1976).

Selenium dioxide, 1, 992–1000; **2,** 360–362; **3,** 245–247; **4,** 422–424; **5,** 575–576; **6,** 509–510.

Review. Oxidations with selenium dioxide have been reviewed (634 references). The reagent is useful for oxidation of carbonyl compounds to 1,2-dicarbonyl compounds, for oxidation of olefins to allylic alcohols, and for benzylic oxidations. It has been used for dehydrogenations, but high-potential quinones are probably more useful for this reaction.[1]

[1] N. Rabjohn, *Org. React.*, **24,** 261 (1976).

Silver carbonate–Celite, 2, 363; **3,** 247–249; **4,** 425–428; **5,** 577–580; **6,** 511–514.

Oxidation of heterocyclic alcohols. This reagent is useful for oxidation of heterocyclic primary alcohols, particularly those containing nitrogen. Note that oxidation of a nonaromatic heterocyclic alcohol can lead to a lactone or a ketone instead of an aldehyde.[1]

Examples:

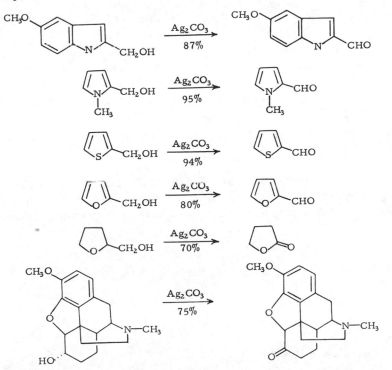

α,β-Epoxy-γ-butyrolactones. These substances are not obtainable by direct epoxidation of Δ²-butenolides with peracids or other usual oxidizing agents. Recently they have been prepared indirectly from *cis*-epoxy diols (2), prepared as shown. The diols can be converted into the lactones (3) with Fetizon's reagent

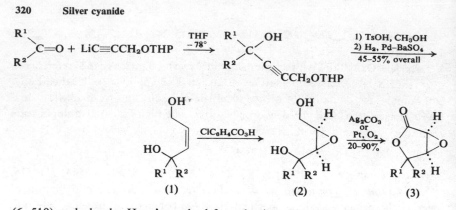

(6, 510) and also by Heyn's method for selective oxidation of primary alcohols (1, 432). The former oxidant is generally more suitable, but there is some variation with the substrate.[2]

[1] M. Fétizon, F. Gomez-Parra, and J.-M. Louis, *J. Heterocyclic Chem.*, **13**, 525 (1976).
[2] R. K. Boeckman, Jr., and E. W. Thomas, *Tetrahedron Letters*, 4045 (1976).

Silver cyanide, 1, 1006; 5, 581.

Hindered esters. Preparation of esters by the reaction of an acyl chloride with an alcohol in the presence of a base (pyridine, dimethylaniline) is improved by use of silver cyanide rather than an organic base. The reaction is carried out in benzene at 20–80° or in HMPT at 80°. Yields are 60–100%.[1]

[1] S. Takimoto, J. Inanaga, T. Katsuki, and M. Yamaguchi, *Bull. Chem. Soc. Japan*, **49**, 2335 (1976).

Silver hexafluoroantimonate, 5, 577. Additional suppliers: Alfa, ROC/RIC.

Isoflavones. 3-Bromoflavanones (1) undergo a 2,3-aryl shift when treated with silver hexafluoroantimonate in methylene chloride to form isoflavones (2).[1]

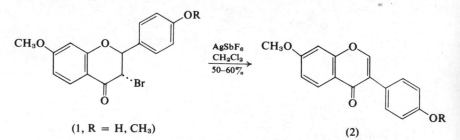

The same reaction occurs, but more slowly, with silver perchlorate. The *cis*-isomers of (1) rearrange more slowly than the *trans*-isomers. This rearrangement may be involved in biosynthesis of isoflavones.[1]

[1] A. Pelter, R. S. Ward, and M. Balasubramanian, *J.C.S. Chem. Comm.*, 151 (1976).

Silver nitrate, 1, 1008–1011; **2,** 366–368; **3,** 252; **4,** 429–430; **5,** 582.

(E)-Alkenes. The reaction of trialkylboranes with alkaline silver nitrate (**2,** 368)[1] to form coupled alkanes has been used to convert 1,2-diorganoboranes from internal alkynes into (E)-alkenes.[2] The reaction is considered to involve an intermediate diradical that couples intramolecularly to the alkene.

Examples:

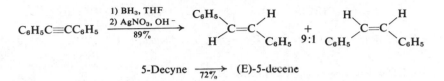

5-Decyne $\xrightarrow[72\%]{}$ (E)-5-decene

Cyclization of dienes. Reaction of dienes with diborane in THF at 0° and then with alkaline silver nitrate at 20° results in cyclic products. This reaction is an extension of the coupling of olefins via hydroboration (**2,** 368).[3]

Examples:

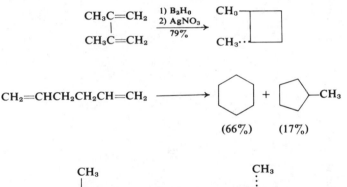

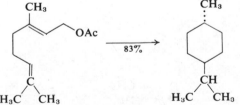

[1] H. C. Brown, *Organic Synthesis via Boranes,* Wiley, New York, 1975, pp. 123–125.
[2] K. Avasthi, S. S. Ghosh, and D. Devaprabhakara, *Tetrahedron Letters,* 4871 (1976).
[3] R. Murphy and R. H. Prager, *ibid.,* 463 (1976).

Silver(I) oxide, 1, 1011; **2,** 368; **3,** 252–254; **4,** 430–431; **5,** 583–585; **6,** 515–518.

Oxidative coupling.[1] Some carboxylic acid esters undergo oxidative dimerization in DMSO when treated with Ag_2O. Metallic silver is formed.

Examples:

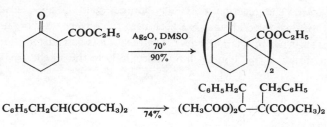

$$C_6H_5CH_2CH(COOCH_3)_2 \xrightarrow[74\%]{} (CH_3COO)_2\overset{\displaystyle C_6H_5H_2C \quad CH_2C_6H_5}{\underset{\displaystyle}{C}}-\overset{}{\underset{}{C}}(COOCH_3)_2$$

[1] Y. Ito, S. Fujii, T. Konoike, and T. Saegusa, *Syn. Comm.*, **6**, 429 (1976).

Silver(II) oxide, 2, 369; **4,** 431–432; **6,** 518.

Oxidative demethylation (**5,** 431–432). 5-Methoxy-1,4,9,10-anthradiquinone (2) has been obtained in 98% yield from 1,4,5-trimethoxyanthraquinone (1).

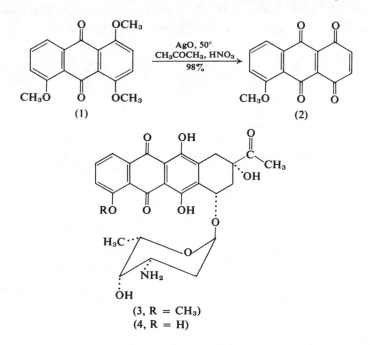

(3, R = CH₃)
(4, R = H)

The product (2) has been converted into the antineoplastic antibiotics (±)-daunomycinone (3) and (±)-carminomycinone (4).[1]

[1] A. S. Kende, Y. Tsay, and J. E. Mills, *Am. Soc.*, **98**, 1967 (1976).

Silver perchlorate, 2, 369–370; **4,** 432–435; **5,** 585–587; **6,** 518–519.

Oxidative dimerization of azirines. The conversion of azirines (1) into pyrazines (2) can be accomplished with $AgClO_4$ in benzene (20°, 4–5 days).[1]

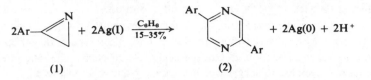

$$2Ar-\text{(1)} + 2Ag(I) \xrightarrow[15-35\%]{C_6H_6} \text{(2)} + 2Ag(0) + 2H^+$$

(1)   (2)

[1] H. Alper and J. E. Prickett, *J.C.S. Chem. Comm.*, 983 (1976).

Silver tosylate, 1, 1018; **2,** 370–371; **3,** 254.

Ring enlargement of **gem-***dibromocyclopropanes* (*cf.* **4,** 432–433). Tosylated medium-sized rings can be obtained in high yield by reaction of *gem*-dibromo-cyclopropanes with this Ag(I) salt. A single product, an (E)-allylic tosylate, is obtained (disrotatory reaction).[1]

Examples:

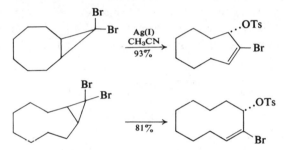

[1] H. J. J. Loozen, W. M. M. Robben, T. L. Richter, and H. M. Buck, *J. Org.*, **41,** 384 (1976).

Silver trifluoroacetate, 1, 1018–1019. Preparation.[1]

$$\text{CCl}_2 \rightarrow \text{C}=\text{O} \quad (1, 1019).$$ Dichloroketene undergoes cycloaddition to the strained triple bond of (1) to form the mono adduct (2, red). Hydrolysis of the

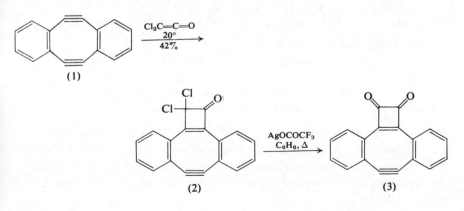

gem-dichloride with H_2SO_4, even under mild conditions, was accompanied with hydration of the triple bond, but the desired conversion to (3, orange-red) was accomplished with silver trifluoroacetate. This "quinone" decomposes on standing at room temperature. Unexpectedly, electrochemical reduction of (3) to a dihydroxycyclobutadienecyclooctatetraene system is almost as easy as reduction of phenanthrene-9,10-quinone.[2]

[1] D. E. Janssen and C. V. Wilson, *Org. Syn. Coll. Vol.*, **4**, 547 (1963).

[2] H. N. C. Wong, F. Sondheimer, R. Goodin, and R. Breslow, *Tetrahedron Letters*, 2715 (1976).

Silver(I) trifluoromethanesulfonate, 6, 520–521.

The triflate can be prepared by reaction of silver oxide with trifluoromethanesulfonic acid (98% yield).[1,2] Protect from light.

Cycloalkanes. Whitesides and Gutowski[2] have described a synthesis of cycloalkanes involving reaction of α,ω-Grignard reagents with this Ag(I) salt, which is definitely superior in this case to tetrakis[iodo(tri-*n*-butylphosphine)-silver(I)].[3] A representative synthesis is outlined in equation (I). The method is

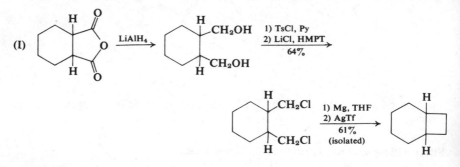

satisfactory for synthesis of four-, five-, and six-membered rings, but as expected, yields are low for medium-sized rings.

[1] R. N. Haszeldine and J. M. Kidd, *J. Chem. Soc.*, 4228 (1954).

[2] G. M. Whitesides and F. D. Gutowski, *J. Org.*, **41**, 2882 (1976).

[3] F. G. Mann, A. F. Wells, and D. Purdie, *J. Chem. Soc.*, 1828 (1937).

Sodium, 1, 1022–1023; 4, 437; 5, 589.

(E)-Disubstituted alkenyl alcohols (4, 437). Full details for the preparation of (E)-4-hexene-1-ol by fragmentation of cyclic β-halo ethers with sodium have been published. References to other examples of this cleavage are cited.[1]

[1] R. Paul, O. Riobé, and M. Maumy, *Org. Syn.*, **55**, 62 (1976).

Sodium–Ammonia, 1, 1041; 2, 374–376; 3, 259; 4, 438; 5, 589–591; 6, 523.

Reduction decyanization. Certain biogenetic-type syntheses of some alkaloids from α-amino acids required elimination of the carboxyl group by conversion to the corresponding nitrile and reductive decyanization. Some α-amino nitriles are decyanated by sodium borohydride, but sodium in liquid ammonia–THF appears to be more generally effective; moreover, no epimerization occurs at chiral centers (last example).[1]

Examples:

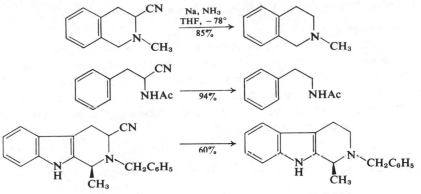

[1] S. Yamada, K. Tomioka, and K. Koga, *Tetrahedron Letters*, 61 (1976).

Sodium–*t*-Butanol–Tetrahydrofurane, 1, 1056; **3,** 378–379; **6,** 523–524.

Benzobarralene (**6**, 523–524). Explicit details for the preparation of this hydrocarbon are available.[1] Batches as large as 10 g. can be prepared in about 3 days. The method is also useful for substituted benzobarralenes.

[1] N. J. Hales, H. Heaney, J. H. Hollinshead, and P. Singh, *Org. Syn.*, submitted (1976).

Sodium acetanilidoborohydride, (1).

The reagent is obtained in a somewhat crude form by reaction of acetanilide with sodium borohydride in pyridine or in α-picoline. On the basis of spectral data it is assigned the structure (1). It is readily soluble in common organic solvents.

$$\underset{CH_3}{\overset{\overset{\displaystyle OBH_3^-\ Na^+}{|}}{C}}=N{-}C_6H_5$$

(1)

Reduction of carbonyl compounds. This borohydride reduces esters at ambient temperatures in good to high yield. Methyl esters are reduced more readily than isopropyl esters. Aldehydes and ketones, as expected, are readily reduced in CH_2Cl_2 at 20° in 30 minutes.[1]

[1] Y. Kikugawa, *Chem. Pharm. Bull. Japan*, **24**, 1059 (1976).

Sodium acetoxyborohydride, CH_3COOBH_3Na. Mol. wt. 95.87.

The reducing agent is prepared from sodium borohydride (50 mmole) and acetic acid (50 mmole) in dioxane (20 ml.). THF and diglyme are also suitable solvents.

Reduction of carboxamides to amines. Primary and secondary amides are reduced by this complex hydride to amines in about 40–90% yield. Tertiary amides can be reduced by sodium trifluoroacetoxyborohydride; N-acetylindoline is converted into N-ethylindoline in 64% yield.[1]

[1] N. Umino, T. Iwakuma, and N. Itoh, *Tetrahedron Letters*, 763 (1976).

Sodium amalgam, 1, 1030–1033; **2,** 373; **3,** 259.

Desulfonylation. Sulfones are useful in synthesis because alkylation α to the SO₂R group is accomplished readily and the sulfone group can be eliminated after the alkylation by various reagents, particularly sodium amalgam (6%). Examples of the synthesis of vinylic halides by alkylation of sulfones with 1,3-dichloropropene and 2,3-dichloropropene are formulated, together with products of hydrolysis (aldehydes and ketones).[1]

(I) $(CH_3)_2C=CHCH_2$
$\overset{\overset{SO_2C_6H_5}{|}}{}$

$\xrightarrow[86\%]{\substack{1)\ \underline{n}\text{-BuLi} \\ 2)\ CH_2ClCCl=CH_2}}$

$(CH_3)_2C=CHCHCH_2CCl=CH_2$
$\overset{\overset{SO_2C_6H_5}{|}}{}$

$\xrightarrow[94\%]{Na(Hg)}$

$(CH_3)_2C=CHCH_2CH_2CCl=CH_2$ $\xrightarrow[88\%]{H_2SO_4}$ $(CH_3)_2C=CHCH_2CH_2COCH_3$

(II) $\underline{n}\text{-}C_4H_9CH_2SO_2C_6H_5$ $\xrightarrow[86\%]{\substack{1)\ \underline{n}\text{-BuLi, CuI} \\ 2)\ CH_2ClCH=CHCl}}$ $\underline{n}\text{-}C_4H_9CHCH_2CH=CHCl$
$\overset{\overset{SO_2C_6H_5}{|}}{}$

$\xrightarrow[80\%]{Na(Hg)}$

$\underline{n}\text{-}C_4H_9CH_2CH_2CH=CHCl$ $\xrightarrow[78\%]{Hg(OAc)_2}$ $\underline{n}\text{-}C_4H_9CH_2CH_2CH_2CHO$

Trost *et al.*[2] also recommend 6% sodium amalgam, but in the presence of disodium hydrogen phosphate, for desulfonylation. Use of the acid phosphate buffer is important to success and allows successful desulfonylation of even allylic sulfones and β-keto sulfides.

Examples:

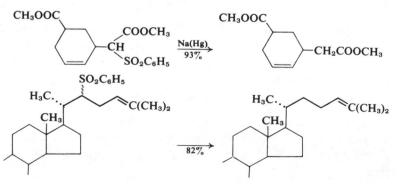

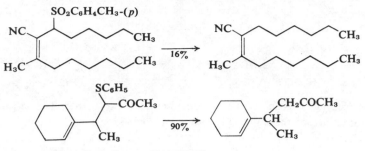

[1] M. Julia and C. Blasioli, *Bull. soc.*, 1941 (1976).
[2] B. M. Trost, H. C. Arndt, P. E. Strege, and T. R. Verhoeven, *Tetrahedron Letters*, 3477 (1976).

Sodium amide–Sodium *t*-butoxide, 4, 439–440; **5,** 593; **6,** 526.

Tetralones, indanones. In the presence of this complex base α,β-unsaturated ketones condense with bromobenzene to form tetralones and/or indanones. Benzyne is the actual reactant.[1]

Examples:

$$C_6H_5CH=CHCOCH_3 + C_6H_5Br \xrightarrow[NaOC(CH_3)_3]{NaNH_2}$$

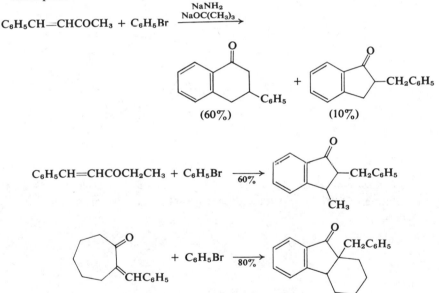

[1] M. Essiz, G. Coudert, G. Guillaumet, and P. Caubère, *Tetrahedron Letters*, 3185 (1976).

Sodium bis(2-methoxyethoxy)aluminum hydride, 3, 260–261; **4,** 441–442; **5,** 596; **6,** 528–529.

Cleavage of aryl benzyl and allyl aryl ethers. Japanese chemists[1] have noted that the benzyl group of (1) is cleaved preferentially on reduction with this hydride (refluxing xylene, 10 hours).

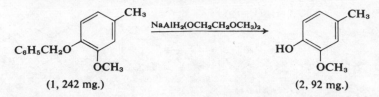

(1, 242 mg.)   (2, 92 mg.)

This cleavage is general for aryl benzyl and allyl aryl ethers and was used in a transformation of the amide (3) into pentazocine (5, a non-narcotic analgesic) from tyrosine.[1]

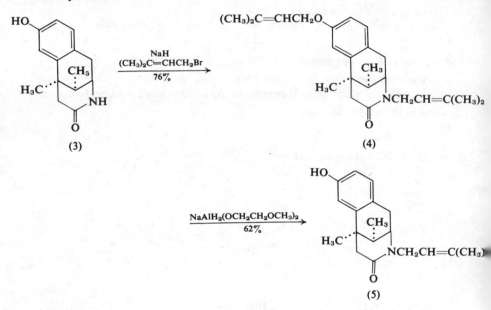

(3)   (4)

(5)

Modified reagent (6, 528–529). Addition of 1 equiv. of absolute ethanol to the solution of SMEAH provides a reagent that reduces lactones to lactols. Examples:

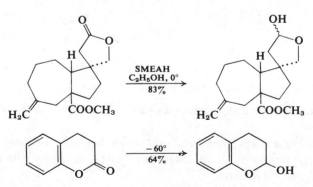

Modification of SMEAH by addition of 1 equiv. of morpholine or N-methyl-piperazine provides a reagent that reduces esters to aldehydes.[2]

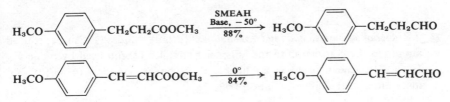

[1] T. Kametani, S.-P. Huang, M. Ihara, and K. Fukumoto, *J. Org.*, **41**, 2545 (1976).
[2] R. Kanazawa and T. Tokoroyama, *Synthesis*, 526 (1976).

Sodium bistrimethylsilylamide, 1, 1046–1047; **3**, 261–262; **4**, 442 443; **6**, 529–530.

Alkylidenetriphenylphosphoranes. This base, soluble in common organic solvents, is very useful for preparation of salt-free solutions of Wittig reactions. Solutions prepared in HMPT or THF can be used for stereoselective preparation of (Z)-alkenes (equation I). No stereoselectivity is observed when the solvent is benzene.[1]

(I) $\qquad R^1CHO + R^2CH{=}P(C_6H_5)_3 \xrightarrow{\text{THF}} R^1CH{=}CHR^2$
$\qquad\qquad\qquad\qquad\qquad\qquad\qquad\qquad (Z/E = \sim 98{:}2)$

[1] H. J. Bestmann, W. Stransky, and O. Vostrowsky, *Ber.*, **109**, 1694 (1976).

Sodium borohydride, 1, 1049–1055; **2**, 377–378; **3**, 262–264; **4**, 443–444; **5**, 597–601; **6**, 530–534.

Reduction of polycyclic quinones. This hydride in DMF is the most efficient of various metal hydrides for reduction of polycyclic quinones to hydroquinones (90–95% yield) with the exception of anthraquinone. With this quinone, lithium aluminum tri-*t*-butoxyhydride is more efficient (75% yield of the hydroquinone diacetate).[1]

Aminoanthraquinones. Nitroanthraquinones are reduced to aminoanthraquinones by $NaBH_4$ in a protic solvent (water, alcohols, aqueous DMF or THF), usually in yields of 80–100%.[2]

Cleavage of α,β-unsaturated p-tosylhydrazones.[3] Reaction of an α,β-unsaturated p-tosylhydrazone with sodium borohydride and methanol affords an allylic or benzylic methyl ether rather than a hydrocarbon as shown in the examples. Actually $NaBH_4$ can be replaced by sodium methoxide or potassium carbonate. The reaction is probably related to the Bamford-Stevens reaction.[4]

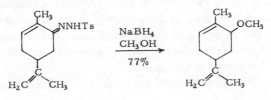

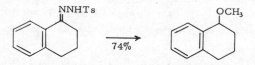

Reduction of amides and imides to amines. Tertiary amides (1) can be reduced to amines (3) by conversion to the Vilsmeier complex (2) followed by reduction with NaBH$_4$. The method is applicable to secondary amides, (4) → (6), but primary amides are converted into nitriles.[5]

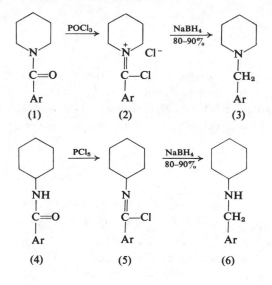

Sodium borohydride in combination with ethanedithiol (but not ethanethiol) or with thiophenol reduces amides to amines in refluxing THF. The reaction proceeds most readily with primary amides; tertiary amides are not reduced to any appreciable extent. Phthalimide (7) is reduced to (8).[6]

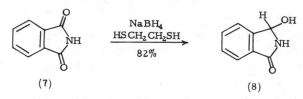

Selective reduction of cyclic imides. Cyclic imides are reduced quantitatively to α-carbinollactams by sodium borohydride in an acidic medium.[7] Regioselectivity has been observed in reduction of succinimides (equation I).[8] The reduction also appears to be stereospecific; only one stereoisomer is obtained when the two alkyl groups at C$_2$ are different.

This reduction was used in a total synthesis of *dl*-mesembrim (4).[9]

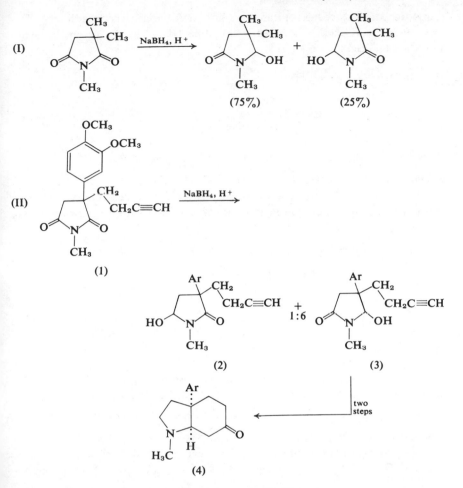

(I)

(75%)   (25%)

(II)

(1)

NaBH₄, H⁺

(2)   +   (3)
         1:6

two
steps

(4)

¹ H. Cho and R. G. Harvey, *J.C.S. Perkin I*, 836 (1976).

² J. O. Morely, *Synthesis*, 528 (1976).

³ R. Grandi, A. Marchesini, U. M. Pagnoni, and R. Trave, *J. Org.*, **41**, 1755 (1976).

⁴ W. R. Bamford and T. S. Stevens, *J. Chem. Soc.*, 4735 (1952).

⁵ Atta-ur-Rahman, A. Basha, N. Waheed, and S. Ahmed, *Tetrahedron Letters*, 219 (1976).

⁶ Y. Maki, K. Kikuchi, H. Sugiyama, and S. Seto, *Chem. Ind.*, 322 (1976).

⁷ J. C. Hubert, W. N. Speckamp, and H. O. Huisman, *Tetrahedron Letters*, 4493 (1972).

⁸ J. B. P. A. Wijnberg, W. N. Speckamp, and H. E. Schoemaker, *ibid.*, 4073 (1974).

⁹ J. B. P. A. Wijnberg and W. N. Speckamp, *ibid.*, 3963 (1975).

Sodium borohydride, sulfurated, 3, 264; 4, 444; 5, 601; 6, 534.

Reduction of nitrogen compounds. This reagent can reduce various organic nitrogen compounds that are inert to sodium borohydride. For example, both azobenzene and azoxybenzene are reduced efficiently to benzidine (equation I).

Phenyl isocyanate is reduced to aniline (62% yield) and N-methylaniline (30% yield). Pyridine N-oxide is reduced to pyridine (100% yield). Aromatic nitro groups are reduced to $-NH_2$.[1]

(I) $C_6H_5N{=}NC_6H_5$ $\xrightarrow{\text{NaBH}_2\text{S}_3}$

H_2N—⬡—⬡—NH_2 $\xleftarrow{\text{NaBH}_2\text{S}_3}$ $C_6H_5N{=}\overset{+}{\underset{\underset{O^-}{|}}{N}}C_6H_5$

[1] J.-R. Brindle, J.-L. Liard, and N. Bérubé, *Canad. J. Chem.*, **54**, 871 (1976).

Sodium borohydride–Nickel chloride, $NaBH_4$–$NiCl_2 \cdot 6H_2O$

Reduction of $\Delta^{16,20(22)}$-cardadienolides. These doubly unsaturated lactones (1) are reduced by sodium borohydride in combination with a transition metal

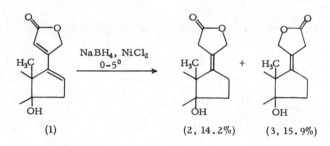

$$\xrightarrow[\text{0–5}^0]{\text{NaBH}_4,\ \text{NiCl}_2}$$

(1)   (2, 14.2%)   (3, 15.9%)

chloride to $\Delta^{17(20)}$-cardenolides, (2) and (3). The product ratio depends markedly on the metal chloride; in the absence of the salt, no reaction occurs.[1]

[1] D. Satoh and T. Hashimoto, *Chem. Pharm. Bull. Japan*, **24**, 1950 (1976).

Sodium carbonate, Na_2CO_3.

Steroid fragmentation. Indian chemists have prepared $\Delta^{5(10)}$-6-ketoestrenes (4) by treatment of 19-acetoxy-5α-bromo-6-ketosteroids (3) with sodium carbonate

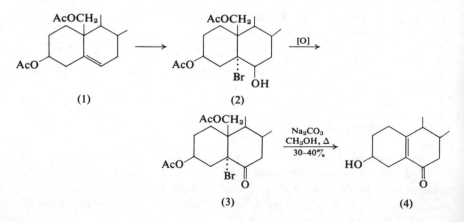

(1)   (2)

$$\xrightarrow[\underset{30–40\%}{\text{CH}_3\text{OH, }\Delta}]{\text{Na}_2\text{CO}_3}$$

(3)   (4)

in refluxing methanol. The fragmentation presumably involves hydrolysis of the 19-acetoxy group followed by deformylative debromination.[1]

[1] S. V. Sunthankar and V. V. Vakatkar, *J.C.S. Chem. Comm.*, 913 (1976).

Sodium cyanide, 4, 446–447; 5, 606–607; 6, 535–536.

Cyanoboration (**4,** 446–447; **5,** 606–607; **6,** 535–536). Detailed instructions have been submitted to *Organic Syntheses* for preparation of di-*n*-octyl ketone from 1-octene by the cyanoborate process.[1]

$$2C_6H_{13}CH{=}CH_2 + \text{[|BH}_2\text{]} \longrightarrow$$

$$\left[\text{[|B(C}_8\text{H}_{17})_2\text{]}\right] \xrightarrow{KCN} K^+ \left[\text{[|}\overset{\underset{|}{CN}}{\bar{B}}(C_8H_{17})_2\text{]}\right] \xrightarrow[\substack{\text{1) (CF}_3\text{CO)}_2\text{O, } -78 \to 20° \\ \text{2) NaOH–H}_2\text{O}_2 \\ 82\%}]{} (C_8H_{17})_2C{=}O$$

Migration of all three primary alkyl groups from boron to the one carbon unit under the influence of TFAA is possible and results in tertiary alcohols. The synthesis of tri-*n*-hexylmethanol is formulated.

$$3C_4H_9CH{=}CH_2 \xrightarrow{BH_3}$$

$$\left[B(C_6H_{13})_3\right] \xrightarrow[\substack{\text{2) (CF}_3\text{CO)}_2\text{O}}]{\text{1) KCN}} \quad \text{[structure]} \quad \xrightarrow[\substack{\text{2) NaOH, H}_2\text{O}_2 \\ 80\%}]{\text{1) (CF}_3\text{CO)}_2\text{O}} (C_6H_{13})_3COH$$

β,γ-*Unsaturated acids* (*esters*). This reaction has been conducted in two cases[2] by reaction of an allylic bromide with sodium cyanide in dry N-methyl-2-pyrrolidone followed by methanolysis. The paper states that copper(I) cyanide[3] was less satisfactory than sodium cyanide.

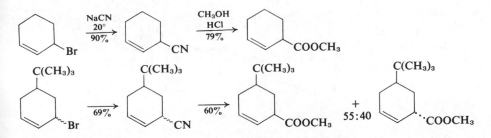

[1] A. Pelter, A. Jones, and K. Smith, *Org. Syn.*, submitted (1976).
[2] S. G. Davies and G. H. Whitham, *J.C.S. Perkin I*, 2279 (1976).
[3] E. Rietz, *Org. Syn. Coll. Vol.*, **III**, 851 (1955).

Sodium cyanoborohydride, 4, 448–451; **5,** 607–609, **6,** 527–538.

Reduction of α,β-unsaturated tosylhydrazones. Hutchins *et al.*[1] reported briefly that reduction of tosylhydrazones of aliphatic α,β-unsaturated ketones produces alkenes resulting from migration of the double bond to the carbon originally bearing the carbonyl group. Taylor and Djerassi[2] have since examined this reduction in detail with steroidal substrates and found that the situation is more complex. Thus reduction of the tosylhydrazone of Δ⁴-cholestenone gives the four products shown in equation (I). Reduction of the tosylhydrazone of Δ⁴,⁶-cholestadienone-3 is even more complex (equation II).

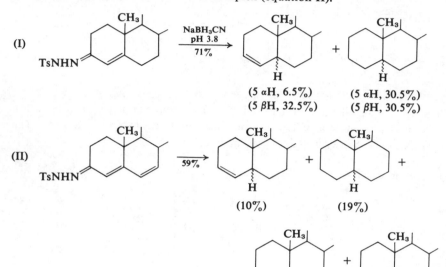

(I) NaBH₃CN pH 3.8 71%

(5 αH, 6.5%)   (5 αH, 30.5%)
(5 βH, 32.5%)   (5 βH, 30.5%)

(II) 59%

(10%)   (19%)

(28%)   (43%)

On the other hand, reduction of *cisoid* substrates (equations III and IV) proceeds cleanly to alkenes with migration of the double bond.

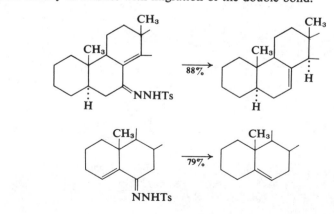

(III) 88%

(IV) 79%

Taylor and Djerassi reason that two different mechanisms are involved: alkenes are obtained by initial hydride reduction of the C=N system, whereas alkanes arise by an initial conjugative addition of hydride. Reductions with NaBD$_3$CN show that deuterium in alkenes is attached to the carbon of the original C=O group.

Reduction of α,β-unsaturated esters, nitriles, and nitro compounds. These substances are reduced to the corresponding saturated compounds[3] in fair to high yield with NaBH$_3$CN in acidic ethanol at 20°.

Deoxygenation of aryl ketones (cf. 5, 607). Aryl ketones can be deoxygenated by a variation of the reduction of aliphatic tosylhydrazones.[4] The method is outlined in equation (I).

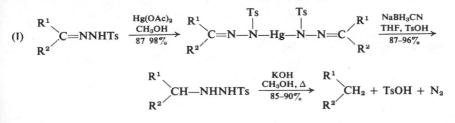

[1] R. O. Hutchins, C. A. Milewski, and B. E. Maryanoff, *Am. Soc.*, **95**, 3662 (1973).
[2] E. J. Taylor and C. Djerassi, *ibid.*, **98**, 2275 (1976).
[3] R. O. Hutchins, D. Rotstein, N. Natale, and J. Fanelli, *J. Org.*, **41**, 3328 (1976).
[4] G. Rosini and A. Medici, *Synthesis*, 530 (1976).

Sodium hydride, 1, 1075–1081; **2,** 380–383; **4,** 452–455; **5,** 610–614; **6,** 541–542.

β-Lactams. An important step in the first stereocontrolled synthesis of a penicillin (3) is the conversion of (1) into a β-lactam (2). The reaction was carried out with sodium hydride in CH$_2$Cl$_2$–DMF at 0° in notably high yield (82%).[1]

$$C_6H_5OC \overset{H_3C \quad CH_3}{\underset{Cl}{\diagdown N \diagup S}} \xrightarrow[82\%]{\text{NaH} \atop \text{CH}_2\text{Cl}_2/\text{DMF}}$$

(1, a_D -39.2°)

(2, a_D -309°)

Several steps →

(3)

[1] J. E. Baldwin, M. A. Christie, S. B. Haber, and L. I. Kruse, *Am. Soc.*, **98**, 3045 (1976).

Sodium hydrosulfite, 1, 1081–1083; **5,** 615–617.

Debromination of **vic-*dibromides*.** This reaction can be carried out with sodium hydrosulfite in DMF (140–145°). Yields are fair to high, but the reaction is not stereospecific. Both *meso-* and *dl*-2,3-dibromobutane give 1:1 mixtures of *cis-* and *trans*-2-butene.[1]

[1] T. Kempe, T. Norin, and R. Caputo, *Acta Chem. Scand.*, **30B,** 366 (1976).

Sodium hydroxide, 1, 1083; **5,** 616–617.

Selenoxide dehydration. Saris and Cava[1] have prepared the unstable benzo-[*c*]selenophene (3) by oxidation of the selenide (1) with neutral H_2O_2 in cold CH_3OH to give a solution of the selenoxide (2). Treatment of (2) with 40% NaOH liberates (3), presumably by dehydration of the ylid (a).

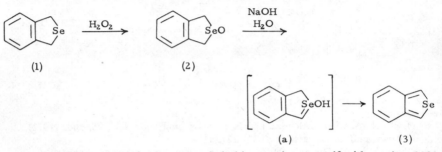

Cava and co-workers[2] have extended this reaction to sulfoxides using 30% aqueous sodium hydroxide at 80–100°, usually in the presence of a phase-transfer catalyst (hexadecyltrimethylammonium bromide).

Examples:

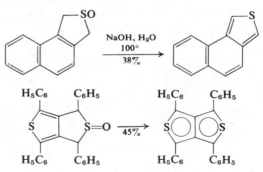

[1] L. E. Saris and M. P. Cava, *Am. Soc.*, **98,** 867 (1976).
[2] C. J. Horner, L. E. Saris, M. V. Lakshmikantham, and M. P. Cava, *Tetrahedron Letters*, 2581 (1976).

Sodium hypobromite, 1, 1083–1084; **2,** 383–384.

Isocyanides.[1] Sodium hypobromite is superior to bromine or lead tetra-acetate–triethylamine for oxidation of 5-aminotetrazoles[2] to isocyanides (equation I). Pentaazafulvenes are probable intermediates.

(I) R(Ar)NH— 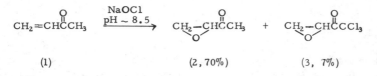 + NaOBr →

[1] G. Höfle and B. Lange, *Org. Syn.*, submitted (1976), *idem., Angew. Ch.*
113 (1976).

[2] Preparations: W. L. Garbrecht and R. M. Herbst, *J. Org.*, **18**, 1269 (1953); R. A. Henry and W. G. Finnegan, *Am. Soc.*, **76**, 926 (1954).

Sodium hypochlorite, **1**, 1084-1087; **2**, 67, **3**, 45, 243; **4**, 456; **5**, 617; **6**, 543.

Epoxidation of methyl vinyl ketone (1).[1] This α,β-unsaturated water-soluble ketone is epoxidized most satisfactorily with 5% aqueous NaOCl (clorox) at 10° at pH $\sim$ 8.5 maintained with HCl. The by-product (3) arises from the halo-form reaction.

$$CH_2=CHCCH_3 \xrightarrow[\text{pH} \sim 8.5]{\text{NaOCl}} CH_2-CHCCH_3 \quad + \quad CH_2-CHCCCl_3$$

(1) (2, 70%) (3, 7%)

Oxidation of alcohols.[2] Alcohols are oxidized in generally high yield by aqueous sodium hypochlorite under phase-transfer catalysis with tetra-*n*-butylammonium bisulfite. Benzene, chloroform, and methylene chloride can serve as the organic phase, but ethyl acetate is the solvent of choice.

Examples:

$$C_6H_5CH_2OH \xrightarrow{76\%} C_6H_5CHO$$

$$(C_6H_5)_2CHOH \xrightarrow{82\%} (C_6H_5)_2CO$$

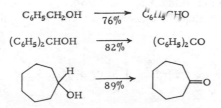

Amines of the type R_2CHNH_2 are oxidized under these conditions to N-chloroimines, $R_2C=NCl$, which are convertible into ketones. Amines of the type RCH_2NH_2 are oxidized predominantly to nitriles.

1,2,4-Oxadiazoles. This heterocyclic system is formed on treatment of amidines or guanidines with sodium hypochlorite and then with sodium carbonate.[3]

Examples:

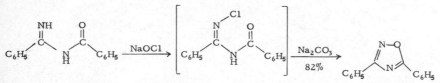

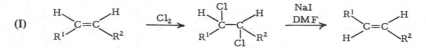

...man, B. Lam, E. L. Anderson, and E. White, V, *Synthesis*, 547 (1976).
G. A. Lee and H. H. Freedman, *Tetrahedron Letters*, 1641 (1976).
[3] N. Götz and B. Zeeh, *Synthesis*, 268 (1976).

Sodium iodide, 1, 1087–1090; **2,** 384; **3,** 267; **4,** 456–457; **6,** 543.

Olefin inversion. This reaction can be accomplished by conversion of the olefin into the *vic*-dichloride (Cl_2 in CH_2Cl_2) or *vic*-bromochloride (HCl–NBS in CH_2Cl_2) followed by sodium iodide reduction in DMF as formulated (equation I). The first reaction involves *trans*-addition, whereas the second is a

(I)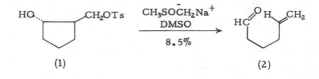

cis-elimination. The method is not useful for *vic*-dibromides, which undergo *trans*-elimination with NaI. It is also only useful for inversion of 1,2-disubstituted olefins.[1]

[1] P. E. Sonnet and J. E. Oliver, *J. Org.*, **41,** 3284 (1976).

Sodium iodide–Zinc.

Reduction of tosylates. Reaction of primary or secondary tosylates or mesylates with sodium iodide and zinc powder in refluxing glyme (or DMSO, HMPT) results in reductive cleavage (equation I). Lithium bromide is less efficient.[1]

(I)

[1] Y. Fujimoto and T. Tatsuno, *Tetrahedron Letters*, 3325 (1976).

Sodium methylsulfinylmethylide (Dimsylsodium), 1, 310–313; **2,** 166–169; **3,** 123–124; **4,** 195–196; **5,** 621; **6,** 546–547.

Secoiridoids. Some years ago Corey and co-workers[1] reported that treatment of the monotosylate of a 1,3-diol with this base could lead to an unsaturated ketone. This reaction as applied recently to the tosylate of the 1,3-diol (1) resulted in fragmentation to the δ,ϵ-unsaturated aldehyde (2), albeit in low yield. Only

polymeric material was obtained with KOC(CH₃)₃ as base. The isomer of (1)
under basic conditions is converted into an oxetane (4) in high yield.

(3)   (4)

The cleavage reaction was used for a synthesis of secoiridoids; for example,
secologanin aglycone O-methyl ether (6) was obtained by fragmentation of (5) in
respectable yield.[2]

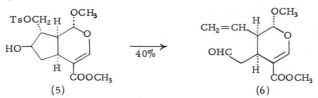

(5)   (6)

β-Acylacrylic esters; α,β-butenolides. Bartlett[3] has published new routes to
these structures from β-keto sulfoxides by reaction of methyl esters with dimsyl-
potassium (method of Corey, **1,** 311). The steps are outlined in equations (I)
and (II). This method was used to prepare the isocardenolide 3β-acetoxy-20-
hydroxy-21-nor-5,22-choladiene-24-oic acid γ-lactone from methyl 3β-hydroxy-
etienate in excellent yield.

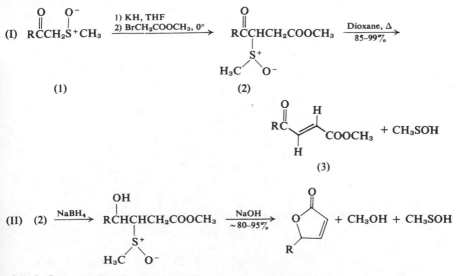

[1] E. J. Corey, R. B. Mitra, and H. Uda, *Am. Soc.,* **86,** 485 (1964).
[2] G. Kinast and L.-F. Tietze, *Ber.,* **109,** 3626 (1976).
[3] P. A. Bartlett, *Am. Soc.,* **98,** 3305 (1976).

Sodium naphthalenide, 1, 711–712; **2,** 289; **4,** 349–350; **5,** 468–470.

Acyloin condensation. The usual preparation of acyloins from aliphatic esters[1] is not applicable to aromatic esters. In this case the reaction with sodium results in formation of the sodium carboxylate and a radical from the R group which undergoes further reactions. However, benzoin can be prepared in moderate yield by treatment of the trimethylsilyl ester of benzoic acid with sodium or by treatment of ethyl benzoate with sodium naphthalenide in THF at room temperature (39% yield).[2]

[1] S. M. McElvain, *Org. React.*, **4**, 256 (1948).
[2] H. Stetter and K. A. Lehmann, *Ann.*, 499 (1973).

Sodium octacarbonylhydridodiiron, $NaHFe_2(CO)_8$. Mol. wt. 359.78, reddish brown solid.

Preparation:

$$Na_2Fe(CO)_4 \cdot 1.5\ O(CH_2CH_2)_2O \xrightarrow{THF} Na_2Fe(CO)_8 \cdot THF \xrightarrow{HOAc} NaHFe_2(CO)_8$$

Reduction of α,β-unsaturated carbonyl compounds. This binuclear cluster compound reduces only the C=C bond of α,β-unsaturated ketones, aldehydes, esters, lactones, amides, and nitriles. Yields are usually >90%. A possible mechanism is suggested in the communication.[1]

Examples:

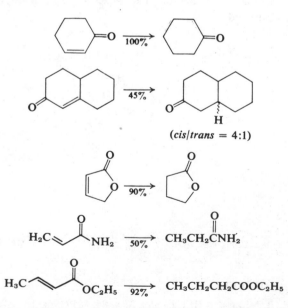

$$\text{(cis/trans = 4:1)}$$

[1] J. P. Collman, R. G. Finke, P. L. Matlock, R. Wahren, and J. I. Brauman, *Am. Soc.*, **98**, 4685 (1976).

... have reported a new synthesis of benz-
alde... ...ion (I). The scheme is successful only when
β-elim... ...impossible. One limitation is that a selenoxide

(I) $ArCH_2$... ...Na^+ $\longrightarrow$ $ArCH_2SeC_6H_5$ $\xrightarrow[\text{Quant.}]{\substack{NaIO_4 \\ 0°}}$

$ArCH_2SeC_6H_5$ $\xrightarrow{110-130°}$ $[ArCH_2OSeC_6H_5]$ $\xrightarrow{60-85\%}$ $ArCHO + [C_6H_5SeH]$

could not be isolated from the oxidation of o-methylbenzyl phenyl selenide
(possibly owing to steric factors).

This rearrangement had been applied previously to benzyl methyl sulfoxides,
but in this case temperatures of 210–280° are required.[2]

[1] I. D. Entwistle, R. A. W. Johnstone, and J. H. Varley, *J.C.S. Chem. Comm.*, 61 (1976).
[2] I. D. Entwistle, R. A. W. Johnstone, and B. J. Millard, *J. Chem. Soc. (C)*, 302 (1967).

**Sodium tetracarbonylferrate(II) (Disodium tetracarbonylferrate), 3, 267–268; 4,
461–468; 5, 624–625; 6, 550–552.**

Amides. Sodium acyltetracarbonylferrates, derived from sodium tetracarbonylferrate and acid chlorides, react with nitro compounds, both aliphatic
and aromatic, to form amides in 70–85% yield.[1]

$$[RCOFe(CO)_4]^- Na^+ + R'NO_2 \xrightarrow[70-85\%]{\substack{1) \ THF, \ 20° \\ 2) \ H_2O}} RCONHR'$$

[1] M. Yamashita, Y. Watanabe, T. Mitsudo, and Y. Takegami, *Tetrahedron Letters*, 1585
(1976).

Sodium trichloroacetate, 1, 1107–1108.

Dichlorocarbene (1, 1107–1108). Dichlorocarbene can be generated from this
precurser at temperatures of 44–63° if 5–10 mole % of a phase-transfer catalyst
is present. Tetra-n-heptylammonium bromide and Aliquat 336 are more effective

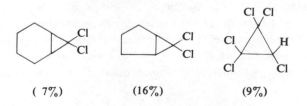

(7%)   (16%)   (9%)

Sodium trifluoroacetoxyborohydride, NaBH$_3$(OCOCF$_3$). Mol. wt. 149.84. The reagent is prepared in THF solution by addition of CF$_3$COOH (1 equiv.) over a period of 10 minutes to a suspension of NaBH$_4$ in THF at 20°.

Reduction of nitriles. This sodium acyloxyborohydride reduces nitriles (aliphatic and aromatic) to primary amines at room temperature in high yield. The reaction appears to be selective; —COOC$_2$H$_5$, —NO$_2$, and —Cl groups are not reduced.[1]

[1] N. Umino, T. Iwakuma, and N. Itoh, *Tetrahedron Letters,* 2875 (1976).

Sodium p-thiocresolate, NaS—⟨⟩—CH$_3$ (1). Mol. wt. 146.18, stable to storage in a dessicator at 20°.

Selective dealkylation of alkyl phenyl ethers (*cf. Sodium thioethoxide,* **4**, 465). This reagent in combination with HMPT cleaves alkyl phenyl ethers under mild conditions. Methoxy groups *ortho* or *para* to a formyl group are cleaved selectively.[1]

Examples:

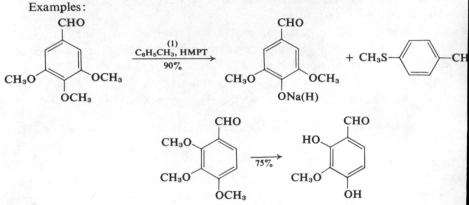

[1] C. Hansson and B. Wickberg, *Synthesis,* 191 (1976).

Stannic chloride, 1, 1111–1113; **3,** 269; **5,** 627–631; **6,** 553–554.

Glycosides.[1] Glycosides can be prepared conveniently by reaction of peracyl derivatives of D-ribofuranose or D-glucopyranose with N,N-dimethylformamide dialkyl acetals under Lewis acid catalysis (equations I and II). Unlike classical methods, this new glycosidation method does not employ the unstable glycosyl halides.

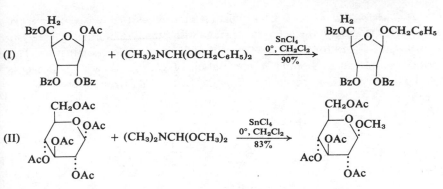

(I)

(II)

The same conditions can be used for preparation of β-D-ribofuranosyl disaccharides (equations III and IV).

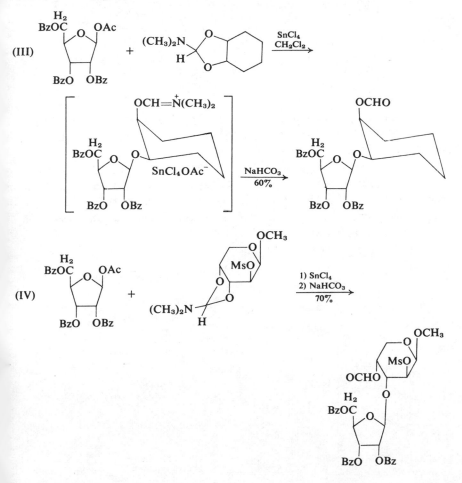

(III)

(IV)

O-Glucuronidation.[2] O-Glucuronidation can be carried out with methyl 1,2,3,4-tetra-O-acetyl-β-D-glucopyranuronate with stannic chloride as catalyst. Both α- and β-glucuronosides are formed, with the latter predominating. The method is particularly useful for preparation of glucuronosides of estrogens, which are obtained in rather low yield by the Koenigs-Knorr method.

Example:

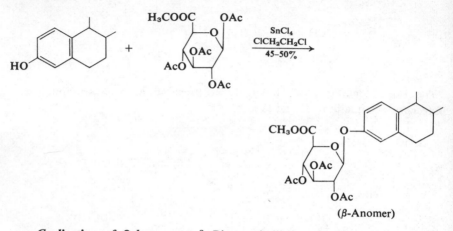

(β-Anomer)

Cyclization of β-keto esters.[3] Biogenetic-like cyclization of polyolefins (**5**, 627–628) has been extended to cyclization of β-keto esters and used in an interesting synthesis of the tricyclic diterpene $\Delta^{8(14)}$-podocarpene-13-one (**5**). Thus treatment of (1), prepared in 61% yield from geranyl bromide and methyl acetoacetate, with stannic chloride in CH_2Cl_2 at $0 \rightarrow 20°$ gave a single cyclic product (2) in 68% yield. This was converted into (5) by several steps as formulated. The intermediate (3) was also converted into a compound used for elaboration of ambreinolide and labdanoid systems.

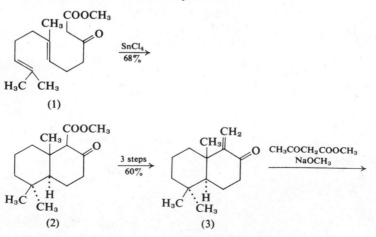

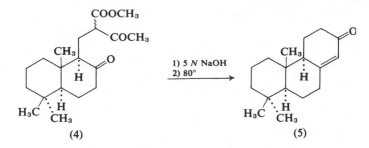

(4)   (5)

Catalysis of a Diels-Alder reaction. The uncatalyzed reaction of methyl *trans*-4-oxobutenoate (1) with *trans*-piperylene (2) results in formation of (3) and (4) in about equal amounts. The reaction catalyzed by a Lewis acid favors

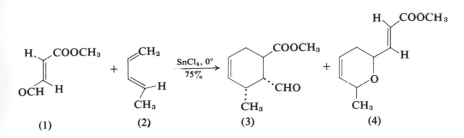

(1)   (2)   (3)   (4)

formation of (3). $SnCl_4$ is a more efficient catalyst than $AlCl_3$ or BF_3 etherate.[4]

Isoquinoline-3-ols.[5] Phenacetyl chlorides react with alkyl or aryl thiocyanates under Friedel-Crafts conditions to form 1-substituted isoquinoline-3-ols (equation I). Stannic chloride is the preferred catalyst and nitrobenzene is the preferred solvent. The reaction fails with acetonitrile or benzonitrile.

(I)

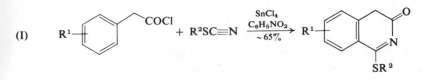

Cyclizations. The biomimetic cyclizations of polyenes have been reviewed by Johnson (86 references).[6]

[1] S. Hanessian and J. Banoub, *Tetrahedron Letters*, 657, 661 (1976).
[2] K. Honma, K. Nakazima, T. Uematsu, and A. Hamada, *Chem. Pharm. Bull. Japan*, **24**, 394 (1976).
[3] R. W. Skeean, G. L. Trammell, and J. D. White, *Tetrahedron Letters*, 525 (1976).
[4] M. Kakushima, J. Espinosa, and Z. Valenta, *Canad. J. Chem.*, **54**, 3304 (1976).
[5] M. A. Ainscough and A. F. Temple, *J.C.S. Chem. Comm.*, 695 (1976).
[6] W. S. Johnson, *Biorganic Chem.*, **5**, 51 (1976).

Stannous octoate (Stannous 2-ethylhexanoate), $Sn[OOCCH(C_2H_5)CH_2CH_2CH_2-CH_3]_2$. Mol. wt. 405.11. Suppliers: Albright and Wilson; Metal and Thermit Corp. (Catalyst T9); ROC/RIC.

Carbamates of tertiary alcohols. The reaction of tertiary alcohols with isocyanates catalyzed by base usually results in dehydration as well as carbamate formation. However, carbamates of tertiary alcohols can be obtained in yields as high as 90% by use of organotin catalysts (stannous octoate, stannous oleate, dibutyltin dilaurate) and an amine.

The same paper reports cyclization of carbamates of unsaturated tertiary alcohols to 2-oxazolidinones (equation I–III).[1]

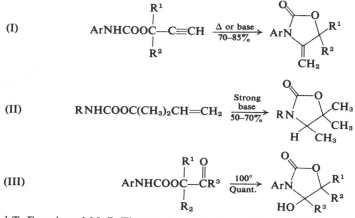

(I) $ArNHCOOC(R^1)(R^2)-C\equiv CH \xrightarrow[70-85\%]{\Delta \text{ or base}}$

(II) $RNHCOOC(CH_3)_2CH=CH_2 \xrightarrow[50-70\%]{\text{Strong base}}$

(III) $ArNHCOOC(R^1)(R_2)-CR^3 \xrightarrow[\text{Quant.}]{100°}$ (with O)

[1] T. Francis and M. P. Thorne, *Canad. J. Chem.*, **54**, 24 (1976).

Sulfur dioxide, 1, 1122; **2**, 292; **4**, 469; **5**, 633; **6**, 558.

Diels-Alder reactions. Heldeweg and Hogeveen[1] have shown that, at least in the case of highly reactive dienes, the initial products of cycloaddition of sulfur dioxide are sulfinic esters rather than sulfones. Thus reaction of the tricyclic diene (1) with SO_2 at room temperature leads to the sulfinic ester (2). This ester is unstable thermally and rearranges mainly to the aromatic ester (3). They also point out the similarity between cycloaddition reactions of sulfur dioxide and selenium dioxide. Thus (1) reacts with the latter reagent to give the seleninic ester (4), probably formed via the intermediate (a).

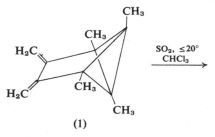

$\xrightarrow[CHCl_3]{SO_2, \leq 20°}$

(1)

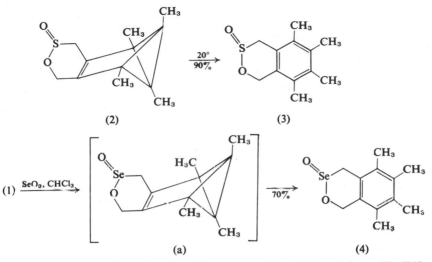

(2)   (3)

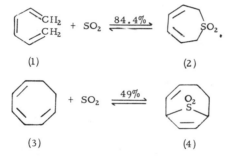

(a)   (4)

1,6-Cyclouaddition.[2] Sulfur dioxide reacts with *cis*-3-hexatriene (1) (20°, several weeks) to form 2,7-dihydrothiepin 1,1-dioxide (2). This cycloaddition appears to be general; indeed substituted *cis*-3-hexatrienes react more readily

than (1). The reaction of SO_2 with 1,3,5-cyclooctatriene (3) involves 1,4-addition to give (4). Both these adducts dissociate into their components when heated and in fact should be useful for isolation and storage of the trienes (1) and (3).

Reaction with ortho esters.[3] Trialkyl(triaryl) ortho esters react with sulfur dioxide to produce esters and dialkyl sulfites (equation I).

(I) $RC(OR^1)_3 + SO_2 \longrightarrow RCOOR^1 + (R^1O)_2SO$

[1] R. F. Heldeweg and H. Hogeveen, *Am. Soc.*, **98**, 2341 (1976).
[2] W. L. Mock and J. H. McCausland, *J. Org.*, **41**, 242 (1976).
[3] M. M. Rogić, K. P. Klein, J. M. Balquist, and B. C. Oxenrider, *ibid.*, **41**, 482 (1975).

Sulfuric acid, 1, 470–472; **5,** 633–639; **6,** 558–560.

Cyclization. A recent synthesis of codeine and morphine utilized the cyclization of (1) to (2) with 80% sulfuric acid in ether.[1] A hydroxyl group was placed

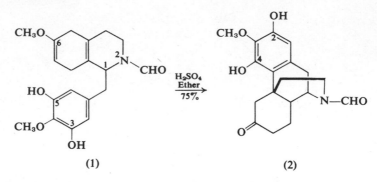

(1)   (2)

at C_3 in the benzyl group of the starting material (1) so that the cyclization could take place in only one way to give (2). The excess hydroxyl group at C_2 was then eliminated[2] by the procedure of Musliner and Gates (**2**, 319–320). The formyl group of the dehydroxylated product was reduced to a methyl group. The resulting compound had previously been converted into both codeine and morphine by Gates and co-workers.[3]

Reduction-rearrangements. The reaction of the carbinol (1) and related $C_{11}C_{18}O$ alcohols with *n*-pentane (hydride donor) in 95% H_2SO_4 leads to formation of 4-homoisotwistane (2) in high yield. Different products, (3) and (4), are formed when phosphoric acid serves as the acid catalyst. The paper suggests several pathways for these hydride-transfer reactions.[4]

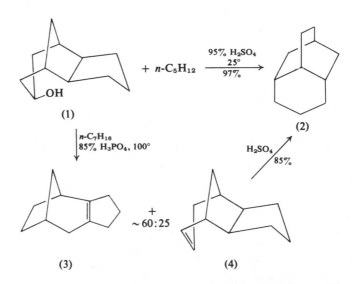

(1)

(2)

(3) $\sim$ 60:25   (4)

Rearrangement of α,β-epoxy ketones.[5] Alkyl-substituted 2,3-epoxycyclo-hexanones are rearranged to 2-hydroxy-2-cyclohexenones by 1.3–2.0% aqueous sulfuric acid. The products usually crystallize as a ketoenolic form. The re-arrangement of 2,3-epoxy-3-methylcyclopentanone to 2-hydroxy-3-methyl-2-cyclopentenone was also reported (last example).

Examples:

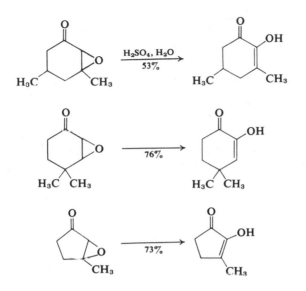

[1] H. C. Beyerman, T. S. Lie, L. Maat, H. H. Bosman, E. Buurman, E. J. M. Bijsterveld, and H. J. M. Sinnige, *Rec. trav.*, **95**, 24 (1976).
[2] H. C. Beyerman, E. Buurman, T. S. Lie, and L. Maat, *ibid.*, **95**, 43 (1976).
[3] M. Gates and G. Tschudi, *Am. Soc.*, **78**, 1380 (1956); M. Gates and M. S. Shepard, *ibid.*, **84**, 4125 (1962).
[4] N. Takaishi, Y. Inamoto, K. Tsuchihashi, K. Aigami, and Y. Fujikura, *J. Org.*, **41**, 771 (1976).
[5] M. T. Langin-Lanteri and J. Huet, *Synthesis*, 541 (1976).

Sulfuryl chloride, 1, 1128–1131; **2,** 394–395; **3,** 276; **4,** 474–475; **5,** 641; **6,** 561.

Chlorination. The chlorination (Cl₂) of *o*-methylphenol (1) results in three products—the *p*-chloro derivative (2), the *o*-chloro derivative (3), and the *o,p*-dichloro derivative (4). Use of sulfuryl chloride results in formation of the first two products, and the reaction requires about 24 hours at 20°. However, if diphenyl sulfide and aluminum chloride are used as catalysts, this chlorination requires only about 2 hours for completion, and the reaction becomes very regioselective for *para*-chlorination to give (2). The paper suggests that sulfuryl chloride converts diphenyl sulfide into diphenyl sulfide dichloride, which then forms a bulky complex with AlCl₃. The same conditions are useful for chlorination of other phenols.[1]

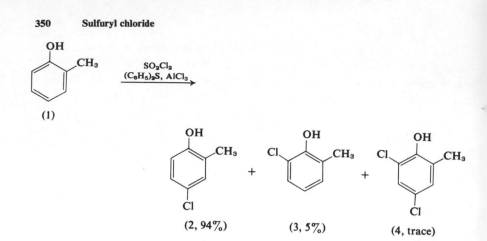

(1)

(2, 94%)   (3, 5%)   (4, trace)

Oxidation of sulfides to sulfoxides. Sulfides are oxidized to sulfoxides in high yield by sulfuryl chloride in the presence of small amounts of wet silica gel (equation I). This oxidation is particularly useful for oxidation of chloromethyl alkyl(aryl) sulfides, which are very susceptible to C—S cleavage (equation II).[2]

$$\text{(I)} \qquad RSR^1 \xrightarrow[\substack{85-100\%}]{\substack{SO_2Cl_2, SiO_2, H_2O \\ CH_2Cl_2, 20^0}} RSOR^1$$

$$\text{(II)} \qquad RSCH_2Cl \xrightarrow[80-100\%]{} RSOCH_2Cl$$

Cleavage of thioacetal groups. Sulfuryl chloride in the presence of wet silica gel oxidatively cleaves thioacetal groups at room temperature. Monosulfoxides are presumably intermediates.[3]

Examples:

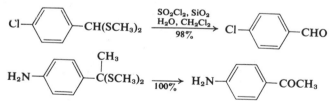

[1] W. D. Watson, *Tetrahedron Letters*, 2591 (1976).
[2] M. Hojo and R. Masuda, *ibid.*, 613 (1976).
[3] *Idem, Synthesis*, 678 (1976).

T

2,4,4,6-Tetrabromocyclohexa-2,5-dienone, (1), **4,** 476–477; **5,** 643–644; **6,** 563.

Reaction with unsaturated alcohols and ketones. This brominating reagent reacts with some unsaturated alcohols, for example (2), to form bromo ethers (3) and (4).[1] This reaction was used in a synthesis of *dl*-incensole (8) from *dl*-mukulol (5) as formulated.[2]

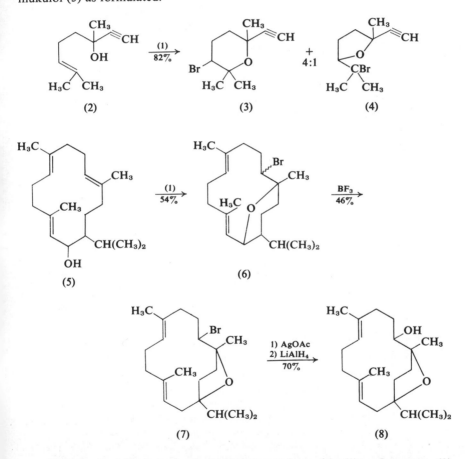

Selective bromination of polyolefins. Farnesyl cyanide (1) and carvone (3) are selectively brominated at the terminal double bond by this tetrabromo compound.[3]

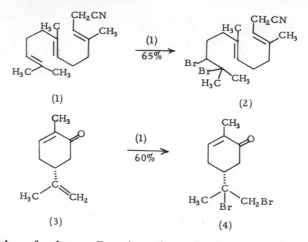

(1)   (2)

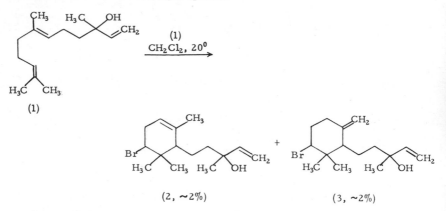

Cyclization of polyenes. Reaction of nerolidol (1) with this brominating agent (1) in CH_2Cl_2 (20°, 3 hours) yields α- and β-snyderol, (2) and (3), in low yield.[4] These bromine-containing monocyclic sesquiterpenes have been isolated recently from a marine red algae species.[5]

(2, ~2%)   (3, ~2%)

p-Bromination of arylamines (**4**, 476). The preparation of 4-bromo-N,N-dimethyl-3-(trifluoromethyl)aniline using this reagent has been published.[6]

[1] T. Kato, I. Ichinose, T. Hosogai, and Y. Kitahara, *Chem. Letters*, 1187 (1976).

[2] T. Kato, C. C. Yen, T. Kobayashi, and Y. Kitahara, *ibid.*, 1191 (1976).

[3] Y. Kitahara, T. Kato, and I. Ichinose, *ibid.*, 283 (1976).

[4] T. Kato, I. Ichinose, A. Kamoshida, and Y. Kitahara, *J.C.S. Chem. Comm.*, 518 (1976).

[5] B. M. Howard and W. Fenical, *Tetrahedron Letters*, 41 (1976).

[6] G. J. Fox, G. Hallas, J. D. Hepworth, and K. N. Paskins, *Org. Syn.*, **55**, 20 (1976).

Tetra-*n*-butylammonium borohydride, 6, 564–565.

Reductions.[1] One advantage of this reagent over other borohydrides is that it is very soluble in methylene chloride. The reactivity is similar to that of sodium borohydride. The reagent reduces acid chlorides rapidly at 20°. Aldehydes and ketones are reduced at convenient rates, but esters are reduced very slowly. Of course the rates also depend on steric bulk of substituents.

[1] D. J. Raber and W. C. Guida, *J. Org.*, **41**, 690 (1976).

Tetra-*n*-butylammonium bromide, 4, 477; 5, 644.

Phosphorylation of primary alcohols and phenols. Zwierzak[1] has found that the Atherton-Todd phosphorylation[2] can be carried out advantageously under phase-transfer catalysis. Carbon tetrachloride and 50% aqueous sodium hydroxide serve as the solvent pair; tetra-*n*-butylammonium bromide or benzyltriethylammonium chloride serve equally well as catalysts. Yields are 35–90% for primary alcohols and phenols; low yields are obtained from other alcohols.

$$R^1O\text{-}P(O)H\text{-}OR^1 + R^2OH \xrightarrow[35-90\%]{\text{cat. } CCl_4, H_2O, NaOH} R^1O\text{-}P(O)OR^2\text{-}OR^1$$

$$R^1 = C_2H_5, \underline{n}\text{-}C_4H_9$$

[1] A. Zwierzak, *Synthesis*, 305 (1976).
[2] F. R. Atherton, H. T. Openshaw, and A. R. Todd, *J. Chem. Soc.*, 660 (1945).

Tetra-*n*-butylammonium chloride, 6, 565.

N-Alkylation of phenylhydrazones. Phenylhydrazones of aldehydes or ketones are alkylated exclusively on nitrogen with phase-transfer catalysis. The products can be hydrolyzed to 1-alkylphenylhydrazines.[1]

$$R^1R^2C=NNHC_6H_5 + R^3X \xrightarrow[50-95\%]{\text{Cat. } NaOH, H_2O} R^1R^2C=NN(C_6H_5)R^3 \xrightarrow[40-70\%]{H_2SO_4, H_2O} C_6H_5NNH_2R^3 + R^1R^2C=O$$

[1] A. Jończyk, J. Włostowska, and M. Mąkosza, *Synthesis*, 795 (1976).

Tetra-*n*-butylammonium fluoride, 4, 477–478; 5, 645.

Additional preparations.[1,2]

Conjugate addition of thiols.[2] This quaternary ammonium fluoride and benzyltrimethylammonium fluoride are effective catalysts for the conjugate addition of thiols to enones. Acetone, acetonitrile, and THF are the solvents of choice. The reaction rate is markedly dependent on the substrates.

Examples:

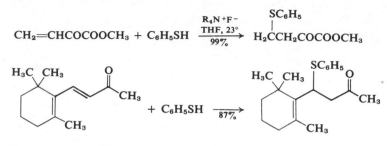

Cleavage of phosphate protecting groups.[3] The three most common protecting groups used in nucleotide triester syntheses are the cyanoethyl ($-CH_2CH_2CN$), trichloroethyl ($-CH_2CCl_3$), and phenyl groups. These groups are all cleaved by 2 equiv. of this fluoride in THF at 20° within 20 minutes. These conditions also cleave alkylsilyl protecting groups. However, these latter groups can be cleaved selectively by 10 equiv. of tetra-*n*-butylammonium fluoride in THF containing acetic acid (20°, 24 hours).

Addition of silylacetylenes to carbonyl compounds.[4] This salt catalyzes the addition of 1-phenyl-2-trimethylsilylacetylene (1) to aldehydes or ketones at room temperature to form adducts which can be isolated in the silylated form (2) or as alcohols (3). The reaction of (a) with α,β-unsaturated carbonyl compounds is generally not useful. See **Ethyl trimethylsilylacetate–Tetra-*n*-butylammonium fluoride,** this volume.

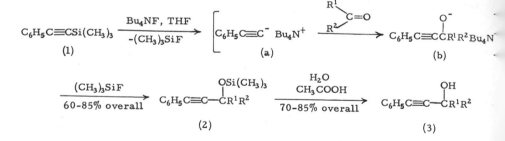

[1] H. B. Henbest and W. R. Jackson, *J. Chem. Soc.*, 954 (1962).
[2] I. Kuwajima, T. T. Murofushi, and E. Nakamura, *Synthesis*, 602 (1976).
[3] K. K. Ogilvie, S. L. Beaucage, and D. W. Entwistle, *Tetrahedron Letters*, 1255 (1976).
[4] E. Nakamura and I. Kuwajima, *Angew. Chem., Int. Ed.*, **15**, 498 (1976).

Tetra-*n*-butylammonium hydrogen sulfate, 6, 565–566.

Dehydrohalogenation. Acetylenes can be prepared by dehydrogenation of compounds of the type RCH=CXR′ or RCHX—CHXR′ by ion-pair extraction (**6,** 565). Yields are generally superior to those obtained by classical methods.[1]

Examples:

$$CH_2BrCHBrCH(OC_6H_5)_2 \xrightarrow[\substack{Pentane, H_2O \\ 80\%}]{Bu_4N^+HSO_4^-, NaOH} HC{\equiv}CCH(OC_6H_5)_2$$

$$C_6H_5CHBrCH_2Br \xrightarrow[87\%]{} C_6H_5C{\equiv}CH$$

$$CCl_2{=}CHCH(OC_2H_5)_2 \xrightarrow[70\%]{} ClC{\equiv}CCH(OC_2H_5)_2$$

[1] A. Gorgues and A. Le Coq, *Tetrahedron Letters*, 4723 (1976).

Tetra-*n*-butylammonium iodide, 5, 646–647; **6,** 566–567.

Wittig-Horner reactions. 1-Aryl-4-phenylbuta-1,3-dienes have been prepared by phase-transfer catalysis of phosphonates (1) and (2) with aromatic aldehydes (equations I and II).[1]

(I) $(C_2H_5O)_2\overset{\overset{O}{\|}}{P}CH_2Ar + C_6H_5CH{=}CHCHO \xrightarrow[70-80\%]{\substack{Cat. \\ C_6H_6, NaOH \\ H_2O}} C_6H_5CH{=}CH{-}CH{=}CHAr$

   (1)

(II) $(C_2H_5O)_2\overset{\overset{O}{\|}}{P}CH_2{-}CH{=}CHC_6H_5 + ArCHO \xrightarrow[55-85\%]{} C_6H_5CH{=}CH{-}CH{=}CHAr$

   (2)

Carbonylation of organic halides. This reaction can be conducted under phase-transfer conditions:

$$RX + CO + 2\,NaOH \xrightarrow[H_2O,\ Xylene]{Pd\left[P(C_6H_5)_3\right]_4,\ cat.} RCOONa + NaX + H_2O$$

Benzyl chloride has been converted into phenylacetic acid (83%) in this way.[2]

[1] C. Piechucki, *Synthesis*, 187 (1976).
[2] L. Cassar, M. Foà, and A. Gardano, *J. Organometal. Chem.*, **121**, C55 (1976).

Tetrachloro-*o*-benzoquinone (*o*-Chloranil), 1, 128–129; **2,** 345; **4,** 76.

Dehydrogenation of pyrazolines. Pyrazoles (2) are obtained in almost quantitative yield by treatment of pyrazolines (1) in ether with this high-potential quinone. The reaction is rapid at 20°.[1]

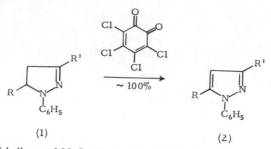

(1)   (2)

[1] N. Latif, N. Mishriky, and N. S. Girgis, *Chem. Ind.*, 28 (1976).

Tetraethylammonium fluoride, 5, 68, 375, 376.

Cleavage of Si—C *bond.* The cleavage of the Si—C bond in (1) and (2) by fluoride ion is relatively facile, but cleavage is difficult or impossible in the case

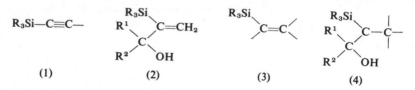

(1)   (2)   (3)   (4)

of (3) and (4). Surprisingly, cleavage of epoxides of type (5) by F⁻ is relatively easy, in fact, easier than cleavage of (2). And indeed epoxides of type (7) are also cleaved. The cleavage occurs with retention of configuration.[1]

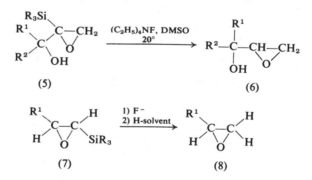

(5)   (6)

(7)   (8)

[1] T. H. Chan, P. W. K. Lau, and M. P. Li, *Tetrahedron Letters*, 2667 (1976).

Tetrahydrofurane–Sulfuryl chloride.

Tetrahydrofuranyl ethers. When sulfuryl chloride is dissolved in THF at room temperature, an exothermic reaction occurs. Addition of this mixture to a solution of an alcohol and triethylamine in THF results in formation of THF ethers in 85–98% yield. The NMR spectra indicate that sulfuryl chloride reacts with THF to form 2-chlorotetrahydrofurane. Thiols, phenols, azoles, and acids can replace ROH.

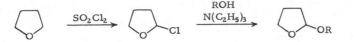

THF ethers are hydrolyzed by very mild acidic conditions; they are hydrolyzed appreciably faster than THP ethers.

The ethers are stable to base, to Grignard and Wittig reagents, and to lithium aluminum hydride.[1]

[1] C. G. Kruse, N. L. J. M. Broekhof, and A. van der Gen, *Tetrahedron Letters*, 1725 (1976).

Tetrakis(triphenylphosphine)nickel(0), $Ni[P(C_6H_5)_3]_4$, **6, 570.**

The complex can also be prepared by reaction of $Ni(acac)_2$, $HAl[CH_2CH-(CH_3)_2]_2$, and $P(C_6H_5)_3$ in THF.

Alkenyl–Aryl coupling. *trans*-Alkenylalanes can be activated for coupling with aryl bromides or iodides with this nickel(0) complex to form arylated *trans*-alkenes in 65–95% yield (equation I). $Pd[P(C_6H_5)_3]_4$ is a less effective catalyst.[1]

(I)

[1] E. Negishi and S. Baba, *J.C.S. Chem. Comm.*, 596 (1976).

Tetrakis(triphenylphosphine)palladium(0), 6, 571–573.

Allylation of amines. Trost and Genêt[1] have synthesized several ring systems characteristic of alkaloids by intramolecular allylation (cyclization) of amines catalyzed by Pd(0).

Examples:

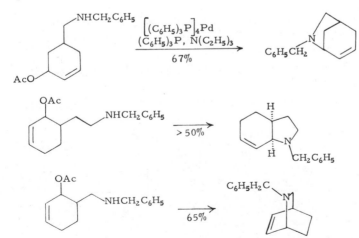

This reaction has been used for a very ready entry into the ibogamine skeleton (1) from tryptamine. The synthesis involves two cyclizations promoted by π-palladium complexes.

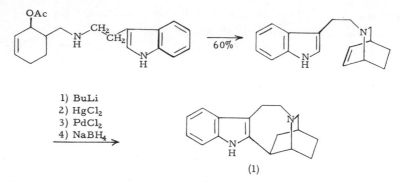

1) BuLi
2) HgCl₂
3) PdCl₂
4) NaBH₄

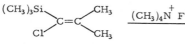

(1)

[1] B. M. Trost and J. P. Genêt, *Am. Soc.*, **98**, 8516 (1976).

Tetramethylammonium fluoride, $(CH_3)_4 N^+ F^-$. Mol. wt. 93.14.

Isopropylidenecarbene.[1] When treated with anhydrous tetramethylammonium fluoride, 1-chloro-1-trimethylsilyl-2-methylpropene undergoes *gem*-dechlorosilylation[2] to form a carbene that reacts with alkenes to form isopropylidenecyclopropanes.

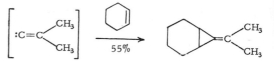

[1] R. F. Cunico and Y.-K. Han, *J. Organometal. Chem.*, **105**, C29 (1976).
[2] Prepared according to G. Urban and R. Dötzer, U.S. Patent 3, 388, 131 (1968); *C.A.*, **63**, P 14909d (1965).

N,N,N′,N′-Tetramethylethylenediamine (TMEDA), 2, 403; **3,** 284–285; **4,** 485–489; **5,** 652; **6,** 576–577.

Metallation of cumene.[1] Cumene (1) is metallated by *n*-pentylsodium in the presence of TMEDA[2] in octane on the ring and is then isomerized to α-cumylsodium (2) in a thermodynamically controlled process.[1]

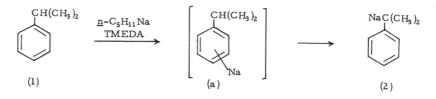

[1] T. F. Crimmins and C. M. Chan, *J. Org.*, **41**, 1870 (1976).
[2] Purified by distillation from BaO or CaH₂.

N,N,N′,N′,-Tetramethylmethanediamine [Bis(dimethylamino)methane], 3, 21–22.

Modified Mannich reagent. deSolms[1] found this reagent superior to the Mannich base for the conversion of (1) into (2).

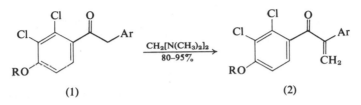

(1)   (2)

[1] S. J. deSolms, *J. Org.*, **41**, 2650 (1976).

Tetramethylthiourea, 1, 1145.

Supplier: Aldrich. This reagent can be purified by vacuum sublimation onto a cold finger (−78°); white, m.p. 78.5°.[1]

Decomposition of selenonium ylides.[2] Tetramethylthiourea catalyzes the decomposition of stabilized selenonium ylides to selenides and olefins.

Examples:

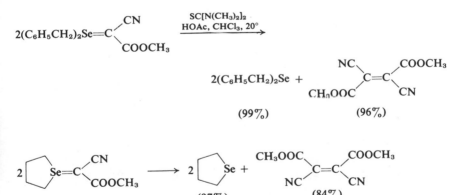

[1] J. S. Hartman, G. J. Schrobilgen, and P. Stilbs, *Canad. J. Chem.*, **54**, 1121 (1976).
[2] S. Tamagaki and I. Hatanaka, *Chem. Letters*, 301 (1976).

Thallium(I) acetate–Iodine

Selective iodination of phenols. Iodination of phenols with iodine results mainly in *para*-substitution; iodination with thallium(I) acetate and iodine in wet acetic acid or in methylene chloride results mainly in *ortho*-substitution. Aromatic amines do not show this variation.[1]

Examples:

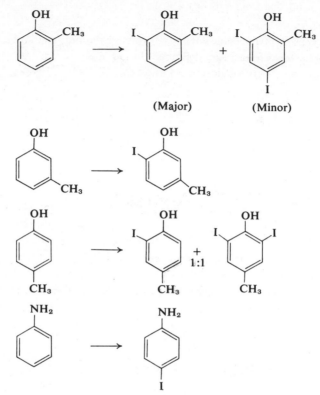

(Major)   (Minor)

[1] R. C. Cambie, P. S. Rutledge, T. Smith-Palmer, and P. D. Woodgate, *J.C.S. Perkin I*, 1161 (1976).

Thallium(III) acetate, 1, 1150–1151; 2, 406; 3, 286; 4, 492; 5, 655.

Reaction with 1-alkynes. Terminal acetylenes react with thallium(III) acetate (2 equiv.) in CHCl$_3$ or HOAc at 0–20° to form the adduct (1) in high yield. When the adduct is heated with an equivalent of Tl(OAc)$_3$, the methyl

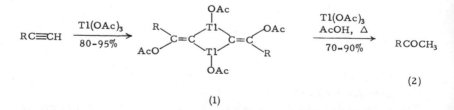

(2)

(1)

ketone (2) is formed.[1] This method for preparation of ketones had been reported previously,[2] but the intermediates were not isolated at that time.

Oxidation of ketosteroids. Oxidation of ketosteroids with this substance involves acetoxylation, dehydrogenation, and Wagner-Meerwein rearrangement.[3]

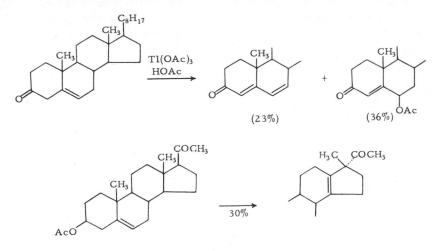

$$(23\%) \qquad (36\%) \quad \text{OAc}$$

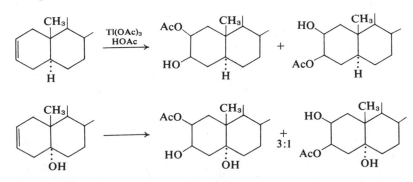

cis-*Hydroxylation.* cis-Hydroxylation of unsaturated steroids has been effected with this reagent in acetic acid.[4]

Examples:

(reaction schemes as shown, 3:1)

[1] S. Uemura, H. Miyoshi, H. Tara, M. Okano, and K. Ichikawa, *J.C.S. Chem., Comm.*, 218 (1976).

[2] S. Uemura, R. Kitoh, K. Fujita, and K. Ichikawa, *Bull. Chem. Soc. Japan*, **40**, 1499 (1967).

[3] G. Ortar and A. Romeo, *J.C.S. Perkin I*, 111 (1976).

[4] E. Glotter and A. Schwartz, *J.C.S. Perkin I*, 1660 (1976).

Thallium(I) bromide, 2, 405–406; 3, 286; 5, 655.

Biphenyls (5, 655). The use of this salt in the synthesis of 4,4′dimethyl-1,1′-biphenyl has been published.[1]

[1] L. F. Elsom, A. McKillop, and E. C. Taylor, *Org. Syn.*, **55**, 48 (1976).

Thallium(I) ethoxide, 2, 407–411; **4**, 501–502; **5**, 656; **6**, 577–578.

Solutions in benzene can be prepared by conversion of thallium(I) acetate in water to thallium(I) hydroxide by reaction with NaOH. This material is dried, suspended in benzene and ethanol (1 equiv.), and stirred for 24 hours under N_2. Thallium(I) ethoxide is obtained in this way in benzene solution in about 95% yield.[1]

[1] M. T. Pizzorno and S. M. Albonico, *Organometal. Chem. Syn.*, **1**, 463 (1972).

Thallium(III) nitrate (TTN), **4**, 492–497; **5**, 656–657; **6**, 578–579.

Oxidations in trimethyl orthoformate. Taylor, McKillop, *et al.*[1] have found that some oxidations with TTN that proceed poorly or fail in the solvents previously employed (dilute HNO_3, acidic CH_3OH, or glyme) are successful in $CH_3OH–HC(OCH_3)_3$ (1:1) or in $HC(OCH_3)_3$ alone. An example is the rearrangement of cinnamaldehydes to arylmalondialdehyde tetramethylacetals (equation I). This reaction was used to convert benzaldehyde into *t*-butylbenzene and should be adaptable to synthesis of $ArC(CH_3)_2R$.

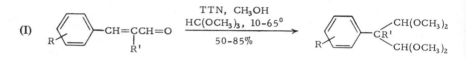

(I)

Cinnamic esters undergo an analogous reaction (equation II). The similar reaction of acetophenones involves two successive methoxythallation reactions (equation III).

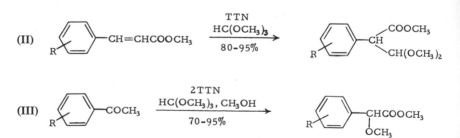

(II)

(III)

Cyclohexanone and cyclopentanone are converted by $TTN–HC(OCH_3)_3$ into the α-methoxyketals.

The exact function of trimethyl orthoformate beyond ketalization in these oxidations is not certain.

α-Alkynoic esters (**4**, 493–494). Detailed directions for the preparation of methyl 2-hexynoate (2) from a 5-pyrazolinone (1) are available. A limitation to this method is that only methyl esters can be prepared; use of alcohols other than methanol result in mixtures of products.[2]

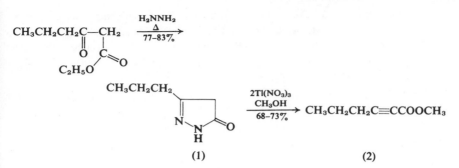

(1)   (2)

Oxidation of phenols to 2,5-cyclohexadienones.[3] Oxidation of hydroquinones with 1 equiv. of thallium(III) nitrate in methanol gives *p*-benzoquinones in high yield (equation I). 2,6-Disubstituted phenols are also oxidized by TTN (2 equiv.) as shown in equation (II). The initial step is believed to be ipso thallation.

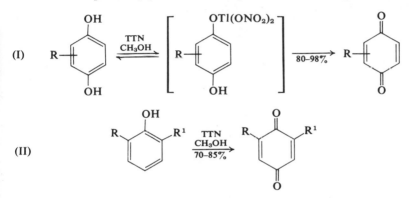

Oxidation of 4-alkylphenols (equation III) and of 4-alkoxyphenols (equation IV) with TTN (1 equiv.) in methanol or trimethyl orthoformate gives 4,4-disubstituted cyclohexa-2,5-dienones in moderate to excellent yield. The last reaction is particularly useful.

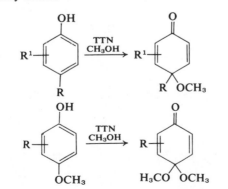

Dethioacetalization. Thioacetals are dethioacetalized to the parent carbonyl compound by thallium(III) nitrate in methanol at 25° in 5–30 minutes; yields are 75–99%. About 2.2 molar equiv. of TTN are required for complete reaction.[4] The present method appears to be preferable to the use of thallium(III) tri-fluoroacetate for this purpose.[5]

Conversion of thioethers into ethers. The thioether (1) is converted into the ether (2) in 92% yield by reaction with TTN (0.6 equiv.) in methanol (5 minutes). Several possible mechanisms for the reaction are suggested in the paper.[6]

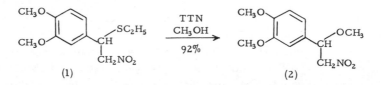

Clay-supported reagent.[7] Thallium(III) nitrate adsorbed on K-10, a mont-morillonite clay,[8] is a remarkably effective reagent for oxidation. Generally oxidations proceed more cleanly than with TTN in methanol. The thallium(I) nitrate formed is also bound to the support. TTN itself is a trihydrate, but the supported reagent appears to be nonhydrated.

Examples:

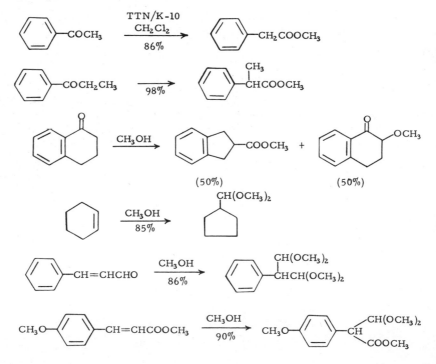

[1] E. C. Taylor, R. L. Robey, K.-T. Liu, B. Favre, H. T. Bozimo, R. A. Conley, C.-S. Chiang, A. McKillop, and M. E. Ford, *Am. Soc.*, **98**, 3037 (1976).

[2] E. C. Taylor, R. L. Robey, D. K. Johnson, and A. McKillop, *Org. Syn.*, **55**, 73 (1976).

[3] A. McKillop, D. H. Perry, M. Edwards, S. Antus, L. Farkas, M. Nógrádi, and E. C. Taylor, *J. Org.*, **41**, 282 (1976).

[4] E. Fujita, Y. Nagao, and K. Kaneko, *Chem. Pharm. Bull. Japan*, **24**, 1115 (1976).

[5] T.-L. Ho and C. M. Wong, *Canad. J. Chem.*, **50**, 3740 (1972).

[6] Y. Nagao, K. Kaneko, M. Ochiai, and E. Fujita, *J.C.S. Chem. Comm.*, 202 (1976).

[7] E. C. Taylor, C.-S. Chiang, A. McKillop, and J. F. White, *Am. Soc.*, **98**, 6750 (1976).

[8] Girdler Catalyst K-10, Girdler Chemicals, Inc., Louisville, Ky.

Thallium(III) perchlorate, 5, 657–658.

   Cyclization of geraniol.[1] Treatment of geraniol (1) with this salt in CH_2Cl_2 at 20° for 2 hours gives three new products: (2), (3), and (4) in a total yield of 91%. The first two products possess the iridoid carbon skeleton. Cyclization to

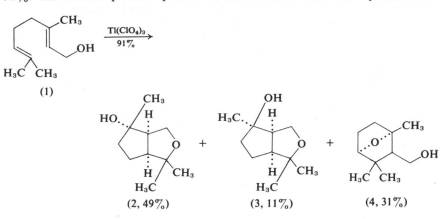

(2, 49%)   (3, 11%)   (4, 31%)

products of type (2) and (3) may be involved in biosynthesis of five-membered terpenes.

[1] Y. Yamada, H. Sanjoh, and K. Iguchi, *J.C.S. Chem. Comm.*, 997 (1976).

Thallium(III) trifluoroacetate, 3, 286–289; **4,** 498–500; **5,** 658–659; **6,** 579–581.

   Aryl iodides (**3,** 287; **4,** 498). Details for the preparation of 2-iodo-*p*-xylene (equation I) have been published. Use of freshly prepared TTFA rather than commercial material is recommended.[1]

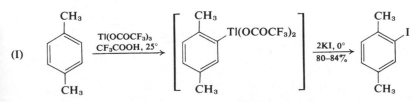

[1] E. C. Taylor, F. Kienzle, and A. McKillop, *Org. Syn.*, **55**, 70 (1976).

Thioglycolic acid, 1, 1153–1154.

Synthesis of chiral peptides. Polish chemists[1] recently reported that α-ferrocenylethanol (2) is converted by thioglycolic acid (1) in the presence of TFA to (3), which is cleaved by aqueous $HgCl_2$ with retention to (2) (equation I).

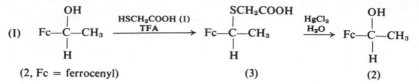

(I)

(2, Fc = ferrocenyl)   (3)   (2)

Eberle and Ugi[2] have made use of this reaction in a method for stereoselective peptide synthesis by four-component condensation using α-ferrocenylalkylamines as the chiral component,[3] as formulated in equation (II). In this way the original chiral α-ferrocenylalkylamine, which is not easily prepared, is recovered.

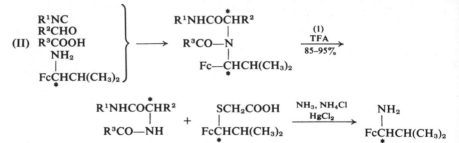

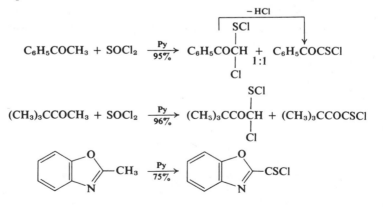

[1] A. Ratajczak and B. Misterkiewicz, *J. Organometal. Chem.*, **91**, 73 (1975).
[2] G. Eberle and I. Ugi, *Angew. Chem., Int. Ed.*, **15**, 492 (1976).
[3] I. Ugi *et al.*, in *Peptides 1974,*, Y. Wolman, Ed., Wiley, New York, 1975.

Thionyl chloride, 1, 1158–1163; **2,** 412; **3,** 290; **4,** 503–505; **5,** 663–667; **6,** 585.

Reaction with active CH_3 compounds. Japanese chemists have reported the oxidation of active methyl compounds to thioacyl chlorides with thionyl chloride.
Examples:

$$C_6H_5COCH_3 + SOCl_2 \xrightarrow[95\%]{Py} C_6H_5COCHCl + C_6H_5COCSCl \quad 1:1$$

$$(CH_3)_3CCOCH_3 + SOCl_2 \xrightarrow[96\%]{Py} (CH_3)_3CCOCHCl + (CH_3)_3CCOCSCl$$

This oxidation is an extension of a reaction reported in 1967 by Simon *et al.*,[2] which is shown in equation (I).

(I) $C_6H_5CH_2COOH + SOCl_2 \xrightarrow[61\%]{Py}$ $C_6H_5\overset{\overset{\displaystyle Cl}{|}}{\underset{\underset{\displaystyle SCl}{|}}{C}}COOCl$

[1] K. Oka and S. Hara, *Tetrahedron Letters*, 2783 (1976).
[2] M. S. Simon, J. B. Rogers, W. Saenger, and J. Z. Gougotas, *Am. Soc.*, **89**, 5838 (1967).

Thionyl chloride–Triethylamine.

Alkene synthesis. The reaction of β-hydroxy selenides with this reagent results in formation of alkenes. This reagent is superior to *p*-TsOH, HClO₄, or TFAA–triethylamine, which have been used previously. The complete synthesis of alkenes from two ketones is formulated in equation (I).[1]

(I)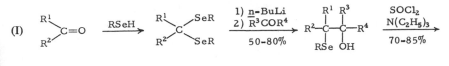

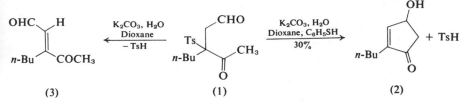

[1] J. Rémion and A. Krief, *Tetrahedron Letters*, 3743 (1976).

Thiophenol, C_6H_5SH. Mol. wt. 110.18, b.p. 169°. Suppliers: Aldrich, Eastman, others.

trans/cis-Equilibration (*cf.* **4**, 505). The final step in a recent synthesis of 4-hydroxycyclopentenones (2) involves treatment of a ketoaldehyde such as (1) with

OHC, H / n-Bu COCH₃ (3) $\xleftarrow[-TsH]{\text{Dioxane}}^{K_2CO_3, H_2O}$ Ts, CHO / n-Bu CH₃ / O (1) $\xrightarrow[30\%]{K_2CO_3, H_2O \atop \text{Dioxane}, C_6H_5SH}$ OH / n-Bu / O (2) + TsH

a base [K_2CO_3 or $N(C_2H_5)_3$] in aqueous dioxane containing some thiophenol. If the thiophenol is omitted, the product is (E)-3-acetyl-2-heptenal (3). The function of the thiophenol is to assist equilibration to the less favored (Z)-form of (3), the form that cyclizes to (2). This reaction has been used in a synthesis of a prostaglandin precursor.[1]

Aldehydes.[2] Terminal alkenes can be converted into aldehydes by free-radical addition[3] of thiophenol, chlorination with NCS, and hydrolysis (equation I). The chloro sulfides can also be converted into alkyl dithianes (equation II).

(I) $RCH=CH_2$ + C_6H_5SH $\xrightarrow[90-100\%]{AIBN}$ $RCH_2CH_2SC_6H_5$ $\xrightarrow[\sim100\%]{NCS \atop CCl_4}$

$$\left[RCH_2CHClSC_6H_5\right] \xrightarrow[50-80\%]{H_2O, (CH_3)_2CO \atop CuCl_2, \; CuO} RCH_2CHO$$

(II) $RCH_2CHClSC_6H_5$ + $HS(CH_2)_3SH$ $\xrightarrow[\sim75\%]{BF_3}$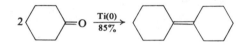

[1] G. K. Cooper and L. J. Dolby, *Tetrahedron Letters*, 4675 (1976).
[2] P. Bakuzis, M. L. Bakuzis, C. C. Fortes, and R. Santos, *J. Org.*, **41**, 2769 (1976).
[3] F. W. Stacey and J. F. Harris, Jr., *Org. React.*, **13**, 150 (1963).

Titanium(0), activated, Ti(0).

An active titanium metal can be prepared by reaction of $TiCl_3$ slurried in THF or DME with potassium[1] or lithium[2] (method of Rieke, **4**, 315).

Reductive coupling of carbonyl compounds to olefins. Reductive coupling of carbonyls to olefins with a low-valent titanium reagent prepared from $TiCl_3$ and $LiAlH_4$ (**6**, 589) tends to give erratic yields. McMurry and Fleming find that use of Ti(0) gives reproducible results. Yields are somewhat higher with metal prepared with potassium, but lithium is easier to handle. 1,2-Diols are also reduced to olefins.[1,2]

Examples:

$$2[(CH_3)_2CH]_2C=O \xrightarrow[40\%]{} [(CH_3)_2CH]_2C=C[CH(CH_3)_2]_2$$

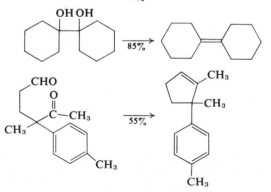

The method can be used for synthesis of unsymmetrical olefins if one component is used in excess. Highest yields are obtained for diaryl ketones; in some cases mixed coupling can occur in a 1:1 ratio.[3]

Examples:

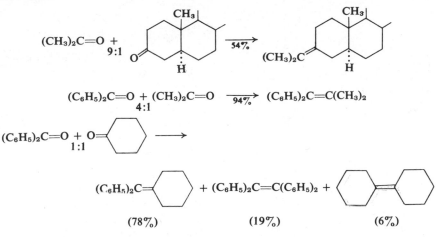

$(CH_3)_2C=O + (C_6H_5)_2C=O \xrightarrow[]{94\%} (C_6H_5)_2C=C(CH_3)_2$
4:1

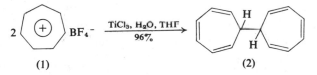

(78%)   (19%)   (6%)

[1] J. E. McMurry and M. P. Fleming, *J. Org.*, **41**, 896 (1976).
[2] M. P. Fleming and J. E. McMurry, *Org. Syn.*, submitted (1976).
[3] J. E. McMurry and L. R. Krepski, *ibid.*, **41**, 3929 (1976).

Titanium(III) chloride, 2, 415; **4,** 506–508; **5,** 669–671; **6,** 587–588.

Reductive dimerization. Bitropyl (2) can be prepared in high yield by reaction of tropylium tetrafluoroborate (1) with $TiCl_3$ or VCl_2 in aqueous THF.[1] The only other satisfactory reagent for such a reduction is zinc dust (tropylium bromide → bitropyl, 97% yield).[2]

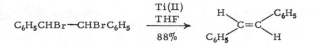

(1)   (2)

[1] G. A. Olah and T.-L. Ho, *Synthesis*, 798 (1976).
[2] W. von E. Doering and L. H. Knox, *Am. Soc.*, **79**, 352 (1957).

Titanium(III) chloride–Lithium aluminum hydride, 6, 588–589, Supplier: Alfa ("McMurry's Reagent").

*Dehalogenation; coupling of allyl or benzyl halides or of aryl **gem-dihalo-alkenes.*** Olah and Prakash[1] have reported further uses of McMurry's reagent to effect these reactions.

Examples:

$$C_6H_5CHBr{-}CHBrC_6H_5 \xrightarrow[88\%]{\substack{Ti(II) \\ THF}} \underset{C_6H_5}{\overset{H}{}}C=C\underset{H}{\overset{C_6H_5}{}}$$

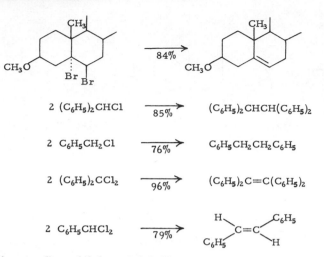

$2 \ (C_6H_5)_2CHCl \xrightarrow[85\%]{} (C_6H_5)_2CHCH(C_6H_5)_2$

$2 \ C_6H_5CH_2Cl \xrightarrow[76\%]{} C_6H_5CH_2CH_2C_6H_5$

$2 \ (C_6H_5)_2CCl_2 \xrightarrow[96\%]{} (C_6H_5)_2C{=}C(C_6H_5)_2$

$2 \ C_6H_5CHCl_2 \xrightarrow[79\%]{}$

Reductive coupling of 1,4- and 1,6-diketones. 1,2-Diphenylcyclobutene and 1,2-diphenylcyclohexene have been prepared in this way (equations I and II).[2]

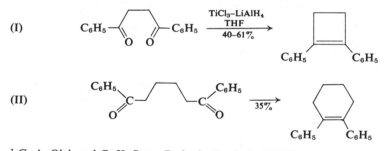

[1] G. A. Olah and G. K. Surya Prakash, *Synthesis*, 607 (1976).
[2] A. L. Baumstark, E. J. H. Bechara, and M. J. Semigran, *Tetrahedron Letters*, 3265 (1976).

Titanium(IV) chloride, 1, 1169–1171; 2, 414–415; 3, 291; 4, 507–508; 5, 671–672; 6, 590–596.

γ,δ-*Unsaturated alcohols.* Allyltrimethylsilane reacts with a wide variety of carbonyl compounds with catalysis with TiCl₄ to form γ,δ-unsaturated alcohols (equation I). The new C—C bond forms exclusively at the γ-carbon of the allysilane (equation II).[1]

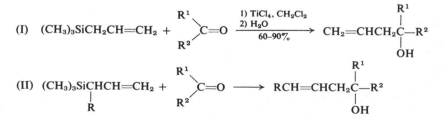

A similar reaction is observed with ketals to form γ,δ-unsaturated ethers (equation III).[2]

(III) $(CH_3)_3SiCH_2CH{=}CH_2 + \underset{R^2}{\overset{R^1}{>}}C(OR)_2 \xrightarrow[\underset{75-95\%}{\text{2) H}_2\text{O}}]{\text{1) TiCl}_4} CH_2{=}CHCH_2\underset{\underset{OR}{|}}{\overset{\overset{R^1}{|}}{C}}{-}R^2$

French chemists[3] have effected the reaction of allylsilanes and aldehydes with $AlCl_3$.

Conjugate allylation of α,β-*enones.*[4] Titanium(IV) chloride is the most effective activator for conjugate addition of allylsilanes to α,β-unsaturated ketones.

(II) $(CH_3)_3SiCH_2CH{=}CH_2 + CH_2{=}CHCOCH_3 \xrightarrow[\underset{-78,\ 3\ \text{hr.}}{\text{TiCl}_4,\ \text{CH}_2\text{Cl}_2}]{} \xrightarrow[59\%]{\text{H}_2\text{O}}$

$$CH_2{=}CH(CH_2)_3COCH_3$$

A useful application of the reaction is introduction of an angular allyl group in fused cyclic enones. An example is formulated in equation (III), which includes transformations of the product (1) into other useful compounds.

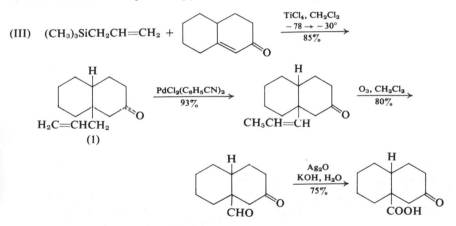

(III) $(CH_3)_3SiCH_2CH{=}CH_2 +$ [fused bicyclic enone] $\xrightarrow[85\%]{\underset{-78\ \rightarrow\ -30°}{\text{TiCl}_4,\ \text{CH}_2\text{Cl}_2}}$

[bicyclic ketone with $H_2C{=}CHCH_2$ group] (1) $\xrightarrow[93\%]{\text{PdCl}_2(\text{C}_6\text{H}_5\text{CN})_2}$ [bicyclic ketone with $CH_3CH{=}CH$ group] $\xrightarrow[80\%]{\text{O}_3,\ \text{CH}_2\text{Cl}_2}$

[bicyclic ketone with CHO group] $\xrightarrow[75\%]{\underset{\text{KOH, H}_2\text{O}}{\text{Ag}_2\text{O}}}$ [bicyclic ketone with COOH group]

δ-Keto esters.[5] δ-Keto esters can be obtained by a Michael-type reaction of O-silylated ketene acetals with α,β-unsaturated carbonyl compounds in the presence of $TiCl_4$ (1 equiv.) (equation I). In the case of methyl vinyl ketone and

(I) $R^1\overset{\overset{O}{\|}}{C}CH{=}C\underset{R^3}{\overset{R^2}{<}} + \underset{R^5}{\overset{R^4}{>}}C{=}C\underset{OSi(CH_3)_3}{\overset{OR^6}{<}} \xrightarrow[\underset{40-98\%}{\text{2) K}_2\text{CO}_3}]{\text{1) TiCl}_4} R^1\overset{\overset{O}{\|}}{C}CH_2{-}\underset{\underset{R^3}{|}}{\overset{\overset{R^2}{|}}{C}}{-}\underset{\underset{R^5}{|}}{\overset{\overset{R^4}{|}}{C}}{-}\overset{\overset{O}{\|}}{C}OR^6$

cyclohexenone improved yields of adducts are obtained by use of the combination $TiCl_4$ and $Ti[OCH(CH_3)_2]_4$ as the promoter.

Claisen rearrangement. The rearrangement of allyl phenyl ethers to *o*-allylphenols can be carried out at 20° (1–24 hours) using TiCl₄ (1–2 equiv.) and N-trimethylsilylacetanilide [CH₃CON(C₆H₅)Si(CH₃)₃, 1 equiv.] as scavenger for hydrogen chloride. Yields are 36–95% (four examples).[6]

[1] A. Hosomi and H. Sakurai, *Tetrahedron Letters*, 1295 (1976).
[2] A. Hosomi, M. Endo, and H. Sakurai, *Chem. Letters*, 941 (1976).
[3] G. Deleris, J. Dunoguès, and R. Calas, *ibid.*, 2449 (1976).
[4] A. Hosomi and H. Sakurai, *Am. Soc.*, **99**, 1673 (1977).
[5] K. Saigo, M. Osaki, and T. Mukaiyama, *Chem. Letters*, 163 (1976).
[6] K. Narasaka, E. Bald, and T. Mukaiyama, *ibid.*, 1041 (1975).

Titanium(IV) chloride–Lithium aluminum hydride, 6, 596.

Reductive dimerization of aldehydes or ketones. Japanese chemists have used this low-valent titanium compound for synthesis of various symmetrical olefins. The yields are improved by the presence of various tertiary amines, particularly 1,8-bis(dimethylamino)naphthalene (3, 22; 4, 35).[1]

Examples:

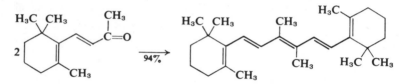

Retinal → β-carotene (50%, pure)

Reduction of C=C and C≡C bonds. Monosubstituted alkenes are reduced by a reagent generated *in situ* from equimolar quantities of TiCl₄ and LiAlH₄ in ether. Disubstituted alkenes are reduced only partially and slowly. This difference can be used for selective reduction. Terminal alkynes are reduced completely to alkanes. Internal alkynes are reduced mainly to alkenes. The *cis*-isomer predominates at low temperatures (−40°). Thus at this temperature 4-octyne is reduced to *cis*-4-octene (73%), *trans*-4-octene (<7%), and *n*-octane (11%).[2]

Hydrogenolysis of allylic alcohols. TiCl₄–LiAlH₄ (1:4 molar ratio) is useful for hydrogenolysis of allylic alcohols, particularly tertiary ones. Yields are somewhat sensitive to reaction conditions.[3] Note that dehydration is the predominant reaction with AlCl₃–LiAlH₄ (1:4).

Examples:

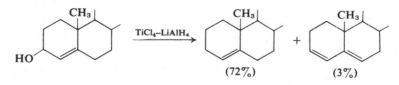

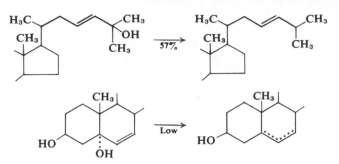

Reductive demethoxylation. Alkenes can be prepared by the reaction of allyl methyl ethers with this reducing agent. The reaction is accompanied by migration of the double bond.[4]

Example:

$$CH_3CH=CHCHCH_2CH_2C_6H_5 \xrightarrow[95\%]{\substack{TiCl_4 \\ 2\ LiAlH_4}} CH_3CH_2CH=CHCH_2CH_2C_6H_5$$

(OCH₃ substituent on the starting material)

$$Z:E = 3:1$$

Reduction of sulfoxides. This titanium reagent reduces sulfoxides to sulfides at $0 \to 20°$ in high yield (80–90%).[5]

[1] A. Ishida and T. Mukaiyama, *Chem. Letters*, 1127 (1976).
[2] P. W. Chum and S. E. Wilson, *Tetrahedron Letters*, 15 (1976).
[3] Y. Fujimoto and N. Ikekawa, *Chem. Pharm. Bull. Japan*, **24**, 825 (1976).
[4] H. Ishikawa and T. Mukaiyama, *Chem. Letters*, 737 (1976).
[5] J. Drabowicz and M. Mikolajczyk, *Synthesis*, 527 (1976).

Titanium(IV) chloride–Magnesium amalgam

Pinacolic coupling of ketones and aldehydes. The reaction of $TiCl_4$ and 70–80 mesh magnesium amalgamated with $HgCl_2$ leads to a Ti(II) species that is effective for intermolecular coupling of aldehydes and ketones as shown in the examples.[1]

Examples:

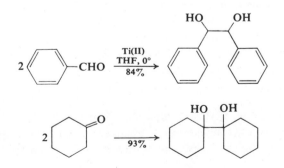

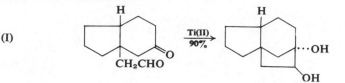

(I)

The reagent can also be used for intramolecular reductive coupling (equation I) but for this reaction a reagent obtained by reduction of cyclopentadienyl-titanium trichloride (Alfa) with lithium aluminum hydride is also effective (equation II). A Ti(II) complex of known structure (1)[2] was also shown to be effective for pinacolic coupling. It is more effective than titanocene.[3]

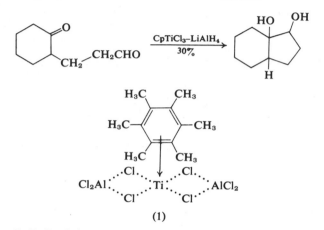

(II)

(1)

[1] E. J. Corey, R. L. Danheiser, and S. Chandrasekaran, *J. Org.*, **41**, 260 (1976).
[2] R. Gieżyński, S. Dzierzogowski, and S. Pasynkiewicz, *J. Organometal. Chem.*, **87**, 295 (1975).
[3] M. D. Rausch and H. Alt, *Am. Soc.*, **96**, 5937 (1974).

p-Toluenesulfinic acid, p-$CH_3C_6H_4SO_2H$. Mol. wt. 156.21, m.p. 85°.

The free acid can be obtained by acidification of an aqueous solution of the commercially available (Aldrich) sodium salt [*Merck Index*, 9th ed., 1976, p. 9227].

cis-trans *Isomerization.* Treatment of a solution of either methyl oleate or methyl elaidate in dioxane (reflux) with 10% *p*-toluenesulfinic acid rapidly gives an equilibrium mixture containing 76% of the *trans*-isomer. No migration of the double bond is observed. Almost any arylsulfinic acid is equally effective. The actual catalytic species is not certain.[1]

[1] T. W. Gibson and P. Strassburger, *J. Org.*, **41**, 791 (1976).

p-Toluenesulfonic acid, 1, 1172–1178; **4,** 508–510; **5,** 673–675; **6,** 597.

Allyl sulfides. Allyl sulfides can be obtained by acid-catalyzed rearrangement of β-hydroxyalkyl phenyl sulfides with migration of the phenylthio group. The rearrangement was first observed in substances of type (I), in which case the C_6H_5S migrates from a tertiary to a secondary position (equation I).[1]

(I)

In more recent work[2] the rearrangement has been extended to a wider range of substrates. Thus migration of the C_6H_5S group from a tertiary center to a primary center is shown in the example formulated in equation (II). No rearrangements

(II)

are observed when both the site of migration and terminus are secondary, or when the site of migration is secondary and the terminus tertiary. But these rearrangements can occur when assisted by a suitably placed $(CH_3)_3Si$ group (equations III and IV). The product in these cases can be rearranged by light to allyl sulfides.

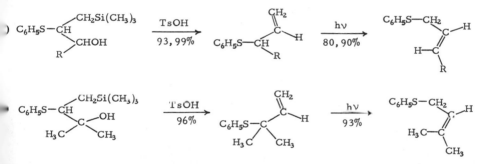

[1] P. Brownbridge and S. Warren, *J.C.S. Chem. Comm.*, 820 (1975).
[2] P. Brownbridge, I. Fleming, A. Pearce, and S. Warren, *ibid.*, 751 (1976).

p-**Toluenesulfonylhydrazine, 1,** 1185–1187; **2,** 417–418; **3,** 293; **4,** 511–512; **5,** 678–681; **6,** 598–600.

   Shapiro-Heath olefin synthesis (**2,** 403; **3,** 284; **4,** 485–489; **5,** 678–680).[1] Dauben and co-workers[2] have reported that the stereochemistry of the hydrazone group affects the course of the reaction. Thus the isomeric tosylhydrazones of pulegone, (1a) and (1b), behave differently: (1a) is converted exclusively or predominantly under these conditions to the diene (2), whereas (1b) always gives a mixture of (2) and (3), with the latter diene predominating. It appears that at least two reaction pathways are involved.

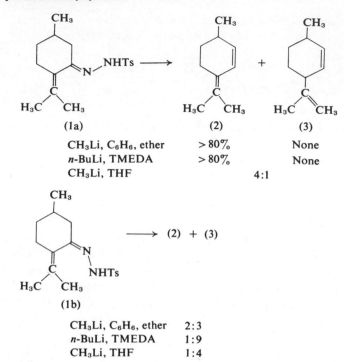

(1a)   (2)   (3)

	(2)	(3)
CH₃Li, C₆H₆, ether	> 80%	None
n-BuLi, TMEDA	> 80%	None
CH₃Li, THF	4:1	

CH_3Li, C_6H_6, ether   > 80%   None
n-BuLi, TMEDA   > 80%   None
CH_3Li, THF   4:1

⟶ (2) + (3)

(1b)

CH₃Li, C₆H₆, ether	2:3
n-BuLi, TMEDA	1:9
CH₃Li, THF	1:4

A vinylogous Shapiro reaction was used in a four-step synthesis of bicyclic diene (7) from α-pinene (4). In place of the usual base (methyllithium), 4 equiv.

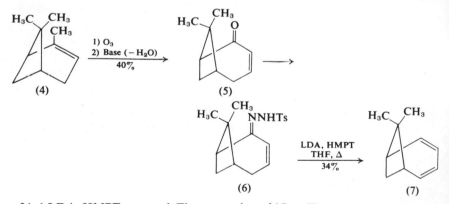

(4)

1) O₃
2) Base (– H₂O)
40%

(5)

LDA, HMPT
THF, Δ
34%

(6)   (7)

of 1:1 LDA–HMPT was used. The conversion of (6) to (7) represents an extension of the Shapiro reaction in that the proton in the γ-position is abstracted rather than that in the α'-position (a bridgehead).[3]

[1] R. H. Shapiro, *Org. React.*, **23**, 405 (1976).
[2] W. G. Dauben, G. T. Rivers, and W. T. Zimmerman, *Tetrahedron Letters*, 2951 (1976).
[3] S. D. Young and W. T. Borden, *ibid.*, 4019 (1976).

p-**Toluenesulfonyl isocyanate,** $CH_3C_6H_4SO_2NCO$. Mol. wt. 197.21, b.p. 91–92°/ 0.5 mm. Suppliers: Aldrich, Eastman, Fluka.

Protection of amino groups. Amino acids are converted into *p*-tosylamino carbonyl derivatives by reaction with this isocyanate (20–80% yield). The derivatives are stable to dilute base and cold TFA. They are deblocked by hot ethanol.[1]

[1] B. Weinstein, T. N.-S. Ho, R. T. Fukure, and E. C. Angell, *Syn. Comm.*, **6**, 17 (1976).

p-**Tolylthiomethyl isocyanide, 4,** 515–516.

4(5)-Thioimidazoles. The anion of this reagent undergoes cycloaddition with nitriles to form these hetcrocycles. The products can be desulfurized with Raney nickel (60% yield).[1]

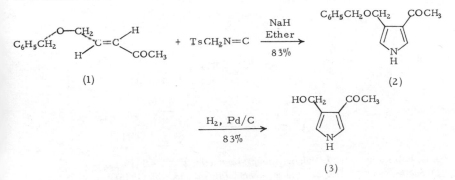

[1] A. M. van Leusen and J. Schut, *Tetrahedron Letters*, 285 (1976).

Tosylmethyl isocyanide, 4, 514–516; **5,** 684–685; **6,** 600.

Pyrrole synthesis. Swiss chemists[1] have synthesized verracarin (3), a secondary metabolite of the soil fungus *Myrothecium verrucaria*, by condensation of (E)-5-benzyloxy-3-pentene-2-one (1) with tosylmethylisocyanide (NaH) followed by hydrogenolysis of the benzyl group. This method is convenient for preparation of C-labeled material.

[1] A. Gossauer and K. Suhl, *Helv.*, **50**, 1698 (1976).

Tri-*n*-butyltin azide, $(n\text{-}C_4H_9)_3SnN_3$. Mol. wt. 332.05, b.p. 118–120°/0.2 torr.

This relatively stable azide is prepared by the reaction of tri-*n*-butyltin chloride with a concentrated aqueous solution of sodium azide at 0°; 91% yield,[1,2]

Isocyanates, carbamates, ureas.[2] These nitrogen-containing compounds can be prepared as shown in the equations. The preparation of the CBZ-derivative of β-alanine is also formulated (equation IV).

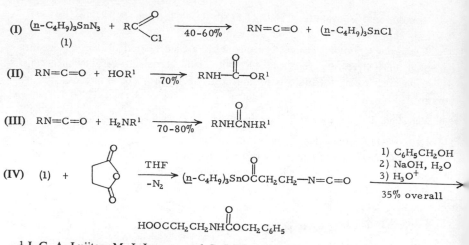

(I) $(\underline{n}\text{-}C_4H_9)_3SnN_3$ + $RC\overset{O}{\underset{Cl}{\diagdown}}$ $\xrightarrow{40\text{-}60\%}$ $RN{=}C{=}O$ + $(\underline{n}\text{-}C_4H_9)_3SnCl$
      (1)

(II) $RN{=}C{=}O$ + HOR^1 $\xrightarrow{70\%}$ $RNH{-}\overset{O}{\overset{\|}{C}}{-}OR^1$

(III) $RN{=}C{=}O$ + H_2NR^1 $\xrightarrow{70\text{-}80\%}$ $RNH\overset{O}{\overset{\|}{C}}NHR^1$

(IV) (1) + [succinic anhydride] $\xrightarrow[-N_2]{THF}$ $(\underline{n}\text{-}C_4H_9)_3SnO\overset{O}{\overset{\|}{C}}CH_2CH_2{-}N{=}C{=}O$ $\xrightarrow[\;35\%\text{ overall}\;]{\begin{array}{l}1)\ C_6H_5CH_2OH\\2)\ NaOH,\ H_2O\\3)\ H_3O^+\end{array}}$

$HOOCCH_2CH_2NH\overset{O}{\overset{\|}{C}}OCH_2C_6H_5$

[1] J. G. A. Luijten, M. J. Janssen, and G. J. M. van der Kerk, *Rec. trav.*, **81**, 202 (1962).
[2] H. R. Kricheldorf and E. Leppert, *Synthesis*, 329 (1976).

Tri-*n*-butyltincarbinol, $(n\text{-}C_4H_9)_3SnCH_2OH$ (1). Mol. wt. 321.06.

This reagent is prepared from $(n\text{-}C_4H_9)_3SnMgCl$ and HCHO (41% yield).[1]

Hydroxymethylation.[2] On treatment with 2 equiv. of *n*-butyllithium (1) is converted into the dianion (a), equivalent to doubly deprotonated methanol. Reaction of (a) with electrophiles yields products of hydroxymethylation. Examples:

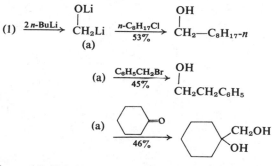

(1) $\xrightarrow{2\,n\text{-}BuLi}$ $\underset{(a)}{\overset{OLi}{\underset{|}{CH_2Li}}}$ $\xrightarrow[53\%]{n\text{-}C_8H_{17}Cl}$ $\overset{OH}{\underset{|}{CH_2{-}C_8H_{17}\text{-}n}}$

(a) $\xrightarrow[45\%]{C_6H_5CH_2Br}$ $\overset{OH}{\underset{|}{CH_2CH_2C_6H_5}}$

(a) $\xrightarrow{46\%}$ [cyclohexanone → 1-(hydroxymethyl)cyclohexanol]

[1] J. C. Lahournère and J. Valade, *Compt. rend.* (*C*), **270**, 2080 (1970).
[2] D. Seebach and N. Meyer, *Angew. Chem., Int. Ed.*, **15**, 438 (1976).

Tri-*n*-butyltin chloride, 6, 604.

Di- and trisubstituted alkenes.[1] Tri-*n*-butyltin chloride reacts with lithium trialkylalkynylborates (1) stereoselectively to form *cis*-vinylboranes (a). Hydrolysis of the boranes gives a single *cis*-alkene (2) in good yield. Other electrophiles such as $(C_2H_5)_2AlCl$ or C_6H_5SCl react with (1) nonstereospecifically. Unfortunately the bulk of tri-*n*-butyltin chloride limits this reaction to unhindered borates.

$$R_3\bar{B}-C\equiv C-R'Li^+ + (C_4H_9)_3SnCl \longrightarrow$$
(1)

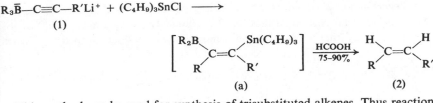

(a)   (2)

This method can be used for synthesis of trisubstituted alkenes. Thus reaction of (a) with thiomethoxymethyllithium–TMEDA[2] gives (3) in ~50% yield.

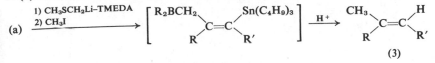

(3)

[1] J. Hooz and R. Mortimer, *Tetrahedron Letters*, 805 (1976).
[2] E. Negishi, T. Yoshida, A. Silveira, Jr., and B. L. Chiou, *J. Org., Chem.*, **40**, 814 (1975).

Tri-*n*-butyltin hydride, 1, 1192–1193; **2,** 424; **3,** 294; **4,** 518–520; **5,** 685–686; **6,** 604.

Deoxy sugars. Radical reduction of the diol thiocarbonate (1) with this tin hydride with an initiator (azobisisobutyronitrile) in refluxing toluene gives, after alkaline hydrolysis, the 5-deoxy sugar (2).[1] The isomeric 6-deoxy sugar (4) can

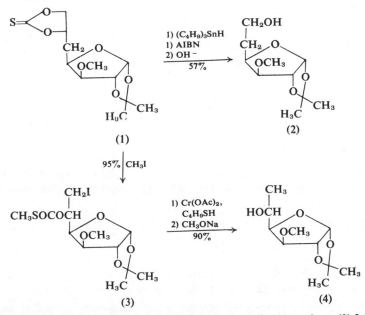

be obtained in good yield by reaction of (1) with methyl iodide to form (3) followed by reduction with chromous acetate in the presence of a thiol (**1,** 148).[2] The reduction of diol thiocarbonates can also be used to obtain 2- and 3-deoxyribofuranose derivatives, as well as 2′- and 3′-deoxynucleosides.

Reduction of vinylcyclopropanes. The ring of vinylcyclopropanes can be selectively reduced without attack of the double bond by radical addition of an organotin hydride followed by protonolysis of the adduct. Both tri-*n*-butyltin

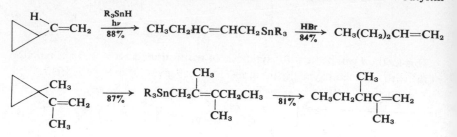

hydride and trimethyltin hydride were used; the former hydride is easier to prepare.[3]

[1] D. H. R. Barton and R. Subramanian, *J.C.S. Chem. Comm.*, 867 (1976).
[2] D. H. R. Barton and R. V. Stick, *J.C.S. Perkin I*, 1773 (1975).
[3] M. Ratier and M. Pereyre, *Tetrahedron Letters*, 2273 (1976).

Tricaprylylmethylammonium chloride (Aliquat 336), 4, 28; 5, 460; 6, 404–406.

1,3-Dithietane.[1] Bis(chloromethyl) sulfoxide (1) reacts with aqueous sodium sulfide[2] in the presence of 1 equiv. of this phase-transfer catalyst to form 1,3-thietane 1-oxide (2) in 36% yield. Bis(chloromethyl) sulfide does not undergo this reaction. However, the product (3) that would have been formed can be obtained by reduction of (2) with borane–tetrahydrofurane (5, 48). The microwave spectra

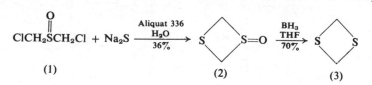

of (2) and eight isotopic modifications indicate that (2) is puckered with the oxygen in an equatorial position. The ring of (3), however, is apparently nearly planar.

[1] E. Block, E. R. Corey, R. E. Penn, T. L. Renken, and P. F. Sherwin, *Am. Soc.*, **98**, 5715 (1976).
[2] J. Lal, *J. Org.*, **26**, 971 (1961).

Trichloroacetic anhydride, 1, 1194; 2, 425.

Polonovski reaction (2, 7; 5, 3). The first synthesis of the dimeric indole alkaloid vinblastine (4), used in the treatment of lymphomas, involved coupling of two indole units using a Polonovski reaction. Thus reaction of the N-oxide (1) and vindoline (2) with trichloroacetic anhydride afforded the quaternary immonium compound (3), which was converted into (4) by standard methods (reduction with NaBH₄, reacetylation).[1]

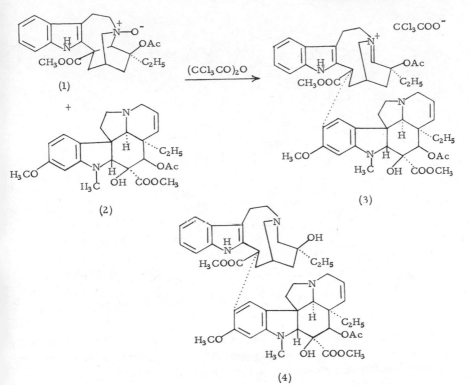

(1)

+

(2)

(3)

(4)

[1] Atta-ur-Rahman, A. Basha, and M. Ghazala, *Tetrahedron Letters*, 2351 (1976).

Trichloroacetonitrile, 1, 1194–1195; **5,** 686; **6,** 604–605.

Synthesis of amines by [3,3]sigmatropic rearrangement.[1] Trichloroacetimidic esters (2) of allylic alcohols (1) can be prepared in 80–100% yield (crude) by reaction with trichloroacetonitrile catalyzed by the corresponding sodium or potassium alkoxide (0°, ether). Inverse addition is preferred for secondary or tertiary alcohols. These esters undergo Claisen-type rearrangement at 25–140° to give allylic trichloroacetamides (3), in which the *trans*-isomer usually predominates.

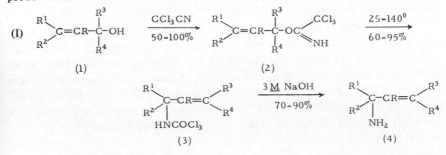

Examples:

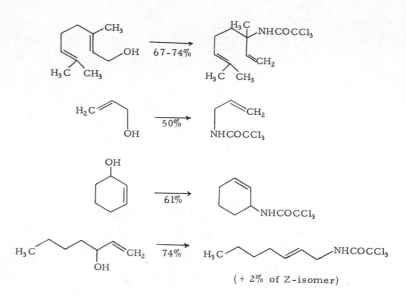

(+ 2% of Z-isomer)

The trichloroacetimidates of 2-alkene-1-ols can also be rearranged with mercuric trifluoroacetate or nitrate (equation II). This rearrangement is believed to involve iminomercuration followed by deoxymercuration.

A similar rearrangement of allyl pseudoureas was found to be less useful.[2]

[1] L. E. Overman, *Am. Soc.*, **98**, 2901 (1976).
[2] S. Tsuboi, P. Stromquist, and L. E. Overman, *Tetrahedron Letters*, 1145 (1976).

$$\text{Trichloroacetyl isocyanate, } Cl_3CCN{=}C{=}O$$

Trichloroacetyl isocyanate, $Cl_3C\overset{O}{\overset{\|}{C}}N{=}C{=}O$, **2**, 425–426.

Reaction with strained olefins. Monsanto chemists[1] reported a few years ago that this isocyanate reacts with norbornene (1) in refluxing toluene (b.p. 110°) to form a 1:1 adduct (2); that is, the isocyanate behaves as a Diels-Alder diene under these conditions. More recently, Agosta *et al.*[2] found that if this reaction is carried out in refluxing *p*-xylene (b.p. 138°) the α,β-unsaturated carbonitrile (3) is obtained in high yield and that (2) cannot be converted into (3) under the same conditions. They suggest that (3) is formed via a [2 + 2]-adduct (a).

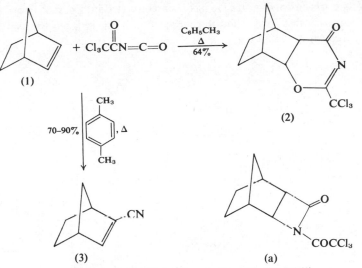

(1)

(2)

(3)   (a)

¹ L. R. Smith, A. J. Speziale, and J. E. Fedder, *J. Org.*, **34**, 633 (1969).
² B. Byrne, C. A. Wilson, II, and W. C. Agosta, *Tetrahedron Letters*, 2189 (1976).

β,β,β-Trichloroethyl chloroformate, 2, 426; **3,** 296–297; **4,** 522; **5,** 686–687.

Amino nitroxides. The β,β,β-trichloroethoxycarbonyl group was used for protection of the less hindered amino group of (1) during oxidation of the more hindered amino group to a nitroxide.¹

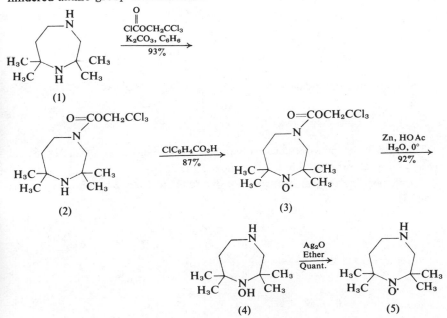

(1)

(2)   (3)

(4)   (5)

N-Demethylation. Cocaine (1) has been converted into norcocaine (3) by use of this formate ester. Vinyl chloroformate can also be used, but the overall yield is somewhat lower (55%).[2]

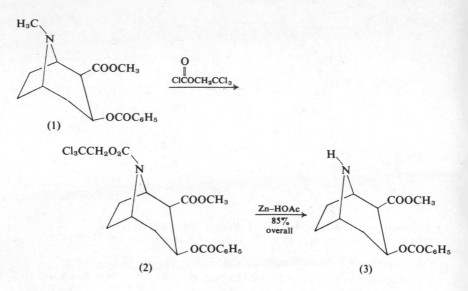

[1] R. Ramasseul, A. Rassat, and P. Rey, *J.C.S. Chem. Comm.*, 83 (1976).
[2] S. W. Baldwin, P. W. Jeffs, and S. Natarajan, *Syn. Comm.*, **7**, 79 (1977).

2,4,6-Trichlorophenylcyanoacetate, 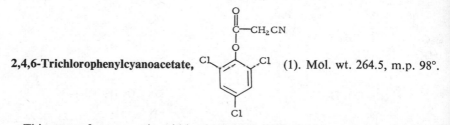 (1). Mol. wt. 264.5, m.p. 98°.

This ester of cyanoacetic acid is obtained in 75% yield by reaction of the acid with 2,4,6-trichlorophenol in the presence of $POCl_3$.

Pyrones. The reagent reacts with β-diketones to form 2-pyrones (equations I and II).[1]

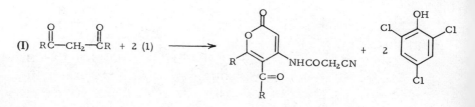

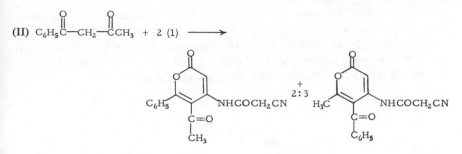

(II) $C_6H_5\overset{O}{\overset{\|}{C}}-CH_2-\overset{O}{\overset{\|}{C}}CH_3$ + 2 (1) ⟶

$\overset{+}{2:3}$

¹ E. Ziegler, F. Raninger, and A. K. Müller, *Ann.*, 250 (1976).

Triethylamine, 1, 1198–1203; **2,** 427–429; **4,** 526; **5,** 689–690.

Hofmann-Loeffler photocyclization. This reaction[1] is commonly carried out in the presence of a strong acid. Japanese chemists[2] have reported a cyclization that proceeds in only 16% yield when sulfuric acid is present but in 50% yield in the presence of triethylamine (equation I).

(I)

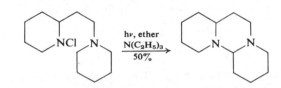

Thio-Claisen rearrangement. When heated in *o*-dichlorobenzene with triethylamine, α-arylthiomethylacrylic acids (1) rearrange to the formerly unknown thiochromane-3-carboxylic acids (2).[3] The transformation probably involves a

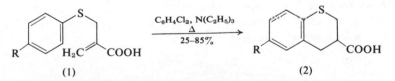

(1)   (2)

thio-Claisen rearrangement followed by a Michael addition. No reaction occurs in the absence of the base.

¹ M. E. Wolff, *Chem. Revs.*, **63**, 55 (1963).
² M. Kimura and Y. Ban, *Synthesis*, 201 (1976).
³ B. Gopalan, K. Rajagopalan, S. Swaminathan, and K. K. Balasubramanian, *Synthesis*, 409 (1976).

Triethyl orthoformate, 1, 1204–1209; **2,** 430; **4,** 527; **5,** 690–691; **6,** 610–611.

Conversion of 1,2-diols to olefins. Vicinal 1,2-diols undergo exchange with triethyl orthoformate to give cyclic orthoformates, which when heated in the presence of a carboxylic acid catalyst undergo a stereospecific *cis*-elimination to form the corresponding olefins, carbon dioxide, and ethanol.[1] Burgstahler *et al.*[2]

have employed this sequence, which was carried out in one flask, for the preparation of the optically active strained olefins (−)-2-bornene (3, R = H) and (+)-2,3-dimethyl-2-bornene (3, R = CH₃). Efforts to prepare (3) by the standard Corey-Winter olefin synthesis (1, 1233–1234; 2, 439–441; 3, 315–316; 4, 269–270, 541–542; 5, 34, 661; 6, 334, 583–584) were less satisfactory.

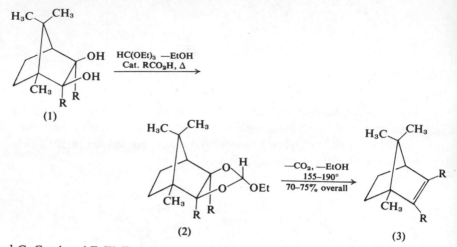

(1)   (2)   (3)

¹ G. Crank and F. W. Eastwood, *Aust. J. Chem.*, **17**, 1392 (1964); J. S. Josan and F. W. Eastwood, *ibid.*, **21**, 2013 (1968).

² A. W. Burgstahler, D. L. Boger, and N. C. Naik, *Tetrahedron*, **32**, 309 (1976).

Triethyloxonium tetrafluoroborate, 1, 1210–1212; **2,** 430–431; **3,** 303; **4,** 527–529; **5,** 691–693.

O-Alkylation of hydroxypyrimidines. Hydroxypyrimidines (1a), which exist in equilibrium with the tautomeric forms (1b) and (1c), undergo selective O-alkylation with Meerwein's reagent.¹

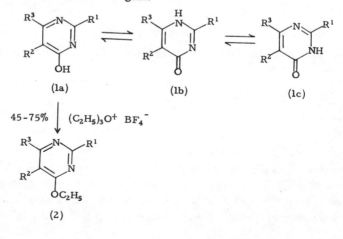

(1a)   (1b)   (1c)

(2)

Rearrangement of cyclobutanones.[2] Treatment of the angularly fused cyclo-butanone (1) with triethyloxonium tetrafluoroborate in CH_2Cl_2 (N_2) results in stereospecific rearrangement to the ketone (2) in high yield. Acid-catalyzed

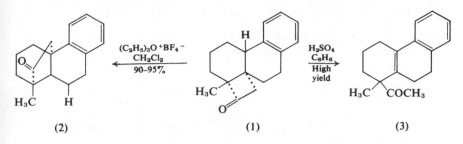

(2)   (1)   (3)

rearrangement results in formation of (3). This rearrangement (1) → (2) is useful in the synthesis of some diterpene alkaloids.[3]

[1] D. G. McMinn, *Synthesis*, 824 (1976).
[2] U. R. Ghatak, B. Sanyal, and S. Ghosh, *Am. Soc.*, **98**, 3721 (1976).
[3] U. R. Ghatak and S. Chakrabarty, *J. Org.*, **41**, 1089 (1976).

Triethyl phosphite, 1, 1212–1216; **2,** 432–433; **3,** 304; **4,** 529–530; **5,** 693; **6,** 612.

Tetrathiafulvalenes. The synthesis of the first known unsymmetrical tetra-thiafulvalene (3) has been carried out by mixed phosphite coupling of (1) and (2). The yield is low (3%) because the main products result from symmetrical coupling of (1) and of (2). The two carbomethoxy groups of (3) can be removed by treatment with LiBr in moist DMSO to give the yellow benzotetrathiaful-valene (4).

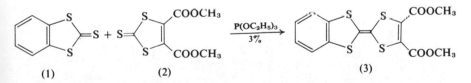

(1)   (2)   (3)

The tetrathiafulvalenes are of interest because of their ability to form charge-transfer salts with metallic properties.[2] Thus (4) forms a charge-transfer salt with tetracyanoquinodimethane.

[1] H. K. Spencer, M. P. Cava, and A. F. Garito, *J.C.S. Chem. Comm.*, 966 (1976).
[2] A. F. Garito and A. J. Heeger, *Accts. Chem. Res.*, **7**, 232 (1974).

Triethylsilane–Boron trifluoride

Reduction of alcohols. Fry *et al.*[1] report that the combination of triethylsilane and BF_3 is more satisfactory than triethylsilane and TFA (**5,** 695) for reduction of alcohols to hydrocarbons. For example, the alcohol (1), which is prone to skeletal rearrangement, is reduced almost entirely to (2) (equation I). Even

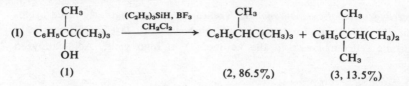

aliphatic secondary alcohols can be reduced: 2-octanol → octane (50%) and 2-adamantanol → adamantane (86%). In some cases triethylsilane was replaced by phenylneopentylmethylsilane.[2]

[1] M. G. Adlington, M. Orfanopoulos, and J. L. Fry, *Tetrahedron Letters*, 2955 (1976).
[2] L. H. Sommer, K. W. Michael, and W. D. Korte, *Am. Soc.*, **89**, 868 (1967).

Trifluoroacetic acid (TFA), 1, 1219–1221; **2,** 433–434; **3,** 305–308; **4,** 530–532; **5,** 695–700; **6,** 613–615.

Cyclization of bicyclic alcohols to tricyclic hydrocarbons. TFA has been found to be excellent for the acid-catalyzed cyclization of the alcohol (1) to the benzobicyclo[3.2.1]octene (2)[1] and of the alcohol (3) to the benzobicyclo[3.3.1]-nonene-2 (4).[2] Sulfuric acid is less satisfactory.

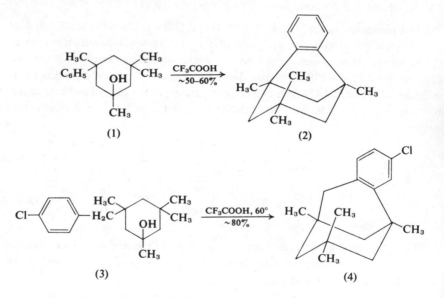

Sulfinamides → sulfinates. Reaction of sulfinamides with an alcohol in the presence of 2 equiv. of a strong acid results in formation of sulfinates. The reaction proceeds with inversion of configuration at the S=O center. Stereo-

(I)
$$R^1SN\begin{array}{c}O\\||\\\end{array}\begin{array}{c}R^2\\ \diagdown\\ R^3\end{array} + R^4OH \xrightarrow[88-91\%]{H^+} R^1SOR^4$$

specificity depends in part on the structure of the alcohol and also on the acid catalyst. Trifluoroacetic acid (in some cases with $AgClO_4$) appears to be superior to TsOH, CF_3SO_3H, and $HSbF_6$.[3]

Olefin cyclization (3, 305–307; 4, 531–532; 5, 696–697). Johnson et al.[4] have obtained 11α-methyl- and 11α-hydroxyprogesterone by total synthesis, in which the key step involves biomimetic polyene cyclization.

[1] B. L. Shapiro, M. J. Gattuso, and G. R. Sullivan, *Tetrahedron Letters*, 223 (1971).
[2] B. L. Shapiro and M. J. Shapiro, *J. Org.*, **41**, 1522 (1976).
[3] M. Mikołajczyk, J. Drabowicz, and B. Bujnicki, *J.C.S. Chem. Comm.*, 568 (1976).
[4] W. S. Johnson and G. E. DuBois, *Am. Soc.*, **98**, 1038 (1976); W. S. Johnson, S. Escher, and B. W. Metcalf, *ibid.*, **98**, 1039 (1976).

Trifluoroacetic anhydride, 1, 1221–1226; 3, 308; 5, 701; 6, 616–617.

Polonovski reaction (3, 7; 5, 3). French chemists[1] have modified this reaction by use of trifluoroacetic anhydride rather than acetic anhydride and have applied this version to a useful synthesis of compounds related to antitumor alkaloids of *Catharanthus roseus*. Thus treatment of the N-oxide (1) of catharanthine, an ibogane alkaloid, with TFAA in CH_2Cl_2 at $-78°$ in the presence of vindoline (2) and then with $NaBH_4$ leads to three products, one of which is the desired (3),

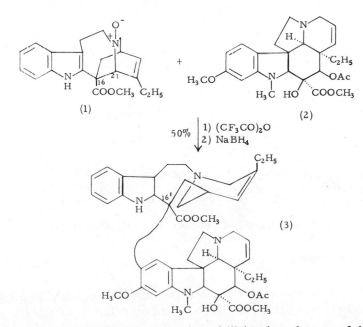

anhydrovinblastine; 50% yield. Formation of (3) involves cleavage of the C_{16}— C_{21} bond of (1) followed by coupling with (2). An attractive feature of this synthesis is that the natural configuration (S) obtains in (3). The biogenetic synthesis follows this *in vitro* synthesis.[2]

[1] N. Langlois, F. Guéritte, Y. Langlois, and P. Potier, *Am. Soc.*, **98**, 7017 (1976).
[2] P. E. Daddona and C. R. Hutchinson, *ibid.*, **96**, 6806 (1974).

Trifluoromethanesulfonic acid trimethylsilyl ester, $F_3CSO_2OSi(CH_3)_3$. Mol. wt. 222.26; b.p. 140°.

The reagent is prepared by the reaction of chlorotrimethylsilane with the silver salt of trifluoromethanesulfonic acid (67% yield).[1]

Silylation of carbonyl compounds.[2] Ketones, diketones, esters, and amides are silylated under mild conditions by this reagent in the presence of triethylamine. Examples:

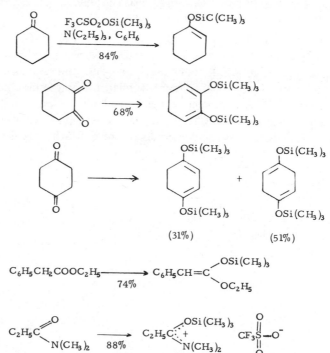

[1] M. Schmeiseer, P. Sartori, and B. Lippsmeier, *Ber.*, **103**, 868 (1970).
[2] G. Simchen and W. Kober, *Synthesis*, 259 (1976).

Trifluoromethanesulfonic anhydride (Triflic anhydride), 4, 533–534; **5,** 702–705; **6,** 618–620.

Nitriles. Aldoximes are dehydrated to nitriles by reaction with triflic anhydride and triethylamine in CH_2Cl_2 at $-78 \rightarrow 20°$ (equation I).[1]

(I) $\underset{H}{\overset{R}{>}}C=NOH \xrightarrow[\underset{88-94\%}{2N(C_2H_5)_3}]{(CF_3SO_2)_2O} RC \equiv N + 2CF_3SO_3^- \overset{+}{H}N(C_2H_5)_3$

[1] J. B. Hendrickson, K. W. Bair, and P. M. Keehn, *Tetrahedron Letters*, 603 (1976).

Trifluoromethylthiocopper, $CuSCF_3$. Mol. wt. 164.61.

The reagent is prepared[1] from the reaction of copper with bis(trifluoromethyl-thio)mercury (*toxic*).[2]

$$Hg(SCF_3)_2 + Cu \longrightarrow CuSCF_3$$
$$(1)$$

Aryl trifluoromethyl sulfides.[1] The reagent, which can be prepared *in situ*, reacts with aryl bromides and iodides to form aryl trifluoromethyl sulfides (DMF, HMPT; 100–160°) in about 80–100% yield.

Examples:

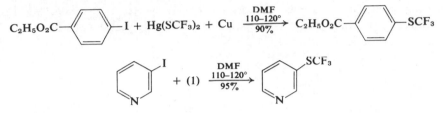

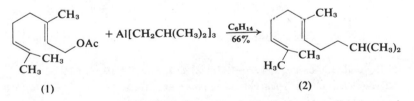

[1] D. C. Remy, K. E. Rittle, C. A. Hunt, and M. B. Freedman, *J. Org.*, **41**, (1976).
[2] E. H. Man, D. D. Coffman, and E. L. Muetterties, *Am. Soc.*, **81**, 3575 (1959).

Triisobutylaluminum, 1, 260; **4,** 535.

Coupling with allylic acetates, formates, and carbonates. The reaction of geranyl acetate in hexane with triisobutylaluminum at $-78 \rightarrow 0°$ results in formation of 2,6,10-trimethyl-2-(E)-6-undecadiene (2) in 66% yield. Other trialkylaluminum reagents can be used; the allylic component can be an acetate,

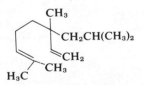

a formate, or a carbonate. In some cases the product is accompanied by the isomer arising from allylic rearrangement, such as (3).[1]

$$\begin{array}{c} CH_3 \\ CH_2CH(CH_3)_2 \\ CH_2 \\ CH_3 \\ H_3C \end{array}$$

[1] S. Hashimoto, Y. Kitagawa, S. Iemura, H. Yamamoto, and H. Nozaki, *Tetrahedron Letters*, 2615 (1976).

2,4,6-Triisopropylbenzenesulfonyl hydrazine, 4, 535.

Diimide. This hydrazine, in the presence of triethylamine, decomposes to diimide at room temperature. It is convenient for hydrogenation of azobenzene to hydrazobenzene (85% yield) and of various olefins (acenaphthylene → acenaphthene, 99% yield).[1]

[1] N. J. Cusack, C. B. Reese, A. C. Risius, and B. Roozpeikar, *Tetrahedron*, **32**, 2157 (1976).

2,2,2-Trimethoxy-4,5-dimethyl-1,3-dioxaphospholene, (1), 2, 97–98.

Branched-chain sugars. A recent synthesis of branched-chain sugars uses the Ramirez dioxaphospholene condensation of a protected D-glyceraldehyde (2) with (1). The products after hydrolysis, (4) and (5), are glycosides of 1-deoxy-3-C-methyl-D-ribohexulose.[1]

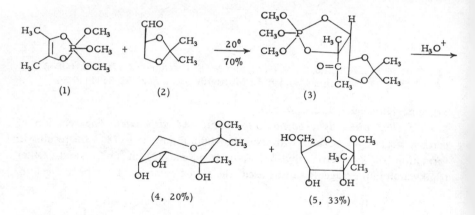

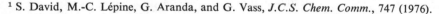

[1] S. David, M.-C. Lépine, G. Aranda, and G. Vass, *J.C.S. Chem. Comm.*, 747 (1976).

Trimethylamine N-oxide, 1, 1230–1231; 2, 434; 3, 309–310; 6, 624–625.

Decomplexation of tricarbonyliron complexes (**6**, 624–625). The intermediate (2) has been isolated from the reaction of trimethylamine oxide with the complex (1). The paper suggests a possible mechanism.[1]

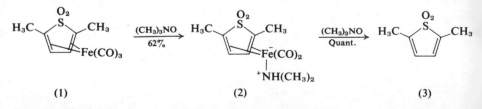

[1] J. H. Eekhof, H. Hogeveen, and R. M. Kellogg, *J.C.S. Chem. Comm.*, 657 (1976).

Trimethyl phosphate, 5, 716–717.

Halogenation; nitration. Trimethyl phosphate is an excellent solvent for halogenation of arenes. The reaction is rapid even at 25° and no hydrogen halide is evolved (methyl halide is formed).

Examples:

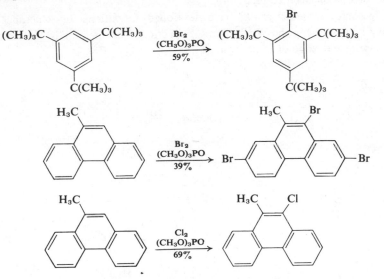

Nitration of reactive arenes can be carried out with sodium nitrate in trimethyl phosphate containing phosphorus pentoxide: anthracene → 9-nitroanthracene (71% yield); 9-methylphenanthrene → 10-nitro-9-methylphenanthrene (37% yield).[1]

[1] D. E. Pearson, M. G. Frazer, V. S. Frazer, and L. C. Washburn, *Synthesis*, 621 (1976).

Trimethyl phosphite, 1, 1233–1235; **2,** 439–441; **3,** 315–316; **4,** 541–542; **5,** 717.

Corey-Winter alkene synthesis (**1,** 1233–1234; **2,** 439–441; **3,** 315–316; **4,** 269–270; 541–542; **5,** 34, 198, 661; **6,** 384). Bauer and Macomber[1] have adapted the Corey-Winter synthesis as a route to alkynes (equation I). The method is

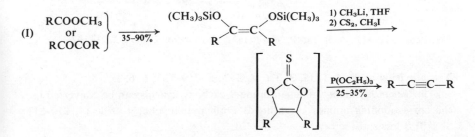

obviously not superior to the route from bishydrazones of α-diketones; however, it offers a useful transformation of esters into alkynes.

[1] D. P. Bauer and R. S. Macomber, *J. Org.*, **41**, 2640 (1976).

Trimethyl phosphonoacetate, 2, 442.

γ-*Lactones*. Bensel *et al.*[1] have developed a synthesis for γ-lactones from epoxy ketones using the Wittig-Horner reaction. The method is illustrated for the case of the oxide of isophorone (1).

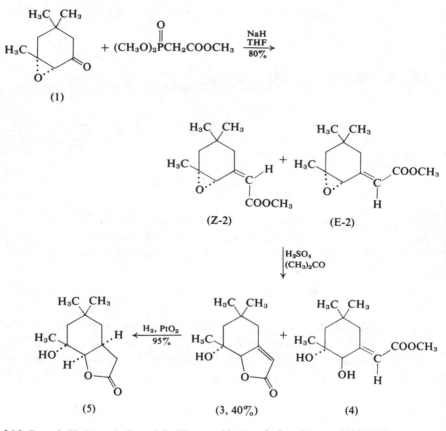

[1] N. Bensel, H. Marschall, and P. Weyerstahl, *Tetrahedron Letters*, 2293 (1976).

Trimethylsilyl azide, 1, 1236; **3,** 316, **4,** 542; **5,** 719–720; **6,** 632.

9-Phenanthrylamines. Phenanthrene-9-carboxylic acids can be converted into the corresponding amines as outlined. Diphenylphosphoryl azide (**4,** 210–211; **5,** 280) is less satisfactory.[1]

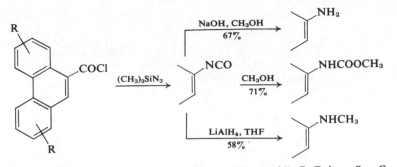

[1] G. Krow, K. M. Damodaran, E. Michener, S. I. Miller, and D. R. Dalton, *Syn. Comm.*, **6**, 261 (1976).

Trimethylsilylbromoketene, $\begin{array}{c}(CH_3)_3Si \\ \diagdown \\ / \\ Br\end{array} C{=}C{=}O$ (1). Mol. wt. 191.13, stable in

solution.

Preparation:

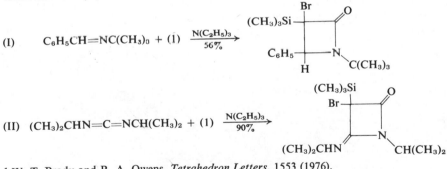

Cycloaddition. This ketene does not react with cyclopentadiene or vinyl ethers, but it does undergo [2 + 2]cycloadditions with imines (equation I) and carbodiimides (II).[1]

(I) $C_6H_5CH{=}NC(CH_3)_3 + (1) \xrightarrow[56\%]{N(C_2H_5)_3}$

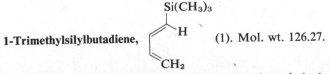

(II) $(CH_3)_2CHN{=}C{=}NCH(CH_3)_2 + (1) \xrightarrow[90\%]{N(C_2H_5)_3}$

[1] W. T. Brady and R. A. Owens, *Tetrahedron Letters*, 1553 (1976).

1-Trimethylsilylbutadiene, $\begin{array}{c}Si(CH_3)_3 \\ \diagup^H \\ \diagdown \\ CH_2\end{array}$ (1). Mol. wt. 126.27.

The diene has been prepared by hydrosilylation of vinylacetylene. Three new methods have been described which are adaptable to synthesis of related dienes (scheme I).[1]

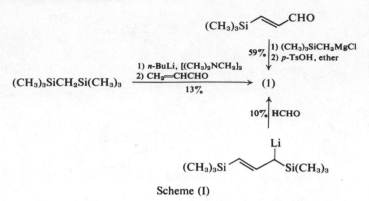

Scheme (I)

Diels-Alder reactions. This diene is not very reactive in cycloadditions, even under BF_3 catalysis, and is less reactive than butadiene itself. In addition the regioselectively is slight.[2] Thus it reacts with methyl acrylate to give both the "ortho" adduct (2) and the "meta" adduct (3) (equation I). A single adduct is

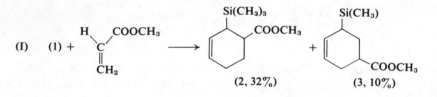

obtained with diethyl oxomalonate (equation II) and with maleic anhydride (presumably *endo*) as shown in equation (III).

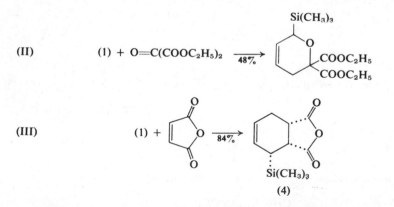

This adduct, 3-trimethylsilyl-4-cyclohexene-1,2-dicarboxylic acid anhydride, and derived allylsilanes (5) and (6) were found to undergo some interesting reactions with electrophiles (scheme II). In each case the substitution is attended by allylic rearrangement.

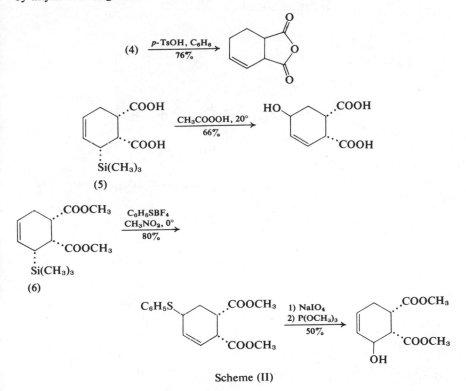

Scheme (II)

[1] M. J. Carter and I. Fleming, *J.C.S. Chem. Comm.*, 679 (1976).
[2] I. Fleming and A. Percival, *ibid.*, 681 (1976).

Trimethylsilyl cyanide, 4, 542–543; 5, 720–722; 6, 632–633.

Isoprenylation of quinones. Evans and Hoffmann[1] have developed a useful route to prenylated quinones, which are natural products involved in various biological processes. The process is illustrated in equation (I) for the synthesis of 2-isopentenylhydroquinone (3). Reaction of the protected quinone (1) with the allylic bromide in the presence of Rieke magnesium (5, 419) affords the epimeric quinols (2). On deprotection the initial product (a) undergoes a facile rearrangement, probably a Cope rearrangement, to (3) in high yield.

The same method was used to synthesize desoxylapachol (4, equation II) and vitamin K$_{2(5)}$ (5, equation III). In the latter synthesis, the [3,3]-sigmatropic rearrangement occurs even at $-22°$.

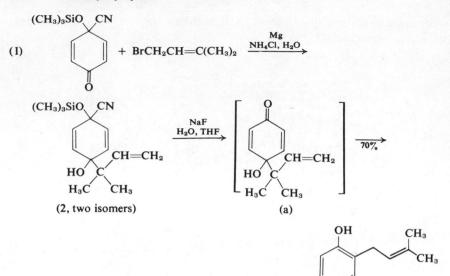

(I)

(2, two isomers)   (a)

(3)

(II)

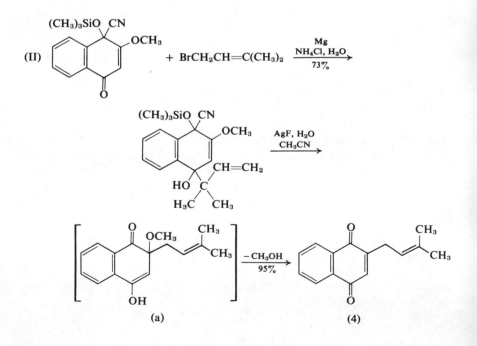

(a)   (4)

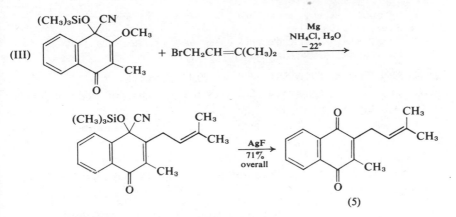

(5)

¹ D. A. Evans and J. M. Hoffman, *Am. Soc.*, **98**, 1983 (1976).

N-(Trimethylsilyl)imidazole, $(CH_3)_3SiN$ ⟨imidazole⟩ (1). Mol. wt. 140.26, b.p. 93–94°/
14 mm. Suppliers: Aldrich, Pearce.

This reagent is prepared from imidazole and hexamethyldisilazane.[1]

Acyl imidazolides. These compounds resemble acyl chlorides in reactivity[2]
and are thus useful in synthesis. Masamune[3] has observed that they can be
prepared by reaction of phenyl and 2,2,2-trifluoroethyl esters with N-(trimethylsilyl)imidazole at 20° in the presence of a trace of sodium phenoxide (I). Alkyl

$$RCOR' + (1) \xrightarrow[>95\%]{C_6H_5ONa} RC(=O)-N\text{(imidazole)}$$

$(R' = C_6H_5, CH_2CF_3)$

esters, S-*t*-butyl thiol esters, and benzenethiol esters do not react under these
conditions. Hence it is possible to activate O-esters selectively.

Silylation. 1,3-Dicarbonyl compounds, even though mainly in the enolic
form, are silylated by standard methods in low yields. The reaction can be effected
in high yield with the combination of hexamethyldisilazane and catalytic amounts
of imidazole.[4]

Examples:

$$CH_3COCH_2COCH_3 \xrightarrow[96\%]{(1),\ \Delta} CH_3\overset{\overset{\displaystyle OSi(CH_3)_3}{|}}{C}=CHCOCH_3$$

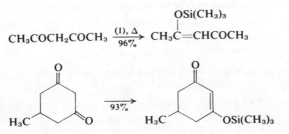

[1] L. Birkofer and A. Ritter, *Angew. Chem., Int. Ed.*, **4**, 417 (1965).
[2] H. A. Staab, *ibid.*, **1**, 351 (1962).
[2] G. S. Bates, J. Diakur, and S. Masamune, *Tetrahedron Letters*, 4423 (1976).
[4] S. Torkelson and C. Ainsworth, *Synthesis*, 722 (1976).

Trimethylsilyllithium, $(CH_3)_3SiLi$. Mol. wt. 80.13, deep red.

This reagent can be prepared readily by reaction at 0° of methyllithium in HMPT with hexamethyldisilane.[1]

Conjugate addition.[2] The reagent undergoes ready 1,4-addition to cyclohexenones to form, after quenching with methanol, 3-silylcyclohexanones. The intermediate can be alkylated cleanly (equation I). The reagent has a strong preference for axial addition (equation II). The reagent thus resembles lithium dialkyl cuprates.

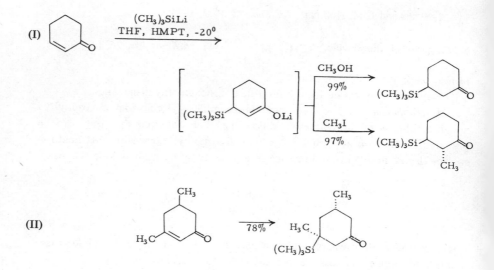

(I)

(II)

[1] H. Sakurai and A. Okada, *J. Organometal. Chem.*, **36**, C13 (1972); Suppliers: PCR and Columbia Organics.
[2] W. C. Still, *J. Org.*, **41**, 3063 (1976).

Trimethylsilylmethylpotassium, 6, 637–638.

Metallation.[1] Trimethylsilylmethylpotassium is the most powerful reagent for metallation of allylic and benzylic positions, although the combination of *n*-butyllithium and potassium *t*-butoxide can often replace it. The combination of *n*-butyllithium and TMEDA is definitely less effective.

Olefinic and cyclopropane hydrogens are metallated most readily by pentyl-sodium,[2] particularly in the presence of potassium *t*-butoxide. For example, camphene (1) can be converted into (2) and (3) as formulated in equation (I). Metallation of nortricyclene is formulated in equation (II). If an activating

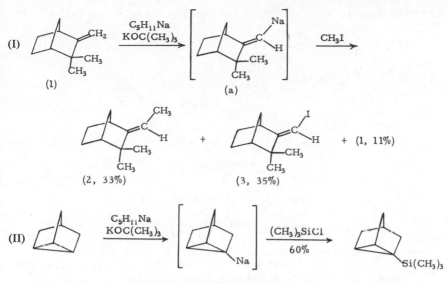

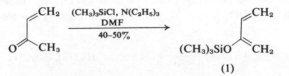

methoxyl group is present, trimethylsilylmethylpotassium can be used in these metallations.

[1] J. Hartmann and M. Schlosser, *Helv.*, **59**, 453 (1976).
[2] J. Hartmann, R. Muthukrishnan, and M. Schlosser, *ibid.*, **57**, 2261 (1974).

2-Trimethylsilyloxy-1,3-butadiene, (1). Mol. wt. 142.27, b.p. 50–55°/50 mm.

The diene can be prepared by reaction of methyl vinyl ketone with chloro-trimethylsilane and triethylamine in DMF (40–50% yield).[1] It can also be prepared, but on a smaller scale, using LDA as base and chlorotrimethylsilane in THF–ether.[2]

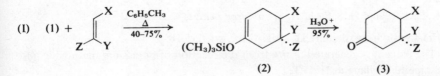

Cyclohexanones. Jung and McCombs[3] have used this substituted diene in place of 2-ethoxy-1,3-butadiene in Diels-Alder reactions. It exhibits similar reactivity, but is much more accessible. The reagent was used to prepare several cyclohexanones in yields of ~40–75% (equation I). In addition the initial adduct can be brominated, hydroxylated, or condensed with an aldehyde (equation II).

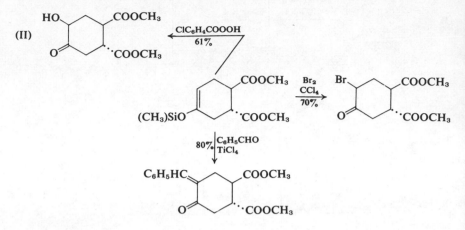

(II)

[1] C. Girard, P. Amice, J. P. Barnier, and J. M. Conia, *Tetrahedron Letters*, 3329 (1974).
[2] M. E. Jung and C. A. McCombs, *Org. Syn.*, submitted (1977).
[3] *Idem*, *Tetrahedron Letters*, 2935 (1976).

Trimethylsilylphenylselenomethyllithium, $(CH_3)_3SiCHLi$ (1). Mol. wt. 249.22.
$\qquad\qquad\qquad\qquad\qquad\qquad\qquad\qquad\quad |$
$\qquad\qquad\qquad\qquad\qquad\qquad\qquad\qquad\ SeC_6H_5$

Preparation:

$$C_6H_5SeSeC_6H_5 \xrightarrow[CH_3OH]{2NaBH_4} [C_6H_5SeNa] \xrightarrow[90\%]{(CH_3)_3SiCH_2Cl}$$

$$(CH_3)_3SiCH_2SeC_6H_5 \xrightarrow[THF,\ -78°]{LDA} (1)$$

Aldehyde synthesis. This reagent can be used to convert halides of the type RCH_2X into RCH_2CHO. A typical synthesis is formulated in equation (I).[1]

(I) $CH_3(CH_2)_3CH_2Br + (1) \xrightarrow[94\%]{THF,\ -78°} (CH_3)_3SiCHCH_2(CH_2)_3CH_3 \xrightarrow[0°]{H_2O_2}$
$\qquad\qquad\qquad\qquad\qquad\qquad\qquad\qquad\qquad\qquad\quad |$
$\qquad\qquad\qquad\qquad\qquad\qquad\qquad\qquad\qquad\quad SeC_6H_5$

$$\left[(CH_3)_3SiCHCH_2(CH_2)_3CH_3 \atop \qquad\quad \underset{O}{\diagup}\!\!SeC_6H_5 \right] \xrightarrow[80\%]{25°} CH_3(CH_2)_4CHO$$

[1] K. Sachdev and H. S. Sachdev, *Tetrahedron Letters*, 4223 (1976).

Trimethylsilylpotassium, $(CH_3)_3SiK$. Mol. wt. 115.18.

The reagent is generated *in situ* from the reaction of potassium methoxide and hexamethyldisilane in HMPT.

Deoxygenation of epoxides. This reagent converts epoxides into olefins with inversion of stereochemistry.[1]

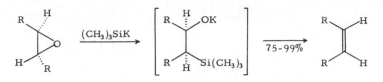

[1] P. B. Dervan and M. A. Shippey, *Am. Soc.*, **98**, 1265 (1976).

Trineopentylneopentylidenetantalum, $Ta[CH_2C(CH_3)_3]_3[CHC(CH_3)_3]$ (1). Mol. wt. 464.49, orange.

Preparation.[1] This "carbene" complex of tantalum is obtained in quantitative yield by the reaction of $Ta[CH_2C(CH_3)_3]_3Cl_2$ [2] with 2 moles of $(CH_3)_3CCH_2Li$.

Reaction with carbonyl groups.[3] This tantalum reagent undergoes Wittig reactions with carbonyl compounds. Notably, it reacts in cases where phosphorus ylides do not (third example).

Examples:

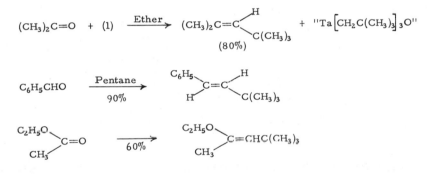

The related niobium compound (2) reacts in the same way, but is more difficult to prepare.

$$Nb[CH_2C(CH_3)_3]_3[CHC(CH_3)_3]$$

(2)

[1] R. R. Schrock, *Am. Soc.*, **96**, 6796 (1974).
[2] Prepared from $TaCl_5$ and 1.5 moles of $Zn[CH_2C(CH_3)_3]_2$.
[3] R. R. Schrock, *Am. Soc.*, **98**, 5399 (1976).

Triphenylphosphine, 1, 1238–1247; **2,** 443–445, **3,** 317–320; **4,** 548–550; **6,** 643–644.

Episulfides. β-Hydroxyethanesulfenyl chlorides react with triphenylphosphine at ambient temperatures or lower to form episulfides. This reaction was observed with (1), derived from a cephalosporin.[1]

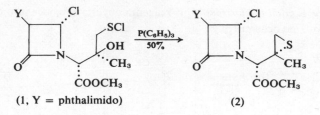

(1, Y = phthalimido)   (2)

[1] J. E. Baldwin and D. P. Hesson, *J.C.S. Chem. Comm.*, 667 (1976).

Triphenylphosphine–Carbon tetrachloride, **1**, 1274; **2**, 445; **3**, 320; **4**, 551–552; **5**, 727; **6**, 644–645.

Dehydration. This reagent is used for conversion of —OH into —Cl. However, if the reaction is conducted in refluxing acetonitrile, secondary alcohols are frequently dehydrated to alkenes.[1]

Examples:

$$CH_3CH_2CHOHC_6H_5 \xrightarrow[\sim 70\%]{\substack{P(C_6H_5)_3, CCl_4 \\ CH_3CN, \Delta}} CH_3CH=CHC_6H_5$$

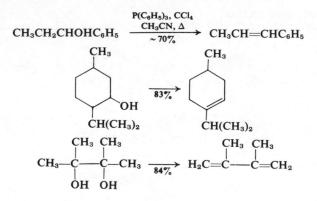

[1] R. Appel and H.-D. Wihler, *Ber.*, **109**, 3446 (1976).

Triphenylphosphine–Diethyl azodicarboxylate, **1**, 245–247; **4**, 553–555; **5**, 727–728; **6**, 645.

Reaction with ketoximes. Ketoximes and an aromatic carboxylic acid react with this system at 0° to form diacylated aromatic amines (equation I).

(I) $(C_6H_5)_2C=NOH + ArCOOH \xrightarrow[C_2H_5OOCN=NCOOC_2H_5]{P(C_6H_5)_3}$

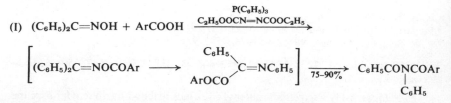

Where the acid is replaced by an alcohol in the system formulated above, the oxime is alkylated (equation II).[1]

(II) $(C_6H_5)_2C=NOH + ROH$ $\xrightarrow[\substack{C_2H_5OOCN=NCOOC_2H_5 \\ 40-70\%}]{P(C_6H_5)_3}$ $(C_6H_5)_2C=NOR$

Esterification of allylic alcohols. Esterification of allylic alcohols with benzoic acid using triphenylphosphine and diethyl azodicarboxylate proceeds with inversion and without allylic rearrangement.[2]

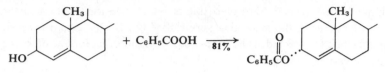

$CH_2=CHCHDOH + C_6H_5COOH \xrightarrow[92\%]{} CH_2=CHCHDOCOC_6H_5$

$OH \rightarrow X$ (X = N_3, CN, CF_3COO, C_6H_5S). The hydroxyl group of 3β-hydroxycholestanol (1) can be replaced with inversion by a number of functional groups by reaction with triphenylphosphine–diethyl azodicarboxylate and a nucleophile as indicated in the formulation. Cholesterol undergoes displacements also with inversion and in comparable yields.[3]

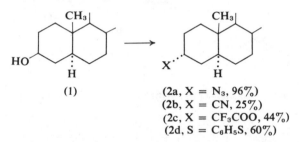

(1)

(2a, X = N_3, 96%)
(2b, X = CN, 25%)
(2c, X = CF_3COO, 44%)
(2d, S = C_6H_5S, 60%)

By this means *cis*-1,2-disubstituted cyclohexanes can be prepared from *trans*-2-substituted cyclohexanols:

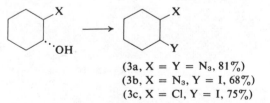

(3a, X = Y = N_3, 81%)
(3b, X = N_3, Y = I, 68%)
(3c, X = Cl, Y = I, 75%)

Lactonization of ω-hydroxy acids. ω-Hydroxy acids, $HO(CH_2)_nCOOH$, are converted mainly into lactones and diolides when treated with this combination of reagents for 2 days. When n = 11, the lactone is the main product (60% yield). When n = 5 or 7, the diolide is the major product.

A related reaction is observed with diols of the type $HO(CH_2)_nOH$ and ethyl cyanoacetate (equation I).[4]

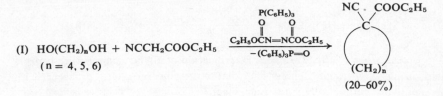

(I) $HO(CH_2)_nOH + NCCH_2COOC_2H_5$ $\xrightarrow[-(C_6H_5)_3P=O]{\substack{P(C_6H_5)_3 \\ C_2H_5OCN=NCOC_2H_5}}$

($n = 4, 5, 6$)

$(20–60\%)$

White and co-workers[5] have used this new lactonization reaction as the last step in a synthesis of the macrolide vermiculine (2). In this case, the sensitivity of the substrate (1) precluded the more usual method of lactonization by activation of the carboxyl group.

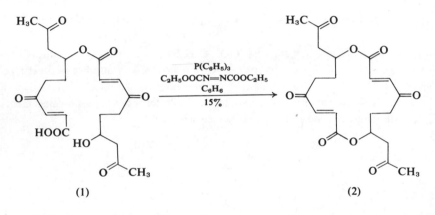

$\xrightarrow[15\%]{\substack{P(C_6H_5)_3 \\ C_2H_5OOCN=NCOOC_2H_5 \\ C_6H_6}}$

(1)   (2)

O-Alkylhydroxylamines. These compounds can be prepared as formulated in equation (I).[6]

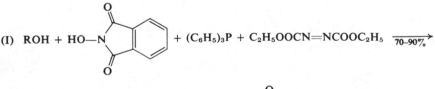

(I) $ROH + HO—N$ $+ (C_6H_5)_3P + C_2H_5OOCN=NCOOC_2H_5$ $\xrightarrow{70-90\%}$

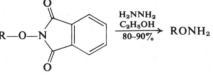

$R—O—N$ $\xrightarrow[80-90\%]{\substack{H_2NNH_2 \\ C_2H_5OH}}$ $RONH_2$

[1] S. Bittner and S. Grinberg, *J.C.S. Perkin I*, 1708 (1976).
[2] G. Grynkiewicz and H. Burzyńska, *Tetrahedron*, **32**, 2109 (1976).
[3] H. Loibner and E. Zbiral, *Helv.*, **59**, 2100 (1976).
[4] T. Kurihara, Y. Nakajima, and O. Mitsunobu, *Tetrahedron Letters*, 2455 (1976).
[5] Y. Fukuyama, C. L. Kirkemo, and J. D. White, *Am. Soc.*, **99**, 646 (1977).
[6] E. Grochowski and J. Jurczak, *Synthesis*, 682 (1976).

Triphenylphosphine–Iodine, $P(C_6H_5)_3-I_2$.

Iododeoxy sugars. Moffatt *et al.*[1] report that in a synthesis of the nucleoside antibiotic angustmycin (3) the combination of triphenylphosphine and iodine[2] is superior to methyltriphenoxyphosphonium iodide (**1,** 1249–1250; **4,** 557–559) for conversion of (1) into (2).

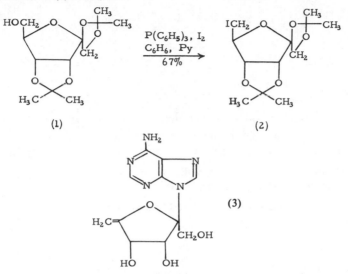

(1)   (2)

(3)

Iodo sugars such as (2) can be dehydrohalogenated with either methanolic sodium methoxide or DBN.

[1] E. J. Prisbe, J. Smejkal, J. P. H. Verheyden, and J. G. Moffatt, *J. Org.*, **41,** 1836 (1976).
[2] A. V. Bayless and N. Zimmer, *Tetrahedron Letters*, 3811 (1968).

Triphenylphosphine dihalides, $(C_6H_5)_3PBr_2$, $(C_6H_5)_3PCl_2$.

Olefin inversion. This inversion can be accomplished by epoxidation, reaction with triphenylphosphine dibromide or dichloride to give a *vic*-dihalide, and then *trans*-elimination of the added halo groups with zinc–acetic acid (equation I). The sequence involves two inversions, one at each carbon joined to oxygen.[1]

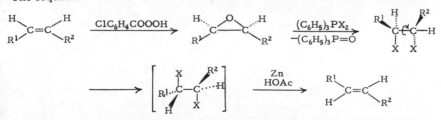

For a related, apparently superior, method for effecting this inversion *see* **Sodium iodide–Dimethylformamide,** this volume.

[1] P. E. Sonnet and J. E. Oliver, *J. Org.*, **41,** 3279 (1976).

Triphenylphosphite ozonide, 3, 323–324; **4**, 559.

Oxidation of alkylidenetriphenylphosphoranes. Ylides of type (1) are oxidized by the triphenylphosphite–ozone adduct to alkenes (2); ylides of type (3) are

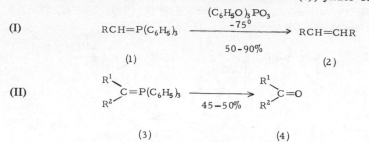

(I)

$$RCH=P(C_6H_5)_3 \quad \xrightarrow[50-90\%]{\substack{(C_6H_5O)_3PO_3 \\ -75°}} \quad RCH=CHR$$

(1)   (2)

(II)

$$\begin{array}{c} R^1 \\ \diagdown \\ R^2 \end{array} C=P(C_6H_5)_3 \quad \xrightarrow[45-50\%]{} \quad \begin{array}{c} R^1 \\ \diagdown \\ R^2 \end{array} C=O$$

(3)   (4)

converted into ketones (4).[1] This oxidation does not involve singlet oxygen since $^1O_2$ is formed from the adduct only at temperatures above $-30°$.

Yields in these oxidations are higher than those obtained by oxygenation.[2]

[1] H. J. Bestmann, L. Kisielowski, and W. Distler, *Angew. Chem., Int. Ed.*, **15**, 298 (1976).
[2] H. J. Bestmann and O. Kratzer, *Ber.*, **96**, 1899 (1963).

2,4,6-Triphenylpyrylium perchlorate, (1). Mol. wt. 408.83, m.p. 290°.
Preparation[1]:

$$C_6H_5CH=CHCOC_6H_5 + C_6H_5COCH_3 + HClO_4 \quad \xrightarrow{100°}$$

(1)

Deamination.[2] Benzylic or allylic primary amines (1) can be deaminated as depicted in equation (I). The pyrolysis step, (4) → (5), may be a retro-ene reaction.

(I) RCH_2NH_2 + (1) $\xrightarrow{85\%}$

  (2, R = $ArCH_2$,
   $H_2C=CHCH_2$)

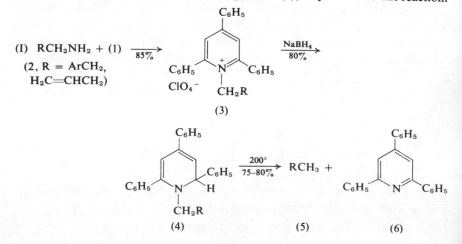

(3)

(4)   (5)   (6)

Isocyanates.[3] Isocyanates can be prepared by reaction of acid hydrazides with (1) to form N-acylimides; on pyrolysis these products decompose to isocyanates (equation II). Isocyanates are also available by a related reaction (equation III). N-Amino-2,4,6-triphenylpyridinium perchlorate (11) is prepared by reaction of (1) with *t*-butyl carbazate (45% yield).

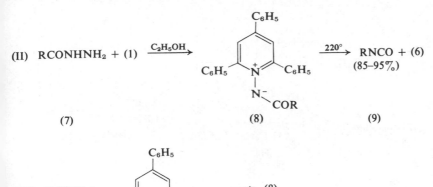

(II) $RCONHNH_2$ + (1) $\xrightarrow{C_2H_5OH}$

$\xrightarrow{220°}$ RNCO + (6)
(85–95%)

(7)   (8)   (9)

(III)   RCOCl + $\xrightarrow[65\ 75\%]{}$ (8)

(10)   (11)

[1] A. T. Balaban and C. Toma, *Tetrahedron Suppl.*, **7**, 1 (1966).
[2] A. J. Boulton, J. Epsztajn, A. R. Katritzky, and P.-L. Nie, *Tetrahedron Letters*, 2689 (1976).
[3] J. R. Bapat, R. J. Blade, A. J. Boulton, J. Epsztajn, A. R. Katritzky, J. Lewis, P. Molina-Buendia, P.-L. Nie, and C. A. Ramsden, *ibid.*, 2691 (1976).

Triphenyl thioborate, $B(SC_6H_5)_3$. Mol. wt. 338.31, m.p. 143–144°, sensitive to moisture.

Preparation.[1] Thiophenol is heated with boron sulfide (B_2S_3, Alpha) in refluxing benzene for 48 hours; yield high.

1,3-Bis(phenylthio)alkenes. α,β-Unsaturated aldehydes and ketones react with triphenyl thioborate at 50° in petroleum ether to form 1,3-bis(phenylthio)-alkenes in 40–80% yield. An example is formulated for conversion of cyclohexenone into (1). This product can be converted into the lithio derivative (a), which is alkylated efficiently to give (2) or which adds to chalcone to give (4). These products can be hydrolyzed to enones, (3) and (5), by mercuric chloride or by cuprous triflate.

These reactions can be applied to acylic enones as shown in equation (I). This example shows that when the anion is unsymmetrical, the less hindered negatively charged carbon atom is alkylated more readily.[2]

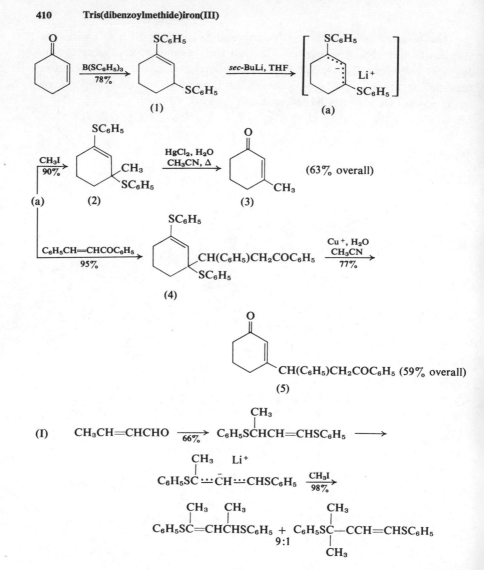

(63% overall)

(59% overall)

¹ J. Brault and J. M. LaLancette, *Canad. J. Chem.*, **42**, 2903 (1964).
² T. Cohen, D. A. Bennett, and A. J. Mura, Jr., *J. Org.*, **41**, 2506 (1976).

Tris(dibenzoylmethide)iron(III), Fe(DBM)$_3$, Fe[CH(COC$_6$H$_5$)$_2$]$_3$. Mol. wt. 725.61, red needles, m.p. 240°.

Preparation.¹ Ferric chloride is added to dibenzoylmethane dissolved in ethanol. A red solid is formed and is then completely precipitated with dilute ammonia. The product is purified by crystallization from benzene–hexane; yield 70%.

Cross coupling of alkenyl halides and Grignard reagents.[2] This iron(III) complex is more effective than ferric chloride from the standpoint of both rate and deactivation of the catalyst. The active catalyst is actually a reduced iron(I) species.

[1] S. M. Neuman and J. K. Kochi, *J. Org.*, **40**, 599 (1975).
[2] R. S. Smith and J. K. Kochi, *ibid.*, **41**, 502 (1976).

Tris(dimethylamino)methane, 5, 653.

α-Alkylidene-γ-butyrolactones. A recent synthesis of these substituted lactones (4) from γ-butyrolactones (1) involves α-*n*-butylthiomethylene-γ-butyrolactones (3). These intermediates can be prepared in high yield by reaction of (1) with

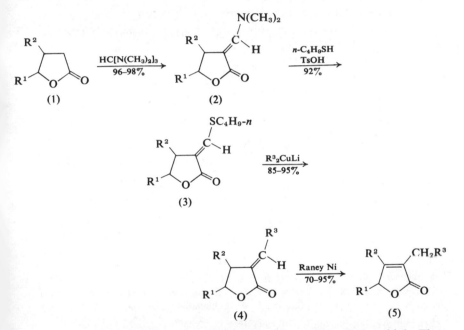

tris(dimethylamino)methane followed by reaction with *n*-butanethiol. These substances react with lithium dialkyl cuprates to form (4) in satisfactory yields. The products (4) can be isomerized to $\Delta^{\alpha,\beta}$-butenolides (5) with deactivated Raney nickel.[1]

[1] S. F. Martin and D. R. Moore, *Tetrahedron Letters*, 4459 (1976).

Tris(dimethylphenylphosphine)norbornadienerhodium(I) hexafluorophosphate, $\{Rh(C_7H_8)[(CH_3)_2C_6H_5P]_3\}^+PF_6^-$. Mol. wt. 754.46, yellow, air-stable. Supplier: Strem.

Preparation.[1]

Hydrogenation of alkynes.[2] This cationic complex of rhodium has been found to be superior to other catalysts, including the Lindlar catalyst, for hydrogenation of alkynes to *cis*-olefins. Thus $C_6H_5C\equiv CCOOC_2H_5$ is hydrogenated in acetone with this catalyst to ethyl *cis*-cinnamate in essentially quantitative yield with no trace of the *trans*-isomer or of the completely reduced acid. In this particular case, use of Lindlar catalyst results primarily in complete reduction.

[1] R. R. Schrock and J. A. Osborn, *Am. Soc.*, **98**, 2134 (1976).
[2] *Idem, ibid.*, **98**, 2134 (1976).

Tris(methylthio)methyllithium, $(CH_3S)_3CLi$, (1) Mol. wt. 160.25. This anion was first reported by Seebach,[1] who prepared it by reaction of *n*-butyllithium with tris(methylthio)methane.

Conjugate addition-acylation. Damon and Schlessinger[2] have reported a stereospecific synthesis of ($\pm$)-protolichesterinic acid (5) from the butenolide (1).[3]

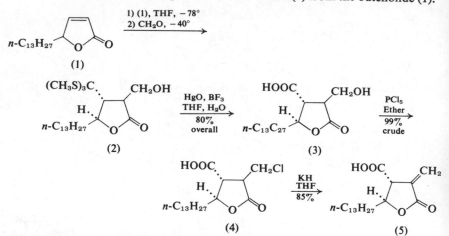

[1] D. Seebach, *Angew. Chem., Int. Ed.*, **6**, 442 (1967).
[2] R. E. Damon and R. H. Schlessinger, *Tetrahedron Letters*, 1561 (1976).
[3] J. L. Herrmann, M. H. Berger, and R. H. Schlessinger, *Am. Soc.*, **95**, 7923 (1973).

Tris(trimethylsilyl) phosphite, $P[OSi(CH_3)_3]_3$.

The phosphite is prepared[1] by the reaction of chlorotrimethylsilane with phosphorous acid in the presence of excess triethylamine.

Lipid phosphonates. The phosphite has been used as an Arbusov reagent for the preparation of lipid phosphonates (equation I).[2]

$$(I) \qquad RX + P[OSi(CH_3)_3]_3 \longrightarrow R\overset{\displaystyle O}{\overset{\|}{P}}[OSi(CH_3)_3]_2 \xrightarrow[\sim 85\%]{H_2O} R\overset{\displaystyle O}{\overset{\|}{P}}(OH)_2$$

The related reagent bis(trimethylsilyl) trimethylsilyloxymethylphosphonite $(CH_3)_3SiOCH_2P[OSi(CH_3)_3]_2$, has been used to prepare substances of the structure (1).[3]

(1)

[1] N. F. Orlov, B. L. Kaufman, L. Sukhi, L. N. Slesar, and E. V. Sudakova, *C.A.* **72**, 21738y (1970).

[2] A. F. Rosenthal, L. A. Vargas, Y. A. Isaacson, and R. Bittman, *Tetrahedron Letters*, 977 (1975).

[3] A. F. Rosenthal, A. Gringauz, and Luis A. Vargas, *J.C.S. Chem. Comm.*, 384 (1976).

Tris(triphenylsilyl)vanadate, 6, 655–656.

Rearrangement of α-acetylenic alcohols. The key step in a stereospecific synthesis of vitamin A from 2,2,6-trimethylcyclohexanone is the rearrangement of (1) to (2).[1] This reaction was carried out with the vanadate in the presence of additional triphenylsilanol. The yield of crude (2) was 96%; purification was accompanied by considerable decomposition, but still the yield of pure (2) was 40%. This vanadate-catalyzed rearrangement appears to be general.

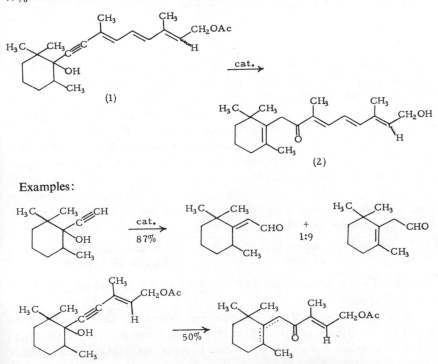

Examples:

The paper includes a study of the activity of related vanadates. Nitro-substituted vanadates are very reactive but somewhat unstable.

This reaction actually was first developed by Pauling *et al.*,[2] also at Hoffmann-La Roche. This publication includes preparation of numerous trialkyl(aryl)-silanols and the corresponding vanadates.

[1] G. L. Olson, H.-C. Cheung, K. D. Morgan, R. Borer, and G. Saucy, *Helv.*, **59**, 567 (1976).
[2] H. Pauling, D. A. Andrews, and N. C. Hindley, *ibid.*, **59**, 1233 (1976).

2-(*p*-Tritylphenyl)thioethanol (TPTE), $(C_6H_5)_3C$—⟨⟩—SCH_2CH_2OH (1).   Mol. wt. 396.53, m.p. 167–168°
Preparation:

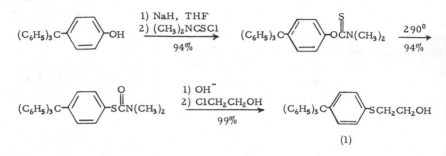

(1)

Protection of 5'-phosphate group in synthesis of dinucleotides. The 5'-phosphate group of mononucleotides is converted into 5'-esters by reaction with (1) in the presence of triisopropylbenzenesulfonyl chloride (TPS; **1**, 1228–1229; **3**, 308) or DCC. The esters are then condensed with a 3'-O-acetylated mononucleotide in pyridine with TPS to form TPTE-dinucleotides in 50–60% yield. The value of (1) is that it confers lipophilic properties on the protected dinucleotides so that they can be extracted into ethyl acetate–butanol or methylene chloride–butanol. The protecting group is removed by oxidation to the corresponding sulfone (NCS) and β-elimination. 2-(*p*-Tritylphenyl)sulfonylethanol (TPSE) can be used in place of (1); in this event, the oxidation step is not necessary. A number of tri- and tetranucleotides were also prepared by this new method.[1]

[1] K. L. Agarwal, Y. A. Berlin, H.-J. Fritz, M. J. Gait, D. G. Kleid, R. G. Lees, K. E. Norris, B. Ramamoorthy, and H. G. Khorana, *Am. Soc.*, **98**, 1065 (1976).

Trityl tetrafluoroborate, 1, 1256–1258; **2,** 454; **4,** 565–567; **6,** 657. Suppliers: Aldrich, Cationics.

Oxidation of alcohols. Trimethylsilyl ethers of secondary alcohols are oxidized in CH_2Cl_2 to ketones in 95–100% yield by trityl tetrafluoroborate. Oxidation of ethers of primary alcohols is too slow to be useful. Consequently selective oxidation of ethers of secondary alcohols is possible. Oxidation of bissilyl ethers is not clean, but use of bistrityl ethers or bis-*t*-butyl ethers is a useful method. The latter ethers are useful for selective oxidation of some 1,3-diols, because only the primary alcohol groups are converted into ethers by isobutylene under standard conditions.[2]

Examples:

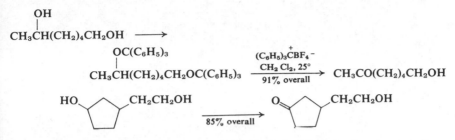

OH
|
CH₃CH(CH₂)₄CH₂OH ⟶

OC(C₆H₅)₃
|
CH₃CH(CH₂)₄CH₂OC(C₆H₅)₃ $\xrightarrow[\text{91\% overall}]{\underset{\text{CH}_2\text{Cl}_2,\ 25°}{(C_6H_5)_3\overset{+}{C}BF_4^-}}$ CH₃CO(CH₂)₄CH₂OH

HO⟋CH₂CH₂OH $\xrightarrow{\text{85\% overall}}$ O⟋CH₂CH₂OH

This method has the limitation that straight-chain 1,2-diols give poor results.

[1] M. E. Jung, *J. Org.*, **41**, 1479 (1976).
[2] M. E. Jung and L. M. Speltz, *Am. Soc.*, **98**, 7882 (1976).

Tungsten hexachloride–Tetramethyltin, WCl₆–Sn(CH₃)₄. (1).

Terpenoids. This catalyst for olefin metathesis[1] has been used for a synthesis of terpenoids with 1-methylcyclobutene (2) as the isoprene synthon. For example, metathesis of (2) and 3-hexene (3) with this catalyst in C₆H₅Cl gives a mixture of the (E)- and (Z)-isomers of (4) in 20–30% yield. A more interesting reaction is

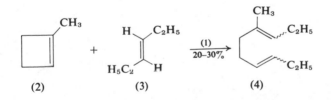

(2)   (3)   (4)

that of geraniol (5) with (2), which leads to farnesol acetate (6, E and Z), but in only 1–2% yields.[2]

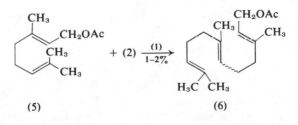

(5)   (6)

[1] P. B. vanDam, M. C. Mittelmeijer, and C. Boelhouwer, *J.C.S. Chem. Comm.*, 1221 (1972).
[2] S. R. Wilson and D. E. Schalk, *J. Org.*, **41**, 3928 (1976).

Tungsten hexafluoride, WF_6. Mol. wt. 297.84, m.p. 2.5°, b.p. 17.1°.

Oxidative cleavage of hydrazones. This salt converts dimethylhydrazones or tosylhydrazones into the carbonyl compound under mild conditions ($CHCl_3$ or Freon 113, 75–95% yield).[1]

[1] G. A. Olah and J. Welch, *Synthesis*, 809 (1976).

U

Uranium hexafluoride, UF_6. Mol. wt. 352.03. Supplier: ROC/RIC. Available from U.S. Energy R and D Administration. Soluble in Freons, CH_2Cl_2, and $CHCl_3$; can be handled in glassware.

Oxidation.[1] This reagent is useful for oxidative cleavage of ethers, as shown in the examples.

Examples:

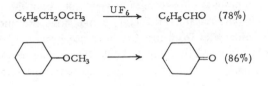

$$C_6H_5CH_2OCH_3 \xrightarrow{UF_6} C_6H_5CHO \quad (78\%)$$

$$C_6H_5CH_2OC_6H_{13}\text{-}\underline{n} \xrightarrow{44\%} \underline{n}\text{-}C_6H_{13}OH$$

The reagent cleaves hydrazones (50–95% yield). Aldehydes are converted into acyl fluorides:

$$RCHO \xrightarrow{UF_6} RCOF \ (30\text{–}40\%)$$

[1] G. A. Olah, J. Welch, and T.-L. Ho, *Am. Soc.*, **98**, 6717 (1976).

Urushibara catalysts, 4, 571; **5,** 743; **6,** 659.

Selective hydrogenations. Japanese chemists have prepared (3) by hydrogenation with Urushibara nickel A catalyst of either (1) or (2).[1]

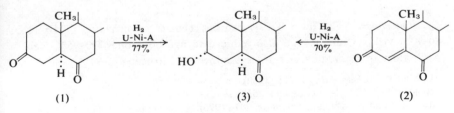

[1] M. Ishige, M. Shiota, and F. Suzuki, *Canad. J. Chem.*, **54**, 2581 (1976).

V

Vanadium(II) chloride, VCl_2. Mol. wt. 121.85, green, deliquescent.

Preparation.[1] Ammonium metavanadate, NH_4VO_3, is dissolved in aqueous NaOH and then added to hydrochloric acid. The solution is then stirred with heavily amalgamated zinc. The solution of VCl_2 thus obtained is stable on storage for several months.

Ketones from nitro compounds.[2] Vanadium(II) chloride can be used to convert secondary nitro compounds into ketones in yields of about 60–70%. However, an acidic medium is required for satisfactory conversions and under such conditions the yields of aldehydes from primary nitro compounds is rather low. The procedure of McMurry using titanium(III) chloride is more useful.

Hydrodehalogenation. α-Halo ketones can be reduced to ketones by VCl_2 in aqueous THF.[3]

Examples:

$$C_6H_5COCH_2Br \xrightarrow[\substack{THF \\ 92\%}]{2VCl_2, H_2O} C_6H_5COCH_3$$

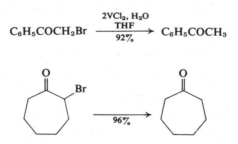

Reductions. VCl_2 (also $TiCl_3$) reduces benzils to benzoins (THF–H_2O, 80–90% yield). Quinones are also reduced to hydroquinones (90–95% yield).[4] Aryl azides are cleaved by VCl_2 to arylamines and nitrogen (70–95% yield).[5]

[1] L. Meites, *J. Chem. Ed.*, **27**, 458 (1950).
[2] R. Kirchhoff, *Tetrahedron Letters*, 2533 (1976).
[3] T.-L. Ho and G. A. Olah, *Synthesis*, 807 (1976).
[4] *Idem, ibid.*, 815 (1976).
[5] T.-L. Ho, M. Henninger, and G. A. Olah, *ibid.*, 815 (1976).

Vanadium oxytrifluoride, 5, 745–746; **6,** 660.

Oxidative coupling. Oxidation of the derivative (1) of norprotosinomine with VOF_3 under controlled conditions gives the dienone (2), which was converted

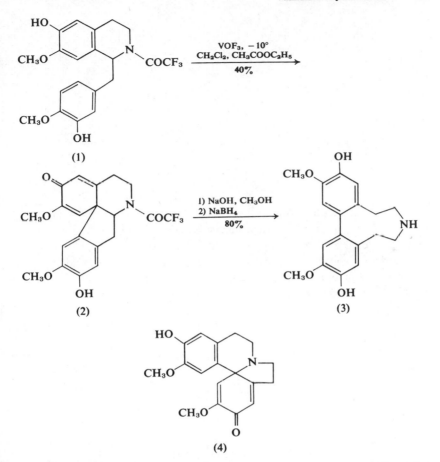

(1)

(2)

(3)

(4)

as shown into the dibenzazonine (3).[1] This substance was shown to be a key intermediate in the biosynthesis of erysodienone (4).[2]

Oxidative coupling of phenethylisoquinolines. Treatment of (±)-N-trifluoro-acetylhomonorlaudanosine (1) with VOF_3 and trifluoroacetic acid in CH_2Cl_2

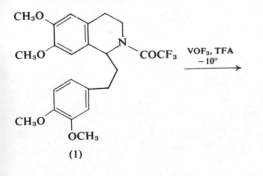

(1)

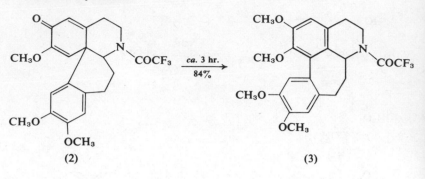

(2)   (3)

at $-10°$ for 10 minutes leads to the dienone (2) in low yield (5%) and (3), also in low yield, together with other products. If the reaction is continued for several hours, the homoaporphine (3) is obtained in high yield.[3]

    Oxidative coupling of benzyltetrahydroisoquinolines. Application of this coupling reaction to benzyltetrahydroisoquinolines such as (1) leads to aporphines (2) also in remarkably high yields.[4]

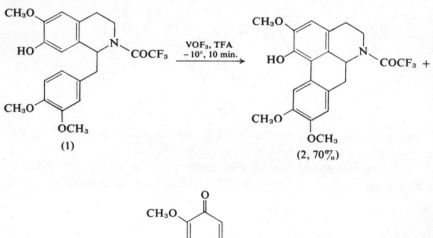

(1)   (2, 70%)

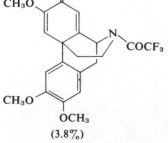

(3.8%)

Nonphenolic oxidative coupling. The final step in a synthesis of (±)-iso-stegnane, (2), a dibenzocyclooctadiene derivative, involved oxidative coupling of (1).[5]

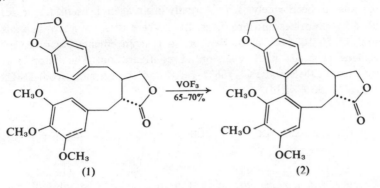

(1)   (2)

This reaction was a key step in the synthesis of the (±)-form of the dibenzo-cyclooctadiene lignan lactone steganacin (5).[6]

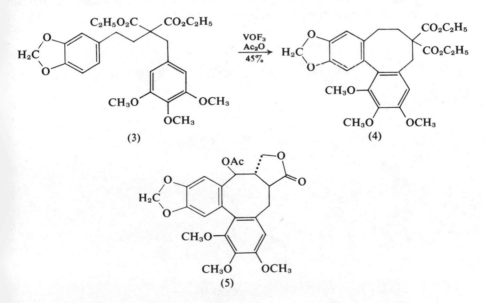

(3)   (4)

(5)

[1] S. M. Kupchan, C.-K. Kim, and J. T. Lynn, *J.C.S. Chem. Comm.*, 86 (1976).

[2] D. H. R. Barton, R. B. Boar, and D. A. Widdowson, *J. Chem. Soc. (C)*, 1213 (1970).

[3] S. M. Kupchan, O. P. Dhingra, C.-K. Kim, and V. Kameswaran, *J. Org.*, **41**, 4047 (1976).

[4] S. M. Kupchan, O. P. Dhingra, and C.-K. Kim, *ibid.*, **41**, 4049 (1976).

[5] R. E. Damon, R. H. Schlessinger, and J. F. Blount, *ibid.*, **41**, 3772 (1976).

[6] A. S. Kende and L. S. Liebeskind, *Am. Soc.*, **98**, 267 (1976).

Vilsmeier reagent, 1, 284–289; **2,** 154; **3,** 116; **4,** 186; **5,** 251; **6,** 220.

Alkyl halides. The Vilsmeier reagent N,N-dimethylchloroformiminium chloride has been reported to convert alcohols into alkyl chlorides, but this reaction has not been widely used.[1] Actually this reagent is useful for preparation of both alkyl chlorides and bromides. In the case of *sec-* or *tert*-alcohols temperatures of 50–100° are used. Few or no rearrangements are observed. The replacement proceeds with inversion of configuration. The chloride is best prepared from DMF and PCl_5 (88% yield) and the bromide from DMF, Br_2, and $P(C_6H_5)_3$ (90% yield).[2]

(I) $ROH + \left[(CH_3)_2N{=}CHX^+\right]X^- \longrightarrow \left[(CH_3)_2\overset{+}{N}{=}CHOR\right]X^- \longrightarrow RX + (CH_3)_2NCHC$

$X = Cl, Br$

Reaction with homophthalic acid.[3] If the reaction of homophthalic acid with $POCl_3$ and DMF is carried out at 0°, the major product is the isochromane-1,3-dione (2). If the reaction is conducted at 100°, the major product is the iso-quinolone (3).[4] Treatment of (2) with hydrochloric acid in methanol results in formation of the methyl ester of isocoumarin-4-carboxylic acid (4).

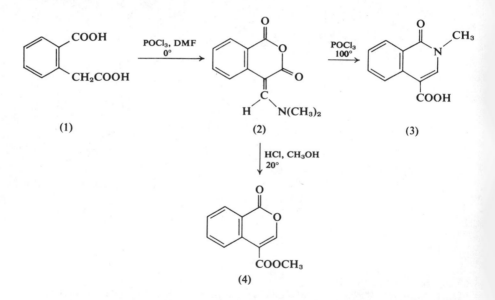

2-Acyl-6-aminofulvenes. Treatment of the anion of acetylcyclopentadiene (1) with the Vilsmeier reagent and sodium methoxide (to neutralize the system) gives 2-acetyl-6-dimethylaminofulvene (2) in 93% yield. The product is hydrolyzed by base to 2-acetyl-6-hydroxyfulvene (3).

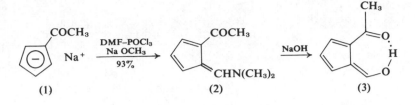

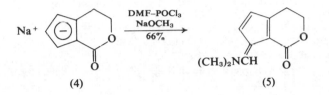
(1)   (2)   (3)

The reaction appears to be general; a further example is the synthesis of (5).[5]

(4)   (5)

Reaction with a purine nitrile. The Vilsmeier reagent reacts with (1) to give a hygroscopic gum considered to be a ketenimine (2), which reacts with water or

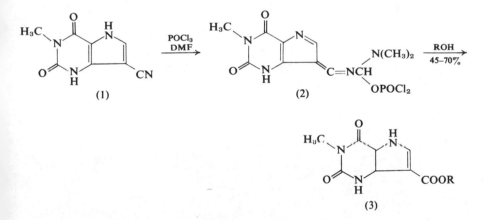
(1)   (2)

(3)

an alcohol to form the acid (3, R = H) or an ester. Reaction of (2) with an amine leads to an amide (55–65% yield).[6]

Chromones.[7] 3-Substituted chromones, including isoflavones (3-arylchromones), can be prepared in fair to excellent yields by treatment of 2-hydroxyphenyl alkyl (arylalkyl) ketones (1)[8] with BF_3 etherate (exothermic reaction) in DMF. Methanesulfonyl chloride in DMF (Vilsmeier reagent) is then added at 52° and the reaction is warmed on a steam bath (90 minutes). Chromones (2) are obtained in 65–96% yield. The function of BF_3 etherate is to deactivate the hydroxylated aromatic ring by complex formation and thus prevent ring formylation.

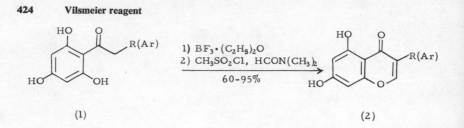

(1)   (2)

[1] H. Eilingsfeld, M. Seefelder, and H. Weidinger, *Ber.*, **96**, 2671 (1963).
[2] D. R. Hepburn and H. R. Hudson, *J.C.S. Perkin I*, 754 (1976).
[3] V. H. Belgaonkar, and R. N. Usgaonkar, *Chem. Ind.*, 954 (1976).
[4] *Idem, Tetrahedron Letters*, 3849 (1975).
[5] T. Fujisawa and K. Sakai, *Tetrahedron Letters*, 3331 (1976).
[6] Y. Okamoto and T. Ueda, *ibid.*, 2317 (1976).
[7] R. J. Bass, *J.C.S. Chem. Comm.*, 78 (1976).
[8] Available by the Hoesch reaction, P. E. Spoerri and A. S. DuBois, *Org. React.*, **5**, 387 (1949).

X

Xenon hexafluoride, XeF_6.

Intercalate in graphite. Israeli chemists[1] have prepared an intercalate of XeF_6 in graphite of the nominal composition $C_{19.1}XeF_6$. The material is stable at room temperature for at least a week and is not shock-sensitive; however, caution in handling is advisable.

The intercalate can be used for selective fluorination of polycyclic arenes. Yields are about the same as those obtained with XeF_2 (6, 669–670), and the same pattern of substitution is obtained. It is recommended for fluorination at hindered positions.[2]

[1] H. Selig, M. Rabinovitz, I. Agranat, C.-H. Lin, and L. Ebert, *Am. Soc.*, **98**, 1601 (1976).
[2] I. Agranat, M. Rabinovitz, H. Selig, and C.-H. Lin, *Synthesis*, 267 (1977).

Z

Zinc, 1, 1276–1284; 2, 459–462; 3, 334–337; 4, 574–577; 5, 753–756; 6, 672–675.

Clemmensen reduction. Reduction of 10 α,β-unsaturated ketones with amalgamated zinc, hydrogen chloride, and acetic anhydride leads to cyclopropyl acetates, the yields of which depend on the temperature. Cyclopropanols have been suggested as intermediates in such reductions; the function of acetic anhydride is to trap these alcohols.[1]

Examples:

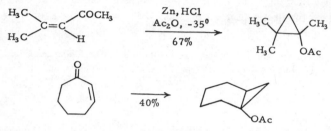

1,2,3,5-Hexatetraene. This new C_6H_6 hydrocarbon (5) has been prepared[2] by reduction of the mixture of dichlorides, (2)–(4), obtained from (1) as shown, with zinc in diethylene glycol dibutyl ether[3] at 70°. The product was identified as the adduct (6) with tetracyanoethylene (TCNE).

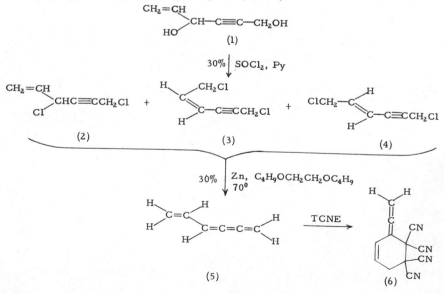

426

Aldehydes from amides. Aromatic tertiary amides (1) can be converted into aryl aldehydes (3) by formation of the Vilsmeier complex (2) followed by reduction with zinc dust (equation I).[4]

(I)

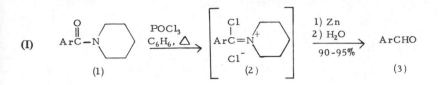

(1)   (2)   (3)

Regiospecific aldol reaction. 2,2,2-Trichloroethyl esters of β-keto acids react with aldehydes in the presence of zinc to form α-substituted β-hydroxy ketones (equation I). The first step probably involves reduction of the ester to form a zinc salt with elimination of dichloroethylene.[5]

(II)
$$R^1\overset{O}{\overset{\|}{C}}\underset{\underset{R^2}{|}}{C}H\overset{O}{\overset{\|}{C}}OCH_2CCl_3 + R^3CHO \xrightarrow[50-90\%]{\underset{25^0}{Zn,\ DMSO}} R^1\overset{O}{\overset{\|}{C}}\underset{\underset{R^2}{|}}{C}H\overset{OH}{\overset{|}{C}}HR^3$$

Dehalogenation (1, 1278–1279; 3, 335; 5, 755). Benzocyclobutadiene (2) has been observed directly by IR spectroscopy by dehalogenation of *cis*-1,2-diiodobenzocyclobutene (2) with zinc under argon in a special apparatus. It is formed as a matrix of the dimer.[6]

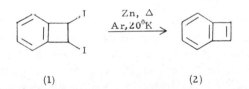

(1)   (2)

14β-Steroids. The unnatural 14β-steroid (3) can be obtained as the major product on reduction of either (1) or (2) with a large excess of zinc dust in methanolic sulfuric acid. Reduction of (1) with zinc dust in acetic acid does not give (3).[7]

Cleavage of 2,2,2-trichloroethyl esters. These esters are usually cleaved by zinc in acetic acid (3, 295–296; 4, 521–522). Just and Grozinger[8] have found that this reaction proceeds smoothly with zinc in aqueous THF buffered at pH 4.2–7.2. Yields of acids are in the range of 75–95%, and the reaction is complete in 10 minutes at 20°. The same conditions can be used to regenerate amines from 2,2,2-trichloroethyl carbamates and to regenerate phenols from 2,2,2-trichloroethyl carbonates.

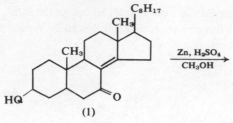

(1)

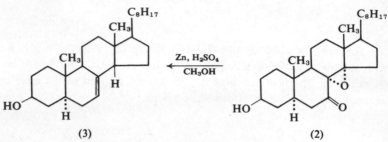

(3)   (2)

[1] C. W. Jefford and A. F. Boschung, *Helv.*, **59**, 962 (1976).
[2] H. Maurer and H. Hopf, *Angew. Chem., Int. Ed.*, **15**, 628 (1976).
[3] E. Kloster-Jensen and J. Wirz, *Helv.*, **58**, 162 (1975).
[4] Atta-Ur-Rahman and A. Basha, *J.C.S. Chem. Comm.*, 594 (1976).
[5] T. Mukaiyama, T. Sato, S. Suzuki, T. Inoue, and H. Nakamura, *Chem. Letters*, 95 (1976).
[6] O. L. Chapman, C. C. Chang, and N. R. Rosenquist, *Am. Soc.*, **98**, 262 (1976).
[7] M. Anastasia, A. Fiecchi, and A. Scala, *J.C.S. Perkin I*, 378 (1976).
[8] G. Just and K. Grozinger, *Synthesis*, 457 (1976).

Zinc–Copper couple, 1, 1292–1293; **5,** 758–760.

Intramolecular Reformatsky-type reactions; α-methylene-γ-lactones.[1] Reaction of the (Z)-bromoaldehyde (1) in a dilute solution of THF with zinc dust at 65° or with zinc–copper couple[2] at 25° leads to the *cis*-fused α-methylene-γ-lactone (2) in 60–62% yields. The reaction involves an intramolecular Reformatsky-type reaction followed by spontaneous lactonization. The same product (2) is also obtained from the (E)-isomer of (1) in 46% yield together with the *trans*-fused isomer of (2). The reaction of (1) with bis(1,5-cyclooctadiene)nickel (4, 33–35; **5,** 34–35) also yields (2) in 52% yield. It is obtained in comparable yield

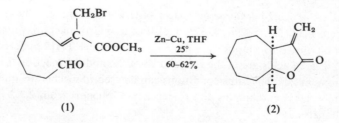

(1)   (2)

from (E)-(1). Thus both metal reagents favor formation of a *cis*-ring fusion in this perhydroazulene derivative.

Cyclic 1,4-diketones from α-bromo ketones (*cf.* **5**, 758). The intramolecular coupling of α-bromo ketones has been extended to synthesis of cyclic 1,4-diketones (equations I and II).[3]

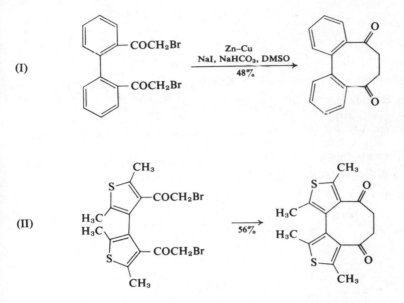

(I)

Zn–Cu
NaI, NaHCO₃, DMSO
48%

(II)

56%

[1] M. F. Semmelhack and E. S. C. Wu, *Am. Soc.*, **98**, 3384 (1976).

[2] Preparation: R. M. Blankenship, K. A. Burdett, and J. S. Swenton, *J. Org.*, **39**, 2300 (1974).

[3] E. Ghera, Y. Gaoni, and S. Shoua, *Am. Soc.*, **98**, 3627 (1976).

Zinc–Chlorotrimethylsilane.

Trimethylsilyl enol ethers. It is relatively easy to obtain trimethylsilyl enol ethers formed under kinetic control from unsymmetrical ketones in high yields by use of LDA as base (**3**, 310–311). However, the more highly substituted enol ethers formed by thermodynamic control have not been available as readily. A new method that leads to the more highly substituted ether uncontaminated with the less highly substituted ether is reduction of α-halo-α-substituted ketones with activated zinc (**1**, 1276) in ether–TMEDA followed by addition of chlorotrimethylsilane in ether. The starting materials can be prepared by chlorination of the α-alkyl ketone with sulfuryl chloride.[1]

Examples:

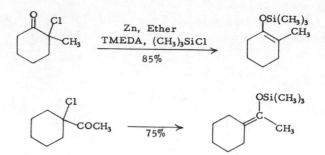

[1] G. M. Rubottom, R. C. Mott, and D. S. Krueger, *Syn. Comm.*, **7**, 327 (1977).

Zinc–Zinc chloride, 6, 675.

Reduction of $\diagup C=C \diagup$. Zinc, zinc chloride, and an alcohol (usually C_2H_5OH) reduce double bonds placed between two carbonyl or cyano groups, as well as the double bonds of indenones and cyclopentadienones.[1]

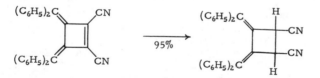

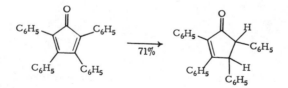

[1] F. Toda and K. Iida, *Chem. Letters*, 695 (1976).

INDEX OF REAGENTS
ACCORDING TO TYPES

ACETOXYLATION: Manganese(III) acetate. Palladium(II) acetate.

ACETOXYMERCURATION: Mercuric acetate.

ACYLOIN CONDENSATION: Sodium naphalenide.

ALDOL CONDENSATION: Di-*n*-butylboryl trifluoromethanesulfonate. Lithium bis(trimethylsilyl)amide. Lithium 1,1-bis(trimethylsilyl)-3-methylbutoxide. Trichloroethanol.

ALLYLIC AMINATION: Selenium–Chloramide-T.

ALLYLIC HYDROPEROXIDATION: Oxygen, singlet.

ALLYLIC OXIDATION: *t*-Butyl perbenzoate.

AMINOALKENYLATION: 1-Chloro-N,N,2-trimethylpropenylamine. Lithio-N,N-dimethylthiopivalamide.

ANNELATION: Chromium hexacarbonyl. 1,3-Dibromo-2-pentene. 3,3-Ethylene-dioxybutylmagnesium bromide. 1-Lithiocyclopropyl phenyl sulfide. 1-Lithio-2-vinylcyclopropane. 2-Methoxy-3-phenylthiobuta-1,3-diene.

ARYLATION: Phenylcopper.

ASYMMETRIC ALKYLATION: (S)-1-Amino-2-methoxymethylpyrrolidine. (Benzylmethoxymethyl)methylamine. (–)-N-Methylephedrine–Lithium tetra-*n*-butylaluminate.

ASYMETRIC CYCLIZATION: Proline.

ASYMETRIC REDUCTION: (–)-N-Dodecyl-N-methylephedrinium bromide. Lithium aluminum hydride–(–)-α-Phenethylamine. L-N-Methyl-N-dodecylephedrinium bromide.

ASYMETRIC SYNTHESIS: (1S,2S,5S)-2-Hydroxypin-3-one. (4S,5S)-(–)-4-Methoxymethyl-2-methyl-5-phenyl-2-oxazoline. (R)-4-Methylcyclohexylidenemethylcopper.

BENZOANNULATION: Dimethyl acetylenedicarboxylate.

BENZOIN CONDENSATION: 3-Benzyl-5-(2-hydroxyethyl)-4-methyl-1,3-thiazolium chloride. Potassium cyanide.

BENZOYLATION: Benzoyltetrazole.

BISANNELATION, Methyl α-bromocrotonate.

BROMINATION: Bromine. Bromotrichloromethane. Magnesium bromide etherate. 2,4,4,6-Tetrabromocyclohexa-2,5-dienone.

BROMINATIVE CYCLIZATION: 2,4,4,6-Tetrabromocyclohexa-2,5-dienone.

BROMOCYCLIZATION: Bromine–Silver tetrafluoroborate.

CANNIZZARO REACTION: Crown ethers.

CHLORINATION: Antimony(V) chloride. Nitrogen dioxide. Sulfuryl chloride.

CLAISEN CONDENSATION: Lithium bis(trimethylsilyl)amide.

CLAISEN REARRANGEMENT: Acetic anhydride-Sodium acetate. Lithium N-isopropylcyclohexylamide. Titanium(IV) chloride.

CLEAVAGE OF:

ALKENES: Ethoxycarbonylformonitril oxide.

ALKYL ARYL ETHERS: Sodium nitrite–Hexamethylphosphoric triamide.

ALLYL ETHERS: Palladium catalysts.

ARYL ESTERS: Cesium carbonate.

CYCLIC ETHERS: Cuprous bromide.

CYCLOPROPANES: Dimethylcopperlithium. Pyridinium chloride.

CYCLOPROPYL KETONES: Hydrogen chloride.
EPOXIDES: Diethylcarbo-t-butoxymethylalane. Dilithioacetate. Hydriodic acid.
ESTERS: Lithium iodide. Potassium superoxide.
ETHERS: Hydrogen iodide.
KETALS: Hydrogen iodide.
METHYL ETHERS: Boron trifluoride etherate–Ethanediol.
OXIMES: Nitrosonium tetrafluoroborate.
TOSYLHYDRAZONES: Boron trifluoride etherate. Sodium peroxide.
2,2,2-TRICHLOROETHYL ESTERS: Zinc.
α,β-UNSATURATED p-TOSYLHYDRAZONES: Sodium borohydride.
CLEMMENSEN REDUCTION: Zinc, Zinc amalgam.
CONJUGATE ADDITION: Diethyl acetylmethylmalonate. Dilithium trimethyl cuprate.
    N,N-Dimethylbenzeneselenenamide. Dimethyl sulfoxide. Lithium di-(2-vinylcyclo-
    propyl)cuprate. Lithium phenylthio(cyclopropyl)cuprate. Lithium phenylthio[(α-
    diethoxymethyl)vinyl]cuprate. 2-(2-Methoxy)-allylidene-1,3-dithiane. 2-Nitropropene.
    Tetra-n-butylammonium fluoride. Titanium(IV) chloride. Tris(methylthio)methyl-
    lithium.
CONJUGATE REDUCTION: Potassium(sodium) decacarbonylhydride chromate.
CROSS-COUPLING: Dichlorobis(triphenylphosphine)palladium(II). Tris(dibenzoylmeth-
    ide)iron(III).
CYANOBORATION: Sodium cyanide.
CYCLIZATION: n-Butyllithium. Diisobutylaluminum hydride. Fluoroantimonic acid. Phos-
    phoric acid–Boron trifluoride complex. Thallium(III) perchlorate. Trifluoroacetic acid.
CYCLOADDITION: Aluminum chloride. Ceric ammonium nitrate. N,N-Diethyl-1-propy-
    nylamine. Dimethyl acetylenedicarboxylate. N,N-Dimethyl-O-ethylphenylpropiol-
    amidium tetrafluoroborate. Ethylaluminum dichloride. Nitrobenzene. Oxygen, singlet.
    Sulfur dioxide. Trichloroacetyl isocyanate. Trimethylsilylbromoketene.
CYCLODEHYDRATION: Boron trifluoride etherate. Polyphosphoric acid.
CYCLOPENTANE ANNELATION: Lithium phenylthio(cyclopropyl)cuprate.
CYCLOPROPANATION: Copper. Copper(II) triflate. Rhodium(II) carboxylates.

DEACYLATION: Hydrazine.
DEAMINATION: 2,4,6-Triphenylpyrylium perchlorate.
DEBENZYLATION: Orthophosphoric acid.
DEBROMINATION: Sodium hydrosulfite.
DECARBOALKOYLATION: Alumina. 1,4-Diazabicyclo[2.2.2]octane. Lithium bromide–
    Hexamethylphosphoric triamide.
DECARBONYLATION: Chlorotris(methyldiphenylphosphine)rhodium(I).
DECARBOXYLATION: Mercuric oxide.
DECHLORINATION: Lithium di-t-butylbiphenyl.
DECHLOROSILYLATION: Tetramethylammonium fluoride.
DEHALOGENATION: Titanium(III) chloride–Lithium aluminum hydride. Zinc.
DEHYDRATION: Alumina. Phosphorus pentoxide. Rhodium(III) chloride–Triphenyl-
    phosphine. Triphenylphosphine–Carbon tetrachloride.
DEHYDROGENATION: Palladium black. Potassium–Graphite. Tetrachloro-o-benzo-
    quinone.
DEHYDROHALOGENATION: Benzyltrimethylammonium hydroxide. Lithium carbonate–
    Lithium halide. Potassium t-butoxide.
N-DEMETHYLATION: Methyl chloroformate. Trichloroethyl chloroformate.
O-DEMETHYLATION: Lithium 2-methylpropane-2-thiolate.

DENITROSATION: Raney nickel.
DEOXYGENATION, AMINE OXIDES: Chromous chloride.
  EPOXIDES: 3-Methyl-2-selenoxobenzothiazole. Trimethylsilylpotassium.
  KETONES: Hydriodic acid–Red phosphorus. Phosphorus (V) sulfide. Sodium cyano-
    borohydride.
  TOSYLATES: Lithium triethylborohydride.
  TOSYLHYDRAZONES: Catecholborane.
DESULFONYLATION: Sodium amalgam.
DESULFURIZATION, EPISULFIDES: 3-methyl-2-selenoxobenzothiazole.
DETHIOACETALIZATION: Glyoxylic acid. Sulfuryl chloride. Thallium(III) nitrate.
DIAZO TRANSFER: Benzyltriethylammonium chloride.
DIELS-ALDER CATALYSTS: Stannic chloride.
DIELS-ALDER REACTIONS: 1,3-Bis(trimethylsilyloxy)-1,3-butadiene. *trans, trans*-1,4-
    Diacetoxybutadiene. 1,2-Dicyanocyclobutene. Ethyl *trans*-1,3-butadiene-1-carbonate.
    Furane. 2-(2-Methoxy)-allylidene-1,3-dithiane. 2-Methoxy-3-phenylthiobuta-1,3-diene.
    1-Methoxy-3-trimethylsilyloxy-1,3-butadiene. Sulfur dioxide. 1-Trimethylsilyl-1,3-
    butadiene. 2-Trimethylsilyloxy-1,3-butadiene.

ENE REACTIONS: Aluminum chloride. *n*-Butyl N-(toluene-*p*-sulfonyl)-iminoacetate.
    Methyl cyanodithioformate. Oxygen, singlet. 4-Phenyl-1,2,4-triazoline-3,5-dione.
EPOXIDATION: Biacetyl. *t*-Butyl hydroperoxide. *t*-Butyl hydroperoxide–Vanadyl acetyl-
    acetate. *m*-Chloroperbenzoic acid. Iodine. Sodium hypochlorite.
ESCHENMOSER FRAGMENTATION: Phenylcopper.
ESTERIFICATION: 1,5-Diazabicyclo[4.3.0]nonene-5. 3,4-Dihydro-2H-pyrido[1,2-*a*]-
    pyrimidine-2-one. Silver cyanide. Triphenylphosphine–Diethyl azodicarboxylate.

FINKELSTEIN REACTION: Dimethyl sulfoxide.
FISHER INDOLE SYNTHESIS: Phenylhydrazine.
FLUORINATION: Fluoroxytrifluoromethane. Perchloryl fluoride. Polyhydrogen fluo-
    ride–Pyridine.
FRAGMENTATION: Hydrazine.
FRIES REARRANGEMENT: Boron trifluoride etherate.

O-GLUCURONIDATION: Stannic chloride.

HALOGENATION: Benzyltriethylammonium chloride. Trimethyl phosphate.
HOFMANN CARBYLAMINE REACTION: Benzyltriethylammonium chloride.
HYDROBORATION: 9-Borabicyclo[3.3.1]nonane.
HYDROFORMYLATION: Hydridocarbonyltris(triphenylphosphine)rhodium(I).
HYDROGENATION, CATALYSTS: Chlorohydridotris(triphenylphosphine)ruthenium(II).
    Dichlorotris(triphenylphosphine)ruthenium(II). Magnesium oxide. Raney nickel.
    Tris(dimethylphenylphosphino)norbornadienerhodium(I) hexafluorophosphate. Urushi-
    barta catalysts.
HYDROGENATION, TRANSFER: Dihydridotetrakis(triphenylphosphine)ruthenium(II).
    Rhodium(III) chloride hydrate.
HYDROXYLATION: Ozone. Ozone–Silica gel. Pertrifluoroacetic acid.
vic-HYDROXYLATION: Osmium tetroxide–*t*-Butyl hydroperoxide. Thallium acetate.
HYDROXYMETHYLATION: Tri-*n*-butyltincarbinol.

IODINATION: Thallium(I) acetate–Iodine.

IODOFLUORINATION: Methyliodine(III) difluoride.
ISOMERIZATION, ALKENES: Polyphosphoric acid. Sodium iodide–Dimethyl-formamide. Triphenylphosphine dihalides.
ALKYNES: Potassium 3-aminopropylamide.
ALKYNOLS: Potassium 3-aminopropylamide.

KNOEVENAGEL CONDENSATIONS: Titanium(IV) chloride.
KOENIGS-KNORR REACTION: Silver(I) oxide.

LACTONIZATION: Boron trifluoride etherate. 2-Chloro-1-methyl-pyridinium iodide. 2,2'-Dipyridyl disulfide–Triphenylphosphine. Phosphorus pentoxide–Methanesulfonic acid. Triphenylphosphine–Diethyl azodicarboxylate.

MANNICH REACTION: N,N-Dimethyl(methylene)ammonium salts. N,N,N'N'-Tetramethylenethanediamine.
METHYLENATION: Methylene chloride–Cesium fluoride.

NITRATION: Trimethyl phosphate.
NITROSATION: Nitrogen trioxide.

OPPENAUER OXIDATION: 2-(Diethylphosphono)propionitrile.
OXIDATION, CATALYSTS: Cuprous chloride.
OXIDATION, REAGENTS: Acetyl nitrate. Bis(tri-n-butyltin)oxide. Bromine–Hexamethylphosphoric triamide. t-Butyl perbenzoate. Ceric ammonium nitrate. N-chlorosuccinimide–Dimethyl sulfide. Chromic acid. Chromic anhydride. Chromic anhydride–Acetic anhydride. Chromic anhydride–Hexamethylphosphoric triamide. 2,3-Dichloro-5,6-dicyano-1,4-benzoquinone. Dimethyl sulfoxide. Dimethyl sulfoxide–Trifluoroacetic anhydride. Diphenylseleninic anhydride. Iodine tris(trifluoroacetate). Lead tetraacetate. N-Methylmorpholine–N–oxide. p-Nitrobenzenesulfonyl peroxide. Oxygen, singlet. Palladium(II) chloride. Peroxybenzimidic acid. Phenylseleninyl chloride. N-Phenyl-1,2,4-triazoline-3,5-dione. Potassium chromate. Potassium superoxide. Pyridinium chlorochromate. Salcomine. Silver carbonate–Celite. Sodium hypochlorite. Sulfuryl chloride–Silica gel. Thallium(III) acetate. Thallium(III) nitrate. Triphenyl phosphite ozonide. Trityl tetrafluoroborate. Uranium hexafluoride.
OXIDATIVE BISEDECARBOXYLATION: Dicarbonylbis(triphenylphosphine)nickel(O).
OXIDATIVE CHLORINATION: Hydrogen peroxide–Hydrochloric acid.
OXIDATIVE CLEAVAGE, ETHERS: Uranium hexafluoride.
  HYDRAZONE, OXIMES: Molybdenum(V) trichloride oxide. Tungsten hexafluoride.
OXIDATIVE COUPLING: Vanadium oxytrifluoride. Vanadium oxyfluoride–Trifluoroacetic acid.
OXIDATIVE CYCLIZATION: Nickel peroxide.
OXIDATIVE DECARBOXYLATION: Bis(triphenylphosphine)nickel dicarbonyl. Lead tetracetate. Oxygen, singlet.
OXIDATIVE DECYANATION: Oxygen.
OXIDATIVE DEMENTHYLATION: Silver(II) oxide.
OXIDATIVE DIMERIZATION: Silver oxide.
OXIDATIVE ESTERIFICATION: N-Bromosuccinimide.
OXIDATIVE SECO REARRANGEMENT: Lead tetracetate.
OXYAMINATION: Osmium tetroxide–Chloramine T.
OXYGENATION: Cuprous chloride. Oxygen. Oxygen, singlet. Ruthenium on alumina.

OXYMERCURATION: Mercuric nitrate.

PHASE-TRANSFER CATALYSTS: Adogen. Benzyldimethyldodecylammonium bromide. Benzyltriethylammonium chloride. N-Dodecylpentamethylphosphoramide. Hexadecyltributylphosphonium bromide. Methyltricaprylylammonium chloride. Tetra-n-butylammonium chloride.

PHENOLIC COUPLING: Potassium ferricyanide.

PHOSPHORYLATION: Phenyl N-phenyl phosphoroamidochloridate. 1-Phosphorylpyrazoles.

PINACOLIC COUPLING: Titanium(IV) chloride–Magnesium amalgam.

POLONOVSKI REACTION: Trichloroacetic anhydride. Trifluoroacetic anhydride.

PRINS REACTION: Formaldehyde.

PROTECTION OF: AMINO GROUPS: 2-Benzoyloxymethylbenzoyl chloride. Di-t-butyldicarbonate. Diphenylphosphinyl chloride. p-Phenylbenzoyl chloride. Phenyl-[2-p-phenylazophenyl)isopropyl]-carbonate. p-Toluenesulfonyl isocyanate.

CARBONYL COMPOUNDS: 2,2-Dibromo-1,3-propanediol.

CARBOXYL GROUPS: p-Chlorobenzyl alcohol. Dimethoxymethane.

HYDROXYL GROUPS: Chloromethyl methyl ether. Dimethoxyethane. Dimethyl sulfoxide–Acetic anhydride. β-Methoxyethoxymethyl chloride. Tetrahydrofurane–Sulfuryl chloride.

PHOSPHATES: 2-(p-Tritylphenyl)thiolethanol.

THIOLS: Diphenyl-4-pyridylmethanol.

TRYPTOPHANE: Crown ethers.

PUMMERER REARRANGEMENT: Acetic anhydride–Trifluoroacetic anhydride.

RACEMIZATION: Carbon, activated.

REARRANGEMENT: α,β-Epoxysilanes: Magnesium bromide etherate.

REDUCTION, REAGENTS: N-Benzyl-1,4-dihydronicotinamide. 9-Borabicyclo[3.3.1]-nonane. 9-Borabicyclo[3.3.1]nonaneate complexes. 2-Chloro-1,3,2-benzodioxaphosphole. Cuprous bromide–Lithium trimethoxyaluminum hydride. Ethylmagnesium bromide–Cuprous iodide. Ferric chloride–Sodium hydride. Lithium–Alkylamines. Lithium–Ammonia. Lithium aluminum hydride. Lithium 9,9-di-n-borabicyclo[3.3.1]-nonate. Lithium trisiamylborohydride. Potassium decacarbonylhydride chromate. Potassium tri-sec-butylborohydride. Sodium acetanilidoborohydride. Sodium acetoxyborohydride. Sodium bis[2-methoxyethoxy]aluminum hydride. Sodium borohydride. Sodium cyanoborohydride. Sodium iodide–Zinc. Sodium octacarbonylhydridoiron. Sodium trifluoroacetoxyborohydride. Tetra-n-butylammonium borohydride. Titanium(IV) chloride–Lithium aluminum hydride. Tri-n-butyltin hydride. Triethylsilane–Boron trifluoride. Vanadium(II) chloride. Zinc–Acetic acid. Zinc–Zinc chloride.

REDUCTIVE CLEAVAGE, OXIMES: Molybdenum(V) trichloride oxide.

REDUCTIVE COUPLING, KETONES: Titanium(0). Titanium(III) chloride–Lithium aluminum hydride.

REDUCTIVE DEMETHOXYLATION: Titanium(IV) chloride–Lithium aluminum hydride.

REDUCTIVE DENITROSATION: Raney nickel.

REDUCTIVE DIMERIZATION: Titanium(III) chloride.

REFORMATSKY REACTION: Ethyl trimethylsilylacetate.

RESOLUTION: Cupric perchlorate. 2,3;4,5-Di-O-isopropylidene-2-keto-L-gulonic acid hydrate. Tartaric acid.

RING EXPANSION: π-Allylpalladium chloride dimer. Benzyltriethylammonium chloride. Dihalomethyllithium. Ethyl azidoformate. Ferric chloride.

ROSENMUND REDUCTION: Palladium catalysts.

SHAPIRO-HEATH OLEFIN SYNTHESIS: *n*-Butyllithium–Tetramethylethylene-diamine. Dimethylformamide. Lithium diisopropylamide. *p*-Toluene-sulfonylhydrazine.
SILYLATION: N,O-Bis(trimethylsilyl)sulfonate. Ethyl trimethylsilylacetate. Trifluoro-methanesulfonic acid trimethylsilyl ester. N-(Trimethylsilyl)imidazole.
SPIROCYCLIZATION: Pyrrolidine.
SYNTHESIS OF:
  α-ACETOXY ALDEHYDES: Lead tetraacetate.
  β-ACYLACRYLIC ESTERS: Sodium methylsulfinylmethylide.
  ALCOHOLS: Bis(phenylthio)methane.
  ALDEHYDES: N-Chlorosuccinimide. *t*-Leucine *t*-butyl ester. Lithium trimethylsilyl-diazomethane. Sodium selenophenolate. Thiphenol. 1-Trimethylsilyl-1-phenylseleno-methyllithium. Zinc.
  ALKENES: Cuprous iodide. Diisobutylaluminum hydride. Dipropenylcopperlithium. Ethyl α-trifluoromethylsulfonyloxy acetate. Iodine. (1-Lithiovinyl)trimethylsilane. Lithium dialkyl cuprates. Lithium phenylethynolate. Methylcopper. Methyllithium. Sodium bistrimethylsilylamide. Thionyl chloride–Triethylamine. Titanium(0). *p*-Toluenesulfonic acid. Tri-*n*-butyltin chloride. Triethyl orthoformate.
  ALKENYL ALCOHOLS: Sodium.
  ALKOXY AMINES: N-Hydroxyphthalimide.
  ALKYLARENES: Copper(I) methyltrialkylborates.
  ALKYL AZIDES: Methyltricaprylylammonium chloride.
  ALKYL CHLORIDES: Hexamethylphosphoric triamide. Phosphorus trichloride–Dimeth-ylformamide.
  α-ALKYLCYCLOHEXENONES: Birch reduction.
  ALKYL FLUORIDES: Ion-exchange resins.
  ALKYL HALIDES: Vilsmeier reagent.
  ALKYL HYDROPEROXIDES: Hydrogen peroxide–Silver trifluoroacetate.
  O-ALKYLHYDROXYLAMINES: Triphenylphosphine–Diethyl azodicarboxylate.
  α-ALKYLIDENE-γ-BUTYROLACTONES: Tris(dimethylamino)methane.
  ALKYL IODIDES: N-Methyl-N,N'-dicyclohexylcarbodiimidium iodide. Triphenylphos-phine–Iodine.
  ALKYNES: Potassium ferricyanide. Tetra-*n*-butylammonium hydrogen sulfate. Tri-methyl phosphate.
  1-ALKYNES: Gringard reagents.
  α-ALKYNOIC ESTERS: Thallium(III) nitrate.
  ALKYNOLS: Tetra-*n*-butylammonium fluoride.
  ALLENE OXIDES: Peracetic acid–Sodium acetate.
  ALLENES: Dimethylcopperlithium. Ferric chloride. Grignard reagents.
  ALLENIC ETHERS: Cuprous bromide.
  ALLENONES: Methyllithium.
  ALLYLIC CHLORIDES: Methanesulfonyl chloride.
  ALLYLIC SULFIDES: *p*-Toluenesulfonic acid.
  AMIDES: 2-Chloro-N-methylbenzothiazolium trifluoromethanesulfonate. Diethyl phos-phorocyanidate. 3,4-Dihydro-2*H*-pyrido[1,2-*a*]pyrimidine-2-one. Potassium hydroxide. Sodium tetracarbonylferrate(II).
  AMINES: Trichloroacetonitrile. Trimethylsilyl azide.
  t-AMINES: Dimethyl(methylene)ammonium salts.
  β-AMINO ALCOHOLS: Lithium tetramethylpiperidide.

AMINO NITROXIDES: β,β,β-Trichloroethyl chloroformate.
ANHYDRONUCLEOSIDES: 1,5-Diazabicyclo[5.4.0]undecene-5.
ARENE OXIDES: 1,5-Diazabicyclo[5.4.0]undecene-5. Oxygen, singlet.
ARENES: 1,5-Diazabicyclo[5.4.0]undecene-5.
ARYL IODIDES: Thallium(III) trifluoroacetate.
ARYL THIOAMIDES: Ethoxy carbonyl isothiocyanate.
AZIDOALKANES: Ferric azide.
AZIRIDINES: Diethyl dibromophosphoroamidate.
AZLACTONES: Polyphosphoric acid.
p-BENZOQUINONES: Ceric ammonium nitrate.
1-BENZOXEPINS: Dimethyloxosulfonium methylide.
BIPHENYLS: Thallium(I) bromide.
α-BROMO ACIDS: N-Bromosuccinimide.
β-BROMOAMINES: Diethyl dibromophosphoramidate.
α-BROMOCARBONYL COMPOUNDS: Bromine. Magnesium bromide.
$\Delta^{\alpha,\beta}$-BUTENOLIDES: Carbon monoxide. Sodium methylsulfinylmethylide.
t-BUTYL KETONES: Lithium phenylthio(t-butyl)cuprate(I).
γ-BUTYROLACTONES: Chromic acid.
CARBAMATES: Stannous octoate. Tri-n-butylstannyl azide.
CARBODIIMIDES: n-Butyllithium.
CARBONYL COMPOUNDS: Sodium amalgam.
CARBOXYLIC ACIDS: Dilithium tetrachlorocuprate. Tetra-n-butylammonium iodide.
CHLOROEPOXIDES: t-Butyl hypochlorite.
β-CHLORO-α,β-UNSATURATED KETONES: Oxalyl chloride.
CHROMONES: Boron trifluoride etherate–Methanesulfonyl chloride–Dimethylformamide.
CYANOHYDRINS: Diperoxo-oxohexamethylphosphoramidomolybdenum(VI).
CYANOFORMATES: 18-Crown-6.
CYCLOALKANES: Silver triflate.
CYCLOALKYNES: Potassium t-butoxide–Dimethyl sulfoxide.
CYCLOBUTENES: Ethylaluminum dichloride.
CYCLOENONES: Piperidine.
CYCLOHEPTENONES: 1-Lithio-2-vinylcyclopropane.
CYCLOHEXANONES: 2-Trimethylsilyloxy-1,3-butadiene.
CYCLOPENTENONES: 2-Nitropropene.
CYCLOPROPANES: Benzyltriethylammonium chloride. Dicobalt octacarbonyl. Lithium diethylamide–Hexamethylphosphoric triamide. Lithium diisopropylamide–Hexamethylphosphoric triamide.
CYCLOPROPANOLS: Lithium 2,2,6,6-tetramethylpiperidide.
CYCLOPROPENYL ETHERS: Lithium 2,2,6,6-tetramethylpiperidide.
CYCLOPROPYL ETHERS: Lithium 2,2,6,6-tetramethylpiperidide.
CYCLOPROPYL KETONES: Cyclopropyltrimethylsilane.
DIACYL PEROXIDES: Potassium superoxides.
1,8-DIARYLNAPHTHALENES: Nickel(II) acetylacetonate.
DIARYL SULFONES: Silver trifluoromethanesulfonate.
1,1-DIBROMOALKENES: Diethyl dibromomethanephosphonate.
gem-DICHLOROCYCLOPROPANES: Benzyltriethylammonium chloride.
1,3-DIENES: Dichlorobis(triphenylphosphine)palladium(II). Manganese(II) chloride. (R)-4-Methylcyclohexylidenemethylcopper.
1,4-DIENES: Diphosphorus tetraiodide. Methylcopper.
1,5-DIENES: Lithium bis(dialkylamines) cuprates.

2,6-DIENIC CARBOXYLIC ACIDS: Trimethylchlorosilane.
1,1-DIFLUORO-1-ALKENES: Chlorodifluoromethane.
N,N-DIFLUOROAMINES: Fluorooxytrifluoromethane.
gem-DIFLUORO COMPOUNDS: Molybdenum hexafluoride.
gem-DIHALIDES: t-Butyl nitrite.
DIHYDROFURANES: Manganic acetate.
2,3-DIHYDROFURANES: Carboethoxycyclopropyltriphenylphosphonium tetrafluoroborate.
1,2-DIKETONES: t-Butoxybis(dimethylamino)methane. Dichlorovinylene carbonate.
1,4-DIKETONES: 2-Nitropropene. Palladium(II) chloride–Copper(I) chloride. Zinccopper couple.
1,5-DIKETONES: Palladium(II) chloride–Copper(II) chloride.
cis-1,2-DIOLS: Osmium tetroxide.
1,3-DIYNES: Dicyclohexylmethylthioborane.
DIYNES: Iodine.
1,4-ENEDIONES: Phenylselenenyl chloride.
ENYNES: Dicobalt octacarbonyl. Palladium catalysts.
EPISULFIDES: Triphenylphosphine.
EPOXIDES: Quinine.
ESTERS: 2-Lithio-2-methylthio-1,3-dithiane.
3-FORMYLCYCLOALKENONES: Bis(methylthio)methyllithium.
FURANES: Methoxyallene.
GLYCIDIC ESTERS: Chloromethyl methyl ether.
GLYCOSIDES: Stannic chloride.
α-HALO-β-LACTAMS: Diethylthallium t-butoxide.
α-HALO-α,β-UNSATURATED CARBONYL COMPOUNDS: Dihalocarbenes.
HETEROCYCLES: Methyl N-tosylmethylthiobenzimidate.
1,2,3,5-HEXATETRAENE: Zinc.
HYDROCRYLATES: Chloromethyl methyl ether.
HYDROPEROXIDES: Hydrogen peroxide–Silver trifluoroacetate. Hydrogen peroxide–Sodium peroxide.
HYDROXAMIC ACIDS: Diborane.
β-HYDROXY ESTERS: Ethyl trimethylsilylacetate–Tetra-n-butylammonium fluoride.
β-HYDROXY ESTERS: Lithium 1,1-bis(trimethylsilyl)-3-methylbutoxide.
α-HYDROXYCARBOXAMIDES: Lithium diisopropylamide.
β-HYDROXY HYDROPEROXIDES: Hydrogen peroxide.
β-HYDROXY KETONES: Phenylcopper.
γ-HYDROXY-α,β-UNSATURATED ESTERS: Methyl methanethiosulfonate.
IMINOAZIRIDINES: Copper(II) trifluoromethanesulfonate.
INDOLES: Bis(acetonitrile)dichloropalladium(II).
1,2-Iodocarboxylates: N-Iodosuccinimide–Carboxylic acids.
IMIDES: Ruthenium tetroxide.
ISOALLOXAZINES: Phenyl isocyanate.
ISOCYANATES: 2,4,6-Triphenylpyrylium perchlorate.
ISOCYANIDES: Benzyltriethylammonium chloride. Sodium hypobromite.
ISOCYANATES: Tri-n-butylstannyl azide.
ISOFLAVONES: Silver hexafluoroantimonate.
ISOXAZOLIDINES: N-Methylhydroxylamine hydrochloride.
KETENE THIOACETALS: Chloramine-T. O,O-Dimethyl formylphosphonate. S,S-Dimethyl thioacetal.

γ-KETO ALDEHYDES: 2-(2-Bromoethyl)-1,3-dioxane.

α-KETO ESTERS: Oxygen.

β-KETO ESTERS: *t*-Butyl trimethylsilylacetate.

δ-KETO ESTERS: Titanium(IV) chloride.

α-KETO HYDROPEROXIDES: Oxygen.

α-KETOLS: 3-Benzyl-4-methylthiazolium chloride.

1,4-KETOLS; 1,4-DIKETONES: Jones reagent.

KETONES: Bis(phenylthio)methane. N,N-Dimethyldithiocarboamoyl acetonitrile. Grignard reagents. Manganese(II) iodide. Phenylsileninyl chloride. Thallium(III) acetate. Vanadium(II) chloride.

β-KETOPHOSPHONATES: Cuprous iodide.

β-LACTAMS: Oxygen. Sodium hydride.

β-LACTONES: Lithium phenylethynolate. Peroxybenzimidic acid.

γ-LACTONES: Trimethyl phosphonacetate.

METHOXYMETHYL ETHERS: Dimethoxymethane.

α-METHOXY-α,β-UNSATURATED ALDEHYDES: 1,2-Dimethoxyvinyllithium.

N-METHYLANILINES: 5,5-Dimethylhydantoin.

α-METHYLENE-γ-BUTYROLACTONES: Dimethyl(methylene)ammonium salts. Lithium diisopropylamide.

α-METHYLENE LACTONES: Formaldehyde. Lithium diisopropylamide. Palladium chloride–Thiourea.

α-METHYLENE-γ-LACTONES: Spiro[2.5]-5,7-dioxa-6,6-dimethylactane-4,8-dione. Zinc-Copper couple.

NITRILES: Acrylonitrile. N-Amino-4,6-diphenylpyridone. *n*-Hexylamine. Hydroxylamine hydrochloride. Lithium diethylamide–Hexamethylphosphoric triamide. Methyl carbazate. Phenyl isocyanate. *p*-Toluenesulfonic acid. Trifluoromethanesulfonic anhydride.

1,2,4-OXADIAZOLES: Sodium hypochlorite.

2-OXAZOLIDINONES: Stannous octoate.

2-OXAZOLIDONES: Carbon dioxide.

OXETANES: Lithium hydroxide.

PEPTIDES: Benzotriazolyloxytris(dimethylamino)-phosphonium hexafluorophosphate. Diethyl phosphorocyanidate. Diphenylphosphinyl chloride. 4,6-Diphenylthieno[3,4-d]-dioxol-2-one-5,5-dioxide. Thioglycolic acid.

PERACIDS: Hydrogen peroxide.

PEROXIDES, CYCLIC: Mercuric nitrate–Hydrogen peroxide.

PHENOLIC ALDEHYDES: Bis(1,3-diphenyl-2-imidazolidinylidene).

2-PIPERIDONES: Ammonium acetate.

POLYNUCLEOTIDES: Benzenesulfonyltetrazole.

PROPARGYLIC ALCOHOLS: Aluminum chloride.

PURINES: N,N-Dimethylhydrazine.

PYRAZINES: Silver perchlorate.

PYRONES: 2,4,6-Trichlorophenylcyanoacetate.

PYRROLES: Lithium diisopropylamide. Tosylmethylisocyanide.

PYRROLIDENES: Lithium diisopropylamide.

PYRROLIDINOENAMINES: Pyrrolidine.

QUATERNARY AMMONIUM HYDROXIDES: Ion-exchange resins.

o-QUINONES: Diphenylsileninic anhydride.

SALICYLALDEHYDES: Oxalyl chloride.

SPIROCYCLOHEXADIENONES: Methyl vinyl ketone.

STYRENES: Formaldehyde.
SULFIDES: Hexamethyldisilazane.
SULFINAMIDES: Dicyclohexylcarbodiimide.
SULFONES: *n*-Butyllithium.
SULFONYL CHLORIDES: Chlorine.
TETRAHYDROPYRIDAZINES: Lithium diisopropylamide.
THIOL ESTERS: 2-Chloro-N-methylbenzothiazolium trifluoromethanesulfonate.
THIOLS: Dicyclohexylcarbodiimide.
TRIMETHYLSILYL ENOL ETHERS: Zinc-Trimethylchlorosilane.
α-TRIMETHYLSILYLVINYL ESTERS: *t*-Butyl bis(trimethylsilyl)lithioacetate.
β,γ-UNSATURATED ACIDS: Sodium cyanide.
γ,δ-UNSATURATED ALCOHOLS: Aluminum chloride. Titanium(IV) chloride.
γ,δ-UNSATURATED ALCOHOLS: Titanium(IV) chloride.
α,β-UNSATURATED ALDEHYDES: Cuprous bromide. Dimethylformamide. (Dicyclo-
    pentadienyl)(chloro)hydridozirconium.
α,β-UNSATURATED ESTERS: Carbon monoxide. Ethyl chloroformate. O-Ethyl-S-
    ethoxycarbonylmethyl dithiocarbonate.
α,β-UNSATURATED KETONES: Aluminum chloride. 1,2-Dichloro-4-phenylthio-2-
    butenes. Lithium diisopropylamide. Oxygen. Phenylselenenyl bromide (chloride).
β,γ-UNSATURATED KETONES: Aluminum chloride.
γ,δ-UNSATURATED KETONES: 9-Borabicyclo[3.3.1]nonane.
α,β-UNSATURATED NITRILES: Potassium hydroxide.
URACIL: Polyphosphoric acid.
UREAS: Tri-*n*-butylstannyl azide.
VINYL CHLORIDES: Chloromethyltriphenylphosphonium chloride.
VINYL ETHERS: Grignard reagents. Methoxymethyldiphenylphosphine oxide.
VINYL HALIDES: Sodium amalgam.
VINYL KETONES: Diethyl oxalate.
VINYLLITHIUM CARBANIONS: n-Butyllithium–Tetramethylethylenediamine.
α-VINYLNITRILES: Benzyltriethylammonium chloride.
VINYLSILANES: Benzenesulfonyl hydrazine. Cuprous iodide.
XANTHINES: N,N-Dimethylhydrazine.

THIOACETALIZATION: Hexacarbonylmolybdenum.
THIO-CLAISEN REARRANGEMENT: Triethylamine.
TRANSESTERIFICATION: Cesium fluoride.
TRANSFER HYDROGENATION: Palladium on carbon.
TRANSTHIOACETALIZATION: Glyoxylic acid.
TRIFLUOROACETYLATION, AMINO ACIDS: Methyl trifluoroacetate.

ULLMANN REACTION: Copper.

VALENCE ISOMERIZATION: Copper.
VON BROWN DEGRADATION: Ruthenium tetroxide.

WITTIG-HORNER REACTION: 2-(Diethylphosphono)propionitrile. O,O-Dimethyl
    formylphosphonate S,S-dimethyl thioacetal. Methoxymethyldiphenylphosphine oxide.
    Tetra-*n*-butylammonium iodide. Trimethyl phosphonoacetate.
WITTIG REACTION: Dichloromethylenetriphenylphosphorane. Formaldehyde. Hexa-
    methylphosphorus triamide. Sodium bistrimethylsilylamide. Trineopentylneopentyli-
    denetantalum.

AUTHOR INDEX

441

SUBJECT INDEX

Page numbers referring to reagents are indicated in **boldface**.